Springer Series in Reliability Engineering

Series Editor

Professor Hoang Pham
Department of Industrial and Systems Engineering
Rutgers, The State University of New Jersey
96 Frelinghuysen Road
Piscataway, NJ 08854-8018
USA

Other titles in this series

The Universal Generating Function in Reliability Analysis and Optimization
Gregory Levitin

Warranty Management and Product Manufacture
D.N.P. Murthy and Wallace R. Blischke

Maintenance Theory of Reliability
Toshio Nakagawa

System Software Reliability
Hoang Pham

Reliability and Optimal Maintenance
Hongzhou Wang and Hoang Pham

Applied Reliability and Quality
B.S. Dhillon

Shock and Damage Models in Reliability Theory
Toshio Nakagawa

Risk Management
Terje Aven and Jan Erik Vinnem

Satisfying Safety Goals by Probabilistic Risk Assessment
Hiromitsu Kumamoto

Offshore Risk Assessment (2nd Edition)
Jan Erik Vinnem

The Maintenance Management Framework
Adolfo Crespo Márquez

Human Reliability and Error in Transportation Systems
B.S. Dhillon

Complex System Maintenance Handbook
D.N.P. Murthy and Khairy A.H. Kobbacy

Recent Advances in Reliability and Quality in Design
Hoang Pham

Product Reliability
D.N.P. Murthy, Marvin Rausand, and Trond Østerås

Mining Equipment Reliability, Maintainability, and Safety
B.S. Dhillon

Advanced Reliability Models and Maintenance Policies
Toshio Nakagawa

Justifying the Dependability of Computer-based Systems
Pierre-Jacques Courtois

Reliability and Risk Issues in Large Scale Safety-critical Digital Control Systems
Poong Hyun Seong

Failure Rate Modeling for Reliability and Risk
Maxim Finkelstein

Javier Faulin · Angel A. Juan · Sebastián Martorell
José-Emmanuel Ramírez-Márquez
(Editors)

Simulation Methods for Reliability and Availability of Complex Systems

Springer

Prof. Javier Faulin
Universidad Pública de Navarra
Depto. Estadística e Investigación Operativa
Campus Arrosadia, Edif. Los Magnolios,
1ª planta
31080 Pamplona
Spain
javier.faulin@unavarra.es

Prof. Sebastián Martorell
Universidad Politécnica de Valencia
Depto. Ingeniería Química y Nuclear
Camino de Vera, s/n
46022 Valencia
Spain
smartore@iqn.upv.es

Assoc. Prof. Angel A. Juan
Open University of Catalonia (UOC)
Computer Science, Multimedia
and Telecommunication Studies
Rambla Poblenou, 156
08015 Barcelona
Spain
ajuanp@uoc.edu

Asst. Prof. José-Emmanuel
Ramírez-Márquez
Stevens Institute of Technology
School of Systems & Enterprises
1 Castle Point on Hudson
Hoboken NJ 07030
USA
jmarquez@stevens.edu

ISSN 1614-7839
ISBN 978-1-4471-2552-5 ISBN 978-1-84882-213-9(eBook)
DOI 10.1007/978-1-84882-213-9
Springer London Dordrecht Heidelberg New York

British Library Cataloguing in Publication Data
A catalogue record for this book is available from the British Library

Cover design: deblik, Berlin, Germany
Typesetting and production: le-tex publishing services GmbH, Leipzig, Germany

Printed on acid-free paper

Springer is part of Springer Science+Business Media (www.springer.com)

Foreword

Satisfying societal needs for energy, communications, transportation, etc. requires complex inter-connected networks and systems that continually and rapidly evolve as technology changes and improves. Furthermore, consumers demand higher and higher levels of reliability and performance; at the same time the complexity of these systems is increasing. Considering this complex and evolving atmosphere, the usage and applicability of some traditional reliability models and methodologies are becoming limited because they do not offer timely results or they require data and assumptions which may no longer be appropriate for complex modern systems. Simulation of system performance and reliability has been available for a long time as an alternative for closed-form analytical and rigorous mathematical models for predicting reliability. However, as systems evolve and become more complex, the attractiveness of simulation modeling becomes more apparent, popular, and useful. Additionally, new simulation models and philosophies are being developed to offer creative and useful enhancements to this modeling approach to study reliability and availability behavior of complex systems. New and advanced simulation models can be more rapidly altered to consider new systems, and they are much less likely to be constrained by limiting and restrictive assumptions. Thus, a more realistic modeling approach can bc cmployed to solve diverse analytical problems.

The editors of this book (Profs. Faulin, Juan, Martorell, and Ramírez-Márquez) have successfully undertaken a remarkable challenge to include topical and interesting chapters and material describing advanced simulation methods to estimate reliability and availability of complex systems. The material included in the book covers many diverse and interesting topics, thereby providing an excellent overview of the field of simulation including both discrete event and Monte Carlo simulation models. Every contributor and author participating in this book is a respected expert in the field, including researchers such as Dr. Lawrence Leemis, Dr. Enrico Zio, and others who are among the most respected and accomplished experts in the field of reliability.

available for some time. In the end, if no effective maintenance policies are taken, any product (component or system, hardware or software) will fail, meaning that it will stop being operative, at least as intended.

Reliability is often defined as the probability that a system or component will perform its intended function, under operating conditions, for a specified period of time. Moreover, availability can be defined as the probability that a system or component will be performing its intended function, at a certain future time, according to some maintenance policy and some operating conditions. During the last few decades, a lot of work has been developed regarding the design and implementation of system maintenance policies. Maintenance policies are applied to many real systems: when one component fails – or there is a high probability that it can fail soon – it is repaired or substituted by a new one, even when the component failure does not necessarily imply the global system failure or status change. For system managers and engineers, it can be very useful to be able to predict the availability function of time-dependent systems in the short, medium, or long run, and how these availability levels can be increased by improving maintenance policies, reliability of individual components or even system structure design. This information can be critical in order to ensure data integrity and safety, quality-of-service, process or service durability, and even human safety. In other words, great benefits can be obtained from efficient methods and software tools that: (1) allow predicting system availability levels at future target times and (2) provide useful information about how to improve these availability levels.

Many authors point out that, when dealing with real complex systems, only simulation techniques, such as MCS and, especially, DES, can be useful to obtain credible predictions for R&A parameters. In fact, simulation has been revealed as a powerful tool in solving many engineering problems. This is due to the fact that simulation methods tend to be simpler to implement than analytic ones and, more importantly, to the fact that simulation methods can model real-systems behavior with great detail. Additionally, simulation methods can provide supplementary information about system internal behavior or about critical components from a reliability/availability point of view. These methods are not perfect either, since they can be computationally intensive and they do not provide exact results, only estimated ones. Applications of simulation techniques in the R&A fields allow modeling details such as multiple-state systems, component dependencies, non-perfect repairs, dysfunctional behavior of components, etc. Simulation-based techniques have also been proposed to study complex systems availability. In fact, during the last few years, several commercial simulators have been developed to study the R&A of complex systems.

Every system built by humans is unreliable in the sense that it degrades with age and/or usage. A system is said to fail when it is no longer capable of delivering the designed outputs. Some failures can be catastrophic in the sense that they can result in serious economic losses, affect humans and do serious damage to the environment. Therefore, the accurate estimation of failures in order to study the R&A of complex systems has revealed as one of the most challenging tasks of research. Taking into account the importance of this type of study and its difficulties, we think that apart from the traditional exact methods in R&A, the use of a very popular tool such as simulation can be a meaningful contribution in the development of new protocols to study complex systems.

Thus, this book deals with both simulation and R&A of complex systems, topics which are not commonly presented together. It is divided into three major parts:

Part I Fundamentals of Simulation in Reliability and Availability Issues;
Part II Simulation Applications in Reliability;
Part III Simulation Applications in Availability and Maintenance.

Each of these three parts covers different contents with the following intentions:

Part I: To describe, in detail, some ways of performing simulation in different theoretical arenas related to R&A.

Part II: To present some meaningful applications of the use of simulation in the study of different scenarios related to reliability decisions.

Part III: To discuss some interesting applications of the use of simulation in the study of different cases related to availability decisions.

Part I presents some new theoretical results setting up the fundamentals of the use of simulation in R&A. This part consists of four chapters. The first, by Zio and Pedroni, describes some interesting uses of MCS to make accurate estimations of Reliability. The second, by K. Durga Rao *et al.*, makes use of simulation to develop a dynamic fault tree analysis providing meaningful examples. Cancela *et al.* develop some improvements of the path-based methods for Monte Carlo reliability evaluation in the third chapter. The fourth, by Leemis, concludes this part by introducing some descriptive simulation methods to generate variates. This part constitutes the core of the book and develops a master view of the use of simulation in the R&A field.

Parts II and III are closely connected. Both of them present simulation applications in two main topics of the book: reliability and availability. Part II is devoted to simulation applications in reliability and Part III presents other simulation applications in availability and maintenance. Nevertheless, this classification cannot be strict because both topics are closely connected.

Part II has five chapters, which present some real applications of simulation in selected cases of reliability. Thus, Chapter 5 (Gosavi and Murray) describes the simulation analysis of the reliability and preventive maintenance of a public infrastructure. Marotta *et al.* discuss reliability models for data integration systems in the following chapter, giving a complementary view of the previous chapter. Chapter 7 makes a comparison between the results given by analytical methods and given by simulation of the power distribution system reliability. This is one of the most meaningful applications of the book. Chapter 8 (Aijaz Shaikh) presents the use of the software Reliasoft to analyse process industries. Chapter 9 (Angel A. Juan *et al.*) concludes this part by explaining some applications of discrete event simulation and fuzzy sets to study structural Reliability in building and civil engineering.

Finally, Part III consists of four chapters. Chapter 10 describes maintenance manpower modeling using simulation. It is a good application of some traditional tools of simulation to describe maintenance problems. Kwang Pil Chang *et al.* present in

Chapter 11 another interesting application in the world of estimating availability in offshore installations. This challenging case is worth reading carefully. Zille *et al.* explain in the twelfth chapter the use of simulation to study the maintained multi-component systems. Last but not least, Farukh Nadeem and Erich Leitgeb describe a simulation model to study availability in optical wireless communication.

The book has been written for a wide audience. This includes practitioners from industry (systems engineers and managers) and researchers investigating various aspects of R&A. Also, it is suitable for use by Ph.D. students who want to look into specialized topics of R&A.

We would like to thank the authors of the chapters for their collaboration and prompt responses to our enquiries which enabled completion of this handbook on time. We gratefully acknowledge the help and encouragement of the editor at Springer, Anthony Doyle. Also, our thanks go to Claire Protherough and the staff involved with the production of the book.

Javier Faulin
Public University of Navarre, Pamplona, Spain

Angel A. Juan
Open University of Catalonia, Barcelona, Spain

Sebastián Martorell
Technical University of Valencia, Valencia, Spain

José-Emmanuel Ramírez-Márquez
Stevens Institute of Technology, Hoboken, New Jersey, USA

Contents

Part I
Fundamentals of Simulation in Reliability and Availability Issues

Chapter 1
Reliability Estimation by Advanced Monte Carlo Simulation

E. Zio and N. Pedroni

Abstract Monte Carlo simulation (MCS) offers a powerful means for evaluating the reliability of a system, due to the modeling flexibility that it offers indifferently of the type and dimension of the problem. The method is based on the repeated sampling of realizations of system configurations, which, however, are seldom of failure so that a large number of realizations must be simulated in order to achieve an acceptable accuracy in the estimated failure probability, with costly large computing times. For this reason, techniques of efficient sampling of system failure realizations are of interest, in order to reduce the computational effort.

In this chapter, the recently developed subset simulation (SS) and line sampling (LS) techniques are considered for improving the MCS efficiency in the estimation of system failure probability. The SS method is founded on the idea that a small failure probability can be expressed as a product of larger conditional probabilities of some intermediate events: with a proper choice of the intermediate events, the conditional probabilities can be made sufficiently large to allow accurate estimation with a small number of samples. The LS method employs lines instead of random points in order to probe the failure domain of interest. An "important direction" is determined, which points towards the failure domain of interest; the high-dimensional reliability problem is then reduced to a number of conditional one-dimensional problems which are solved along the "important direction."

The two methods are applied on two structural reliability models of literature, i.e., the cracked-plate model and the Paris–Erdogan model for thermal-fatigue crack growth. The efficiency of the proposed techniques is evaluated in comparison to other stochastic simulation methods of literature, i.e., standard MCS, importance sampling, dimensionality reduction, and orthogonal axis.

Energy Department, Politecnico di Milano, Via Ponzio 34/3, 20133 Milan, Italy

P. Faulin, A. Juan, S. Martorell, and J.E. Ramírez-Márquez (eds), *Simulation Methods for Reliability and Availability of Complex Systems*.

1.1 Introduction

In the performance-based design and operation of modern engineered systems, the accurate assessment of reliability is of paramount importance, particularly for civil, nuclear, aerospace, and chemical systems and plants which are safety-critical and must be designed and operated within a risk-informed approach (Thunnissen *et al.* 2007; Patalano *et al.* 2008).

The reliability assessment requires the realistic modeling of the structural/mechanical components of the system and the characterization of their material constitutive behavior, loading conditions, and mechanisms of deterioration and failure that are anticipated to occur during the working life of the system (Schueller and Pradlwarter 2007).

In practice, not all the characteristics of the system under analysis can be fully captured in the model. This is due to: (1) the intrinsically random nature of several of the phenomena occurring during the system life; (2) the incomplete knowledge about some of these phenomena. Thus, uncertainty is always present in the hypotheses underpinning the model (model uncertainty) and in the values of its parameters (parameter uncertainty); this leads to uncertainty in the model output, which must be quantified for a realistic assessment of the system (Nutt and Wallis 2004).

In mathematical terms, the probability of system failure can be expressed as a multidimensional integral of the form

$$P(F) = P(\boldsymbol{x} \in F) = \int I_F(\boldsymbol{x}) q(\boldsymbol{x}) \mathrm{d}\boldsymbol{x} \tag{1.1}$$

where $\boldsymbol{x} = \{x_1, x_2, \ldots, x_j, \ldots, x_n\} \in \Re^n$ is the vector of the uncertain input parameters/variables of the model, with multidimensional probability density function (PDF) $q: \Re^n \to [0, \infty)$, $F \subset \Re^n$ is the failure region and $I_F: \Re^n \to \{0, 1\}$ is an indicator function such that $I_F(\boldsymbol{x}) = 1$, if $\boldsymbol{x} \in F$ and $I_F(\boldsymbol{x}) = 0$, otherwise. The failure domain F is commonly defined by a so-called performance function (PF) or limit state function (LSF) $g_x(\boldsymbol{x})$ which is lower than or equal to zero if $\boldsymbol{x} \in F$ and greater than zero, otherwise.

In practical cases, the multidimensional integral in Equation 1.1 cannot be easily evaluated by analytical methods nor by numerical schemes. On the other hand, Monte Carlo simulation (MCS) offers an effective means for estimating the integral, because the method does not suffer from the complexity and dimension of the domain of integration, albeit it implies the nontrivial task of sampling from the multidimensional PDF. The MCS solution to Equation 1.1 entails that a large number of samples of the values of the uncertain parameters $\boldsymbol{x}$ be drawn from $q(\boldsymbol{x})$ and that these be used to compute an unbiased and consistent estimate of the system failure probability as the fraction of the number of samples that lead to failure. However, a large number of samples (inversely proportional to the failure probability) is necessary to achieve an acceptable estimation accuracy: in terms of the integral in Equation 1.1 this can be seen as due to the high dimensionality n of the problem and the large dimension of the relative sample space compared to the failure region of interest (Schueller 2007). This calls for new simulation techniques for performing robust estimations with a limited number of input samples (and associated low computational time).

In this respect, effective approaches are offered by subset simulation (SS) (Au and Beck 2001, 2003b) and line sampling (LS) (Koutsourelakis *et al.* 2004; Pradlwarter *et al.* 2005).

In the SS method, the failure probability is expressed as a product of conditional failure probabilities of some chosen intermediate events, whose evaluation is obtained by simulation of more frequent events. The evaluation of small failure probabilities in the original probability space is thus tackled by a sequence of simulations of more frequent events in the conditional probability spaces. The necessary conditional samples are generated through successive Markov chain Monte Carlo (MCMC) simulations (Metropolis *et al.* 1953; Hastings 1970; Fishman 1996), gradually populating the intermediate conditional regions until the final target failure region is reached.

In the LS method, *lines*, instead of random *points*, are used to probe the failure domain of the high-dimensional problem under analysis (Pradlwarter *et al.* 2005). An "important direction" is optimally determined to point towards the failure domain of interest and a number of conditional, one-dimensional problems are solved along such a direction, in place of the high-dimensional problem (Pradlwarter *et al.* 2005). The approach has been shown to perform always better than standard MCS; furthermore, if the boundaries of the failure domain of interest are not too rough (i.e., almost linear) and the "important direction" is almost perpendicular to them, the variance of the failure probability estimator could be ideally reduced to zero (Koutsourelakis *et al.* 2004).

In this chapter, SS and LS schemes are developed for application to two structural reliability models of literature, i.e., the cracked-plate model (Ardillon and Venturini 1995) and the Paris–Erdogan thermal-fatigue crack growth model (Paris 1961). The problem is rather challenging as it entails estimating failure probabilities of the order of 10^{-7}. The effectiveness of SS and LS is compared to that of other simulation methods, e.g., the importance sampling (IS), dimensionality reduction (DR) and orthogonal axis (OA) methods (Gille 1998, 1999). In the IS method, the PDF $q(\boldsymbol{x})$ in Equation 1.1 is replaced with an importance sampling distribution (ISD) arbitrarily chosen so as to generate samples that lead to failure more frequently (Au and Beck 2003a); in the DR method, the failure event is re-expressed in such a way as to highlight one important variable (say, x_j) and the failure probability is then computed as the expected value of the cumulative distribution function (CDF) of x_j conditional on the remaining $(n-1)$ variables; finally, in the OA method, a sort of importance sampling is performed around the most likely point in the failure domain (Gille 1998, 1999).

The remainder of the chapter is organized as follows. In Section 1.2, a general presentation of the SS and LS schemes implemented for this study is given. In Section 1.3, the IS, DR, and OA methods taken as terms of comparison are briefly summarized. The results of the application of SS and LS to the cracked-plate and thermal-fatigue crack growth models are reported in Sections 1.4 and 1.5, respectively. Based on the results obtained, a critical discussion of the simulation techniques adopted and compared in this work is offered in the last section. For completeness of the contents of the chapter, detailed descriptions of the Markov Chain Monte Carlo (MCMC) simulation method used for the development of the SS and LS algorithms are provided in Appendices 1 and 2, respectively.

1.2 Simulation Methods Implemented in this Study

1.2.1 The Subset Simulation Method

Subset simulation is an adaptive stochastic simulation method originally developed for efficiently computing small failure probabilities in structural reliability analysis (Au and Beck 2001). The underlying idea is to express the (small) failure probability as a product of (larger) probabilities conditional on some intermediate events. This allows converting a rare event simulation into a sequence of simulations of more frequent events. During simulation, the conditional samples are generated by means of a Markov chain designed so that the limiting stationary distribution is the target conditional distribution of some adaptively chosen event; by so doing, the conditional samples gradually populate the successive intermediate regions up to the final target (rare) failure region (Au and Beck 2003b).

1.2.1.1 The Basic Principles

For a given target failure event F of interest, let $F_1 \supset F_2 \supset \ldots \supset F_m = F$ be a sequence of intermediate events, so that $F_k = \cap_{i=1}^{k} F_i$, $k = 1, 2, \ldots, m$. By sequentially conditioning on the event F_i, the failure probability $P(F)$ can be written as

$$P(F) = P(F_m) = P(F_1) \prod_{i=1}^{m-1} P(F_{i+1}|F_i) \tag{1.2}$$

Notice that even if $P(F)$ is small, the conditional probabilities involved in Equation 1.2 can be made sufficiently large by appropriately choosing m and the intermediate events $\{F_i, i = 1, 2, \ldots, m - 1\}$.

The original idea of SS is to estimate the failure probability $P(F)$ by estimating $P(F_1)$ and $\{P(F_{i+1}|F_i): i = 1, 2, \ldots, m - 1\}$. Considering, for example, $P(F) \approx 10^{-5}$ and choosing $m = 4$ intermediate events such that $P(F_1)$ and $\{P(F_{i+1}|F_i): i = 1, 2, 3, 4\} \approx 0.1$, the conditional probabilities can be evaluated efficiently by simulation of the relatively frequent intermediate events (Au and Beck 2001).

Standard MCS can be used to estimate $P(F_1)$. On the contrary, computing the conditional probabilities in Equation 1.2 by MCS entails the nontrivial task of sampling from the conditional distributions of $\boldsymbol{x}$ given that it lies in F_i, $i = 1, 2, \ldots, m-1$, i.e., from $q(\boldsymbol{x}|F_i) = q(\boldsymbol{x})I_{F_i}(\boldsymbol{x})/P(F)$. In this regard, MCMC simulation provides a powerful method for generating samples conditional on the intermediate regions F_i, $i = 1, 2, \ldots, m-1$ (Au and Beck 2001, 2003b). For completeness, the related algorithm is presented in Appendix 1.

1.2.1.2 The Algorithm

In the actual SS implementation, with no loss of generality it is assumed that the failure event of interest can be defined in terms of the value of a critical system response variable Y being lower than a specified threshold level y, i.e., $F = \{Y < y\}$. The sequence of intermediate events $\{F_i: i = 1, 2, \ldots, m\}$ can then be correspondingly defined as $F_i = \{Y < y_i\}$, $i = 1, 2, \ldots, m$ where $y_1 > y_2 > \ldots > y_i > \ldots > y_m = y > 0$ is a decreasing sequence of intermediate threshold values (Au and Beck 2001, 2003b).

The choice of the sequence $\{y_i: i = 1, 2, \ldots, m\}$ affects the values of the conditional probabilities $\{P(F_{i+1}|F_i): i = 1, 2, \ldots, m-1\}$ in Equation 1.2 and hence the efficiency of the SS procedure. In particular, choosing the sequence $\{y_i: i = 1, 2, \ldots, m\}$ *a priori* makes it difficult to control the values of the conditional probabilities $\{P(F_{i+1}|F_i): i = 1, 2, \ldots, m-1\}$. For this reason, in this work, the intermediate threshold values are chosen adaptively in such a way that the estimated conditional probabilities are equal to a fixed value p_0 (Au and Beck 2001; Au and Beck 2003b).

Schematically, the SS algorithm proceeds as follows (Figure 1.1):

1. Sample N vectors $\{\boldsymbol{x}_0^k: k = 1, 2, \ldots, N\}$ by standard MCS, i.e., from the original probability density function $q(\cdot)$. The subscript "0" denotes the fact that these samples correspond to "conditional level 0."
2. Set $i = 0$.
3. Compute the values of the response variable $\{Y(\boldsymbol{x}_i^k): k = 1, 2, \ldots, N\}$.
4. Choose the intermediate threshold value y_{i+1} as the $(1-p_0)N^{\text{th}}$ value in the decreasing list of values $\{Y(\boldsymbol{x}_i^k): k = 1, 2, \ldots, N\}$ (computed at step 3 above) to define $F_{i+1} = \{Y < y_{i+1}\}$. By so doing, the sample estimate of $P(F_{i+1}|F_i) = P(Y < y_{i+1}|Y < y_i)$ is equal to p_0 (note that it has been implicitly assumed that $p_0 N$ is an integer value).
5. If $y_{i+1} \leqslant y_m$, proceed to step 10 below.
6. *vice versa*, i.e., if $y_{i+1} > y_m$, with the choice of y_{i+1} performed at step 4 above, identify the $p_0 N$ samples $\{\boldsymbol{x}_i^u : u = 1, 2, \ldots, p_0 N\}$ among $\{\boldsymbol{x}_i^k: k = 1, 2, \ldots, N\}$ whose response Y lies in $F_{i+1} = \{Y < y_{i+1}\}$: these samples are at "conditional level $i+1$" and distributed as $q(\cdot|F_{i+1})$ and function as seeds of the MCMC simulation (step 7 below).
7. Starting from each one of the samples $\{\boldsymbol{x}_i^u: u = 1, 2, \ldots, p_0 N\}$ (identified at step 6 above), use MCMC simulation to generate $(1-p_0)N$ additional condi-

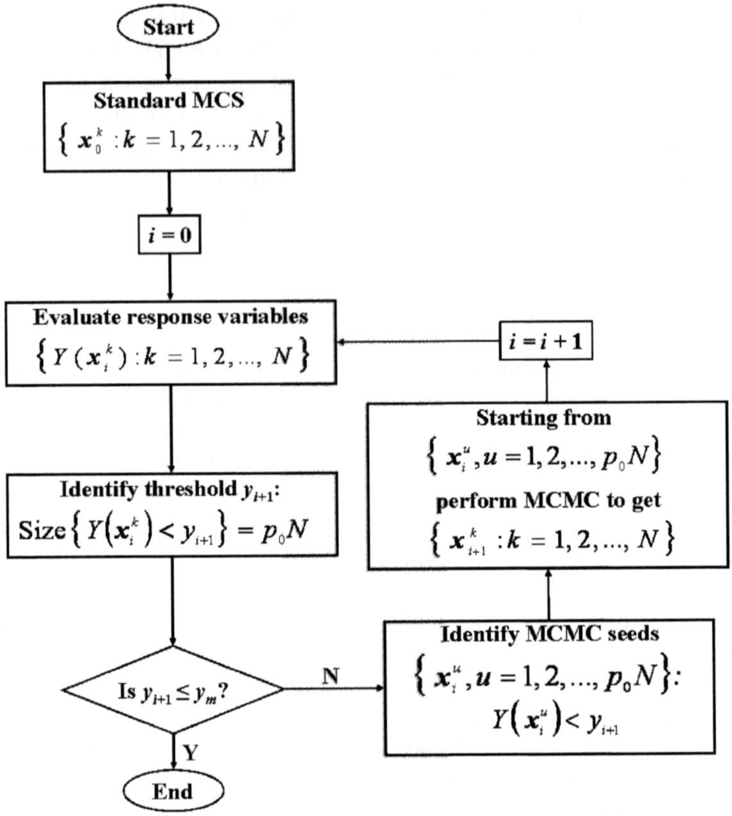

Figure 1.1 The SS algorithm

tional samples distributed as $q(\cdot|F_{i+1})$, so that there are a total of N conditional samples $\{x_{i+1}^k : k = 1, 2, \ldots, N\} \in F_{i+1}$, at "conditional level $i + 1$."

8. Set $i \leftarrow i + 1$.
9. Return to step 3 above.
10. Stop the algorithm.

For the sake of clarity, a step-by-step illustration of the procedure for conditional levels 0 and 1 is provided in Figure 1.2 by way of example.

Notice that the procedure is such that the response values $\{y_i : i = 1, 2, \ldots, m\}$ at the specified probability levels $P(F_1) = p_0$, $P(F_2) = p(F_2|F_1)P(F_1) = p_0^2, \ldots, P(F_m) = p_0^m$ are estimated, rather than the event probabilities $P(F_1)$, $P(F_2|F_1), \ldots, P(F_m|F_{m-1})$, which are *a priori* fixed at p_0. In this view, SS is a method for generating samples whose response values correspond to specified probability levels, rather than for estimating probabilities of specified failure events. As a result, it produces information about $P(Y < y)$ *versus* y at all the simulated values of Y rather than at a single value of y. This feature is important because the

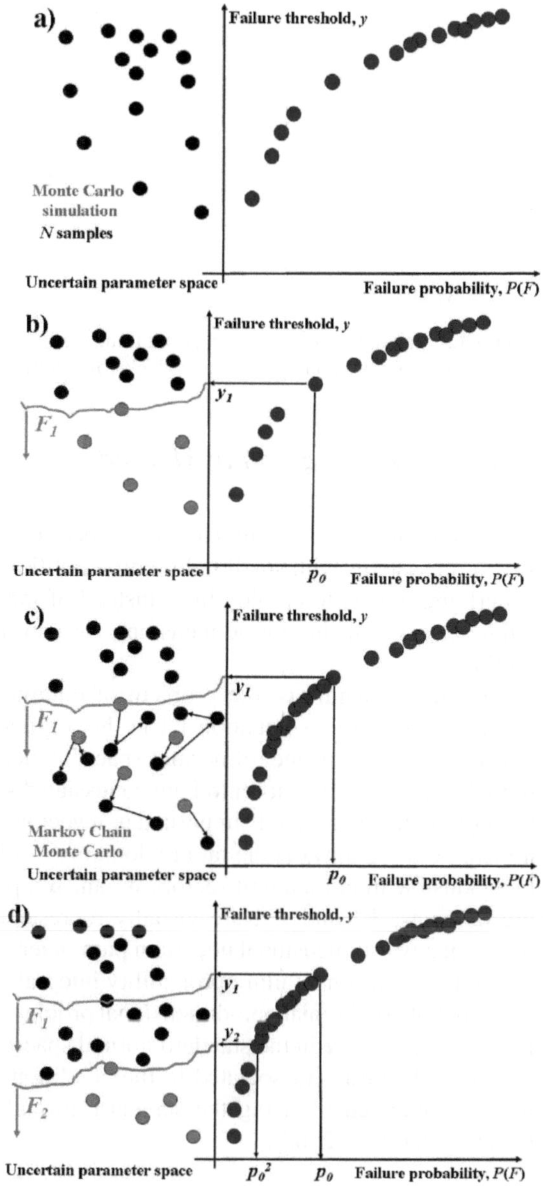

Figure 1.2 The SS procedure: (a) Conditional level 0: standard Monte Carlo simulation; (b) Conditional level 0: adaptive selection of y_1; (c) Conditional level 1: MCMC simulation; (d) Conditional level 1: adaptive selection of y_2 (Au 2005)

whole trend of $P(Y < y)$ *versus* y provides much more information than a point estimate (Au 2005).

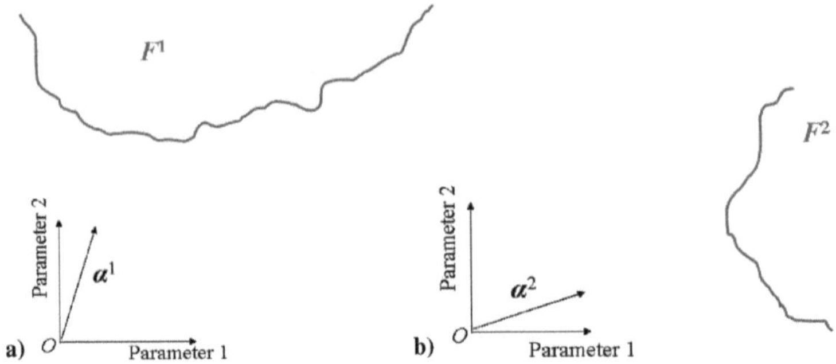

Figure 1.3 Examples of possible important unit vectors $\boldsymbol{\alpha}^1$ (a) and $\boldsymbol{\alpha}^2$ (b) pointing towards the corresponding failure domains F^1 (a) and F^2 (b) in a two-dimensional uncertain parameter space

1.2.2 The Line Sampling Method

Line sampling was also originally developed for the reliability analysis of complex structural systems with small failure probabilities (Koutsourelakis *et al.* 2004). The underlying idea is to employ lines instead of random points in order to probe the failure domain of the high-dimensional system under analysis (Pradlwarter *et al.* 2005).

In extreme synthesis, the problem of computing the multidimensional failure probability integral in Equation 1.1 in the original "physical" space is transformed into the so-called "standard normal space," where each random variable is represented by an independent central unit Gaussian distribution. In this space, a *unit* vector $\boldsymbol{\alpha}$ (hereafter also called "important unit vector" or "important direction") is determined, pointing towards the failure domain F of interest (for illustration purposes, two plausible important unit vectors, $\boldsymbol{\alpha}^1$ and $\boldsymbol{\alpha}^2$, pointing towards two different failure domains, F^1 and F^2, are visually represented in Figure 1.3a and b, respectively, in a two-dimensional uncertain parameter space). The problem of computing the high-dimensional failure probability integral in Equation 1.1 is then reduced to a number of conditional one-dimensional problems, which are solved along the "important direction" $\boldsymbol{\alpha}$ in the standard normal space. The conditional one-dimensional failure probabilities (associated to the conditional one-dimensional problems) are readily computed by using the standard normal cumulative distribution function (Pradlwarter *et al.* 2005).

1.2.2.1 Transformation of the Physical Space into the Standard Normal Space

Let $\boldsymbol{x} = \{x_1, x_2, \ldots, x_j, \ldots, x_n\} \in \Re^n$ be the vector of uncertain parameters defined in the original physical space $\boldsymbol{x} \in \Re^n$. For problems where the dimension n is not so small, the parameter vector $\boldsymbol{x}$ can be transformed into the vector $\boldsymbol{\theta} \in \Re^n$, where each element of the vector $\boldsymbol{\theta}_j$, $j = 1, 2, \ldots, n$, is associated with a central unit Gaussian standard distribution (Schueller *et al.* 2004). The joint PDF of the random parameters $\{\boldsymbol{\theta}_j : j = 1, 2, \ldots, n\}$ is then

$$\varphi(\boldsymbol{\theta}) = \prod_{j=1}^{n} \phi_j\left(\theta_j\right) \tag{1.3}$$

where $\phi_j(\theta_j) = \left(1/\sqrt{2\pi}\right) e^{-\theta_j^2/2}$, $j = 1, 2, \ldots, n$.

The mapping from the original, physical vector of random variables $\boldsymbol{x} \in \Re^n$ to the standard normal vector $\boldsymbol{\theta} \in \Re^n$ is denoted by $T_{x\theta}(\cdot)$ and its inverse by $T_{\theta x}(\cdot)$, i.e.:

$$\boldsymbol{\theta} = T_{x\theta}\left(\boldsymbol{x}\right) \tag{1.4}$$

$$\boldsymbol{x} = T_{\theta x}\left(\boldsymbol{\theta}\right) \tag{1.5}$$

Transformations 1.4 and 1.5 are in general nonlinear and are obtained by applying Rosenblatt's or Nataf's transformations, respectively (Rosenblatt 1952; Nataf 1962; Huang and Du 2006). They are linear only if the random vector $\boldsymbol{x}$ is jointly Gaussian distributed. By transformation 1.4, also the PF or LSF $g_x(\cdot)$ defined in the physical space (Section 1.1) can be transformed into $g_\theta(\cdot)$ in the standard normal space:

$$g_\theta(\boldsymbol{\theta}) = g_x(\boldsymbol{x}) = g_x(T_{\theta x}(\boldsymbol{\theta})) \tag{1.6}$$

Since in most cases of practical interest the function $g_\theta(\boldsymbol{\theta})$ is not known analytically, it can be evaluated only point-wise. According to Equation 1.6, the evaluation of the system performance function $g_\theta(\cdot)$ at a given point $\boldsymbol{\theta}^k, k = 1, 2, \ldots, N_T$, in the standard normal space requires (1) a transformation into the original space, (2) a complete simulation of the system response, and (3) the computation of the system response from the model. The computational cost of evaluating the failure probability is governed by the number of system performance analyses that have to be carried out (Schueller *et al.* 2004).

1.2.2.2 The Important Direction α for Line Sampling

Three methods have been proposed to estimate the important direction $\boldsymbol{\alpha}$ for LS. In (Koutsourelakis *et al.* 2004), the important unit vector $\boldsymbol{\alpha}$ is taken as pointing in the direction of the "design point" in the standard normal space. According to a geometrical interpretation, the "design point" is defined as the vector point $\boldsymbol{\theta}^*$ on the limit state surface $g_\theta(\boldsymbol{\theta}) = 0$ which is closest to the origin in the standard normal space (Schueller *et al.* 2004). It can be demonstrated that $\boldsymbol{\theta}^*$ is also the point of maximum likelihood (Freudenthal 1956; Schueller and Stix 1987). Then, the unit important vector $\boldsymbol{\alpha}$ can be easily obtained by normalizing $\boldsymbol{\theta}^*$, i.e., $\boldsymbol{\alpha} = \boldsymbol{\theta}^*/\|\boldsymbol{\theta}^*\|_2$, where $\|\cdot\|_2$ denotes the usual Euclidean measure of a vector.

However, the design points, and their neighborhood, do not always represent the most important regions of the failure domain, especially in high-dimensional spaces (Schueller *et al.* 2004). Moreover, the computational cost associated with the calculation of the design point can be quite high, in particular if long-running numerical codes are required to simulate the response of the system to its uncertain input parameters (Schueller *et al.* 2004), as it is frequently the case in structural reliability.

In Pradlwarter *et al.* (2005), the direction of $\boldsymbol{\alpha}$ is taken as the normalized gradient of the performance function in the standard normal space. Since the unit vector $\boldsymbol{\alpha} = \{\alpha_1, \alpha_2, \ldots, \alpha_j, \ldots, \alpha_n\}$ points towards the failure domain F, it can be used to draw information about the relative importance of the random parameters $\{\theta_j : j = 1, 2, \ldots, n\}$ with respect to the failure probability $P(F)$: the more relevant a random variable in determining the failure of the system, the larger the corresponding component of the unit vector $\boldsymbol{\alpha}$ will be (Pradlwarter *et al.* 2005). Such quantitative information is obtained from the gradient of the performance function $g_\theta(\boldsymbol{\theta})$ in the standard normal space, $\nabla g_\theta(\boldsymbol{\theta})$:

$$\nabla g_\theta(\boldsymbol{\theta}) = \left[\frac{\partial g_\theta(\boldsymbol{\theta})}{\partial \theta_1}\ \frac{\partial g_\theta(\boldsymbol{\theta})}{\partial \theta_2} \cdots \frac{\partial g_\theta(\boldsymbol{\theta})}{\partial \theta_j} \cdots \frac{\partial g_\theta(\boldsymbol{\theta})}{\partial \theta_n}\right]^T \tag{1.7}$$

The gradient in Equation 1.7 measures in a unique way the relative importance of a particular random variable with respect to the failure probability $P(F)$: the larger the (absolute) value of a component of Equation 1.7, the greater the "impact" of the corresponding random variable on the performance function $g_\theta(\boldsymbol{\theta})$ in the standard normal space. In other words, given a specified finite variation $\Delta\boldsymbol{\theta}$ in the parameter vector $\boldsymbol{\theta}$, the performance function $g_\theta(\boldsymbol{\theta})$ will change most if this variation is taken in the direction of Equation 1.7. Thus, it is reasonable to identify the LS important direction with the direction of the gradient in Equation 1.7 and compute the corresponding unit vector $\boldsymbol{\alpha}$ as the *normalized* gradient of the performance function $g_\theta(\cdot)$ in the standard normal space, i.e., $\boldsymbol{\alpha} = \nabla g_\theta(\boldsymbol{\theta}) / \|\nabla g_\theta(\boldsymbol{\theta})\|_2$ (Pradlwarter *et al.* 2005).

On the other hand, when the performance function is defined on a high-dimensional space, i.e., when many parameters of the system under analysis are random, the computation of the gradient $\nabla g_\theta(\boldsymbol{\theta})$ in Equation 1.7 becomes a numerically challenging task. Actually, as the function $g_\theta(\boldsymbol{\theta})$ is known only implicitly through the response of a numerical code, for a given vector $\boldsymbol{\theta} = \{\theta_1, \theta_2, \ldots, \theta_j, \ldots, \theta_n\}$ at least n system performance analyses are required to determine accurately the gradient at a given point of the performance function $g_\theta(\cdot)$ by straightforward numerical differentiation, e.g., the secant method (Ahammed and Melchers 2006; Fu 2006).

Finally, the important unit vector $\boldsymbol{\alpha}$ can also be computed as the normalized "center of mass" of the failure domain F of interest (Koutsourelakis *et al.* 2004). A point $\boldsymbol{\theta}^0$ is taken in the failure domain F. This can be done by traditional Monte Carlo sampling or by engineering judgment when possible. Subsequently, $\boldsymbol{\theta}^0$ is used as the initial point of a Markov chain which lies entirely in the failure domain F.

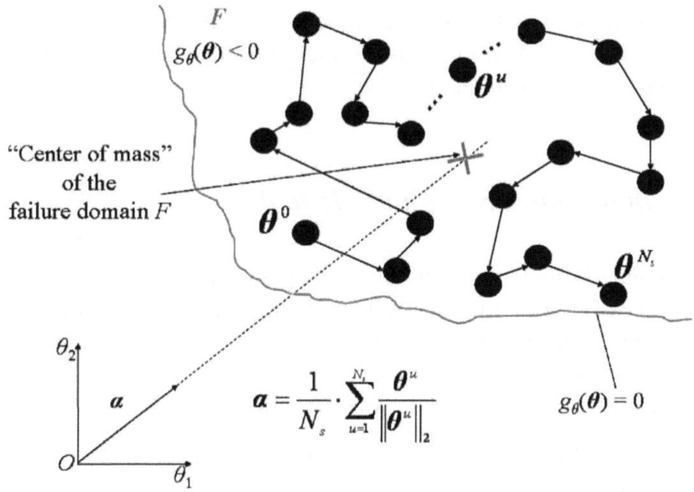

Figure 1.4 Line sampling important unit vector $\boldsymbol{\alpha}$ taken as the normalized "center of mass" of the failure domain F in the standard normal space. The "center of mass" of F is computed as an average of N_s failure points generated by means of a Markov chain starting from an initial failure point $\boldsymbol{\theta}^0$ (Koutsourelakis *et al.* 2004)

For that purpose an MCMC Metropolis–Hastings algorithm is employed to generate a sequence of N_s points $\{\boldsymbol{\theta}^u : u = 1, 2, \ldots, N_s\}$ lying in the failure domain F (Metropolis *et al.* 1956). The unit vectors $\boldsymbol{\theta}^u/\|\boldsymbol{\theta}^u\|_2$, $u = 1, 2, \ldots, N_s$, are then averaged in order to obtain the LS important unit vector as $\boldsymbol{\alpha} = \frac{1}{N_s}\cdot\sum_{u=1}^{N_s}\boldsymbol{\theta}^u/\|\boldsymbol{\theta}^u\|_2$ (Figure 1.4). This direction is not optimal, but it is clear that it provides a good approximation of the important regions of the failure domain (at least as the sample size N_s is large). On the other hand, it should be noticed that the procedure implies N_s additional system analyses by the deterministic model simulating the system, which substantially increase the computational cost associated to the simulation method.

In the implementation of LS for this work, the method based on the normalized "center of mass" of the failure domain F has been employed, because it relies on a "map" approximating the failure domain F under analysis (given by the failure samples generated through a Markov chain) and thus it provides in principle the most realistic and reliable estimate for the LS important direction $\boldsymbol{\alpha}$.

For completeness, a thorough description of the LS algorithm and its practical implementation issues is given in Appendix 2 at the end of the chapter.

1.3 Simulation Methods Considered for Comparison

The performance of SS (Section 1.2.1) and LS (Section 1.2.2) will be compared to that of the IS (Section 1.3.1), DR (Section 1.3.2), and OA (Section 1.3.3) methods; the comparison will be made with respect to the results reported in Gille (1998, 1999) for the two literature case studies considered, of the cracked-plate and thermal-fatigue crack growth models.

1.3.1 The Importance Sampling Method

The concept underlying the IS method is to replace the original PDF $q(\boldsymbol{x})$ with an IS distribution (ISD) $\tilde{q}(\boldsymbol{x})$ arbitrarily chosen by the analyst so as to generate a large number of samples in the "important region" of the sample space, i.e., the failure region F (Au and Beck 2003a; Schueller *et al.* 2004).

The IS algorithm proceeds as follows (Schueller *et al.* 2004):

1. Identify a proper ISD, $\tilde{q}(\cdot)$, in order to increase the probability of occurrence of the failure samples.
2. Express the failure probability $P(F)$ in Equation 1.1 as a function of the ISD $\tilde{q}(\cdot)$:

$$\begin{aligned} P(F) &= \int I_F(\boldsymbol{x})q(\boldsymbol{x})\mathrm{d}\boldsymbol{x} \\ &= \int \left[\frac{I_F(\boldsymbol{x})q(\boldsymbol{x})}{\tilde{q}(\boldsymbol{x})}\right]\tilde{q}(\boldsymbol{x})\mathrm{d}\boldsymbol{x} \\ &= E_{\tilde{q}}\left[\frac{I_F(\boldsymbol{x})q(\boldsymbol{x})}{\tilde{q}(\boldsymbol{x})}\right] \end{aligned} \tag{1.8}$$

3. Draw N_T independent and identically distributed (i.i.d.) samples $\{\boldsymbol{x}^k : k = 1, 2, \ldots, N_T\}$ from the ISD $\tilde{q}(\cdot)$; if a good choice for the ISD $\tilde{q}(\cdot)$ has been made, the samples $\{\boldsymbol{x}^k : k = 1, 2, \ldots, N_T\}$ should be concentrated in the failure region F of interest.
4. Compute an estimate $\hat{P}(F)$ for the failure probability $P(F)$ in Equation 1.1 by resorting to the last expression in Equation 1.8:

$$\hat{P}(F) = \frac{1}{N_T}\sum_{k=1}^{N_T}\frac{I_F\left(\boldsymbol{x}^k\right)q\left(\boldsymbol{x}^k\right)}{\tilde{q}\left(\boldsymbol{x}^k\right)} \tag{1.9}$$

5. The variance $V\left[\hat{P}(F)\right]$ of the estimator $\hat{P}(F)$ in Equation 1.9 is given by

$$\begin{aligned} V\left[\hat{P}(F)\right] &= \frac{1}{N_T}V_{\tilde{q}}\left[\frac{I_F(\boldsymbol{x})q(\boldsymbol{x})}{\tilde{q}(\boldsymbol{x})}\right] \\ &= \frac{1}{N_T}\left(\int \frac{I_F(\boldsymbol{x})q^2(\boldsymbol{x})}{\tilde{q}^2(\boldsymbol{x})}\tilde{q}(\boldsymbol{x})\mathrm{d}\boldsymbol{x} - P(F)^2\right) \end{aligned} \tag{1.10}$$

It is straightforward to verify that the quantity in Equation 1.10 becomes zero when

$$\tilde{q}(\boldsymbol{x}) = \tilde{q}_{\text{opt}}(\boldsymbol{x}) = \frac{I_F(\boldsymbol{x})q(\boldsymbol{x})}{P(F)} \tag{1.11}$$

This represents the optimal choice for the importance sampling density which is practically unfeasible since it requires the *a priori* knowledge of $P(F)$. Several techniques have been developed in order to approximate the optimal sampling density in Equation 1.11 or to at least find one giving small variance of the estimator in Equation 1.9. Recent examples include the use of engineering judgment (Pagani *et al.* 2005), design points (Schueller *et al.* 2004) and kernel density estimators (Au and Beck 2003a).

1.3.2 The Dimensionality Reduction Method

The objective of the DR method is to reduce the variance associated to the failure probability estimates by exploiting the property of conditional expectation (Gille 1998, 1999). In extreme synthesis, the failure event $g_x(\boldsymbol{x}) \leqslant 0$ is re-expressed in such a way as to highlight one of the n uncertain input variables of $\boldsymbol{x}$ (say, x_j); then, the failure probability estimate is computed as the expected value of the CDF of x_j conditional on the remaining $(n-1)$ input variables. By so doing, the zero values contained in the standard MCS estimator (i.e., $I_F(\boldsymbol{x}) = 0$, if $\boldsymbol{x} \in F$) are removed: this allows one to (1) reach any level of probability (even very small) and (2) reduce the variance of the failure probability estimator (Gille 1998, 1999).

The DR algorithm proceeds as follows (Gille 1998, 1999):

1. Write the failure event $g_x(\boldsymbol{x}) = g_x\left(x_1, x_2, \ldots, x_j, \ldots, x_n\right) \leqslant 0$ in such a way as to highlight one of the n uncertain input variables (e.g., x_j):

$$x_j \leqslant h_x\left(\boldsymbol{x}_{-j}\right), \quad j = 1, 2, \ldots, n \tag{1.12}$$

   where $h_x(\cdot)$ is a function defined on $\Re^{n-1}$ which takes values on the set of all (measurable) subsets of $\Re$ and $\boldsymbol{x}_{-j}$ is a vector containing all the uncertain input variables except x_j, i.e., $\boldsymbol{x}_{-j} = \left(x_1, x_2, \ldots, x_{j-1}, x_{j+1}, \ldots, x_n\right)$.
2. Write the failure probability $P(F)$ as follows:

$$\begin{aligned} P(F) &= P\left[g_x(\boldsymbol{x}) \leqslant 0\right] \\ &= P\left[x_j \leqslant h_x(\boldsymbol{x}_{-j})\right] \\ &= E_{\boldsymbol{x}_{-j}}\left\{F_{x_j|\boldsymbol{x}_{-j}}\left[h_x\left(\boldsymbol{x}_{-j}\right)\right]\right\} \end{aligned} \tag{1.13}$$

   where $F_{x_j|\boldsymbol{x}_{-j}}(\cdot)$ is the CDF of x_j conditional on $\boldsymbol{x}_{-j}$, i.e., $\boldsymbol{x}_{-j} = (x_1, x_2, \ldots, x_{j-1}, x_{j+1}, \ldots, x_n)$.

3. Draw N_T samples $\left\{\boldsymbol{x}_{-j}^k : k = 1, 2, \ldots, N_T\right\}$, where $\boldsymbol{x}_{-j}^k = \left(x_1^k, x_2^k, \ldots, x_{j-1}^k, x_{j+1}^k, \ldots, x_n^k\right)$, from the $(n-1)$-dimensional marginal PDF $q_{\mathrm{m}}\left(\boldsymbol{x}_{-j}\right)$, i.e., $q_{\mathrm{m}}\left(\boldsymbol{x}_{-j}\right) = q_{\mathrm{m}}\left(x_1, x_2, \ldots, x_{j-1}, x_{j+1}, \ldots, x_n\right) = \int_{x_j} q\left(x_1, x_2, \ldots, x_j, \ldots, x_n\right) \mathrm{d}x_j$.
4. Using the last expression in Equation 1.13, compute an unbiased and consistent estimate $\hat{P}(F)$ for the failure probability $P(F)$ as follows:

$$\hat{P}(F) = \frac{1}{N_T} \sum_{k=1}^{N_T} F_{x_j|\boldsymbol{x}_{-j}} \left[h_x\left(\boldsymbol{x}_{-j}^k\right)\right] \tag{1.14}$$

It is worth noting that in Equation 1.14 the failure probability estimate is computed as the expected value of the CDF $F_{x_j|\boldsymbol{x}_{-j}}(\cdot)$ of x_j conditional on the remaining $(n-1)$ input variables. Since this quantity takes values *between* 0 and 1, the zero values contained in the standard MCS estimator (i.e., $I_F(\boldsymbol{x}) = 0$, if $\boldsymbol{x} \in F$) are removed: this allows one to (1) reach any level of failure probability (even very small) and (2) reduce the variance of the failure probability estimator. However, such method can not always be applied: first, the performance function $g_x(\cdot)$ must be known analytically; second, it must have the property that one of the uncertain input variables can be separated from the others to allow rewriting the failure condition $g_x(\boldsymbol{x}) \leqslant 0$ in the form of Equation 1.12 (Gille 1998, 1999).

Finally, notice that DR can be considered a very special case of LS (Section 1.2.2) where the performance function $g_x(\cdot)$ is analytically known and the important direction $\boldsymbol{\alpha}$ coincides with the "direction" of the variable x_j, i.e., $\boldsymbol{\alpha} = \left(0, 0, \ldots, x_j, \ldots, 0, 0\right)$.

1.3.3 The Orthogonal Axis Method

The OA method combines the first-order reliability method (FORM) approximation (Der Kiureghian 2000) and MCS in a sort of importance sampling around the "design point" of the problem (see Section 1.2.2.2).

The OA algorithm proceeds as follows (Gille 1998, 1999):

1. Transform $\boldsymbol{x} = \left\{x_1, x_2, \ldots, x_j, \ldots, x_n\right\} \in \Re^n$, i.e., the vector of uncertain parameters defined in the original physical space $\boldsymbol{x} \in \Re^n$, into the vector $\boldsymbol{\theta} \in \Re^n$, where each element of the vector $\boldsymbol{\theta}_j$, $j = 1, 2, \ldots, n$, is associated with a central unit Gaussian standard distribution (Schueller *et al.* 2004) (see Section 1.2.2.2). Thus, the joint probability density function of $\boldsymbol{\theta}$ can simply be written as

$$\varphi_n(\boldsymbol{\theta}) = \prod_{j=1}^{n} \phi\left(\boldsymbol{\theta}_j\right) \tag{1.15}$$

where $\phi\left(\boldsymbol{\theta}_j\right) = \left(1/\sqrt{2\pi}\right) e^{-\left(\boldsymbol{\theta}_j^2/2\right)}$, $j = 1, 2, \ldots, n$.

2. Find the "design point" $\boldsymbol{\theta}^*$ of the problem (see Section 1.2.2.2).
3. Rotate the coordinate system (i.e., by means of a proper rotation matrix $\boldsymbol{R}$) so that the new coordinate θ_n is in the direction of the axis defined by the design point $\boldsymbol{\theta}^*$.
4. Define a new failure function $g_{\text{axis}}(\boldsymbol{\theta})$ as

$$g_{\text{axis}}(\boldsymbol{\theta}) = g(\boldsymbol{R}\boldsymbol{\theta}) \tag{1.16}$$

5. Writing $\boldsymbol{\theta}$ as $\left(\tilde{\boldsymbol{\theta}}, \theta_n\right)$, where $\tilde{\boldsymbol{\theta}} = (\theta_1, \theta_2, \ldots, \theta_{n-1})$, express the failure probability $P(F)$ as follows:

$$\begin{aligned} P(F) &= P\left[g_{\text{axis}}\left(\tilde{\boldsymbol{\theta}}, \theta_n\right) \leqslant 0\right] \\ &= \int P\left[g_{\text{axis}}\left(\tilde{\boldsymbol{\theta}}, \theta_n\right) \leqslant 0 | \tilde{\boldsymbol{\theta}}\right] \varphi_{n-1}\left(\tilde{\boldsymbol{\theta}}\right) \mathrm{d}\tilde{\boldsymbol{\theta}} \\ &= E_{\tilde{\boldsymbol{\theta}}}\left\{P\left[g_{\text{axis}}\left(\tilde{\boldsymbol{\theta}}, \theta_n\right) \leqslant 0\right]\right\} \end{aligned} \tag{1.17}$$

6. Generate N_T i.i.d. $(n-1)$-dimensional samples $\left\{\tilde{\boldsymbol{\theta}}^k : k = 1, 2, \ldots, N_T\right\}$, where $\tilde{\boldsymbol{\theta}}^k = (\theta_1^k, \theta_2^k, \ldots, \theta_{n-1}^k)$.
7. Compute an estimate $\hat{P}(F)$ for the failure probability $P(F)$ as follows:

$$\hat{P}(F) = \frac{1}{N_T} \sum_{k=1}^{N_T} P\left[g_{\text{axis}}\left(\tilde{\boldsymbol{\theta}}^k, \theta_n\right) \leqslant 0\right] \tag{1.18}$$

The terms $P\left[g_{\text{axis}}\left(\tilde{\boldsymbol{\theta}}^k, \theta_n\right) \leqslant 0\right]$, $k = 1, 2, \ldots, N_T$, are evaluated with an iterative algorithm which searches for the roots of the equation $g_{\text{axis}}\left(\tilde{\boldsymbol{\theta}}^k, \theta_n\right) = 0$ (Gille 1998, 1999).

It is worth noting that the idea underlying the OA method is essentially the same as that of LS (Section 1.2.2). However, in OA the "important direction" is forced to coincide with that of the design point of the problem; moreover, OA employs a rotation of the coordinate system which can be difficult to define in very high-dimensional problems.

1.4 Application 1: the Cracked-plate Model

The cracked-plate model is a classical example in fracture mechanics and its relative simplicity allows a detailed and complete study of different simulation techniques. A thorough description of this model can be found in Ardillon and Venturini (1995).

Table 1.1 Names, descriptions, and units of measure of the variables of the cracked-plate model

Name	Description	Unit of measure
K_c	Critical stress intensity factor	$\text{MPa}\sqrt{\text{m}}$
a	Initial length of the defect	m
F	Shape factor of the defect	–
s_∞	Uniform normal loading (stress) to which the plate is subject	MPa

1.4.1 The Mechanical Model

A metal plate of infinite length with a defect of initial length equal to a [m] is considered. The plate is supposed to be subject to a uniform normal loading (i.e., stress) s_∞ [MPa]. The intensity factor K [$\text{MPa}\sqrt{\text{m}}$], determined by the uniform loading in the neighborhood of the defect is defined as follows:

$$K = F s_\infty \sqrt{\pi a} \tag{1.19}$$

where F is the shape factor of the defect. The plate is supposed to break (i.e., fail) when the intensity factor K in Equation 1.19 becomes greater than or equal to a critical value K_c, i.e.:

$$K = F s_\infty \sqrt{\pi a} \geqslant K_c \tag{1.20}$$

The variables of the mechanical model are summarized in Table 1.1.

1.4.2 The Structural Reliability Model

From the point of view of a structural reliability analysis, the cracked-plate mechanical model of Section 1.4.1 is analyzed within a probabilistic framework in which the variables K_c, a, F, and s_∞ are uncertain (for simplicity of illustration with respect to the notation of the previous sections, the four variables are hereafter named x_1, x_2, x_3, and x_4, respectively).

Referring to Equation 1.20, the performance function $g_x(\boldsymbol{x})$ of the system is

$$g_x(\boldsymbol{x}) = g_x(x_1, x_2, x_3, x_4) = x_1 - x_3 x_4 \sqrt{\pi x_2} \tag{1.21}$$

The failure region F is then expressed as

$$F = \{\boldsymbol{x} : g_x(\boldsymbol{x}) \leqslant 0\} = \left\{(x_1, x_2, x_3, x_4) : x_1 \leqslant x_3 x_4 \sqrt{\pi x_2}\right\} \tag{1.22}$$

Finally, the probability of system failure $P(F)$ is written as follows:

$$P(F) = P(\boldsymbol{x} \in F) = P[g_x(\boldsymbol{x}) \leqslant 0] = P\left(x_1 \leqslant x_3 x_4 \sqrt{\pi x_2}\right) \tag{1.23}$$

Table 1.2 Probability distributions and parameters (i.e., means and standard deviations) of the uncertain variables x_1, x_2, x_3, and x_4 of the cracked-plate model of Section 1.4.2 for the four case studies considered; the last row reports the values of the corresponding exact (i.e., analytically computed) failure probabilities, $P(F)$ (Gille 1998, 1999). N = Normal distribution; LG = Lognormal distribution

	Case 0	Case 1	Case 2	Case 3
$x_1(k)$	$N(149.3, 22.2)$	$N(149.3, 22.2)$	$N(160, 18)$	$LG(149.3, 22.2)$
$x_2(a)$	$N(5\times10^{-3}, 10^{-3})$	$N(5\times10^{-3}, 10^{-3})$	$N(5\times10^{-3}, 10^{-3})$	$LG(5 \times 10^{-3}, 10^{-3})$
$x_3(F)$	$N(0.99, 0.01)$	$N(0.99, 0.01)$	$N(0.99, 0.01)$	$LG(0.99, 0.01)$
$x_4(s_\infty)$	$N(600, 60)$	$N(300, 30)$	$N(500, 45)$	$LG(600, 60)$
$P(F)$	1.165×10^{-3}	4.500×10^{-7}	4.400×10^{-7}	3.067×10^{-4}

1.4.3 Case Studies

Four case studies, namely case 0 (reference case), 1, 2, and 3, are considered with respect to the structural reliability model of Section 1.4.2. Each case study is characterized by different PDFs for the uncertain variables x_1, x_2, x_3, and x_4 and by different failure probabilities $P(F)$: these features are summarized in Table 1.2. Notice that in cases 0, 1 and 2 the random variables are independent and normally distributed, whereas in case 3 they are independent and lognormally distributed. Moreover, it is worth noting that the exact (i.e., analytically computed) failure probabilities $P(F)$ approximately range from 10^{-3} to 10^{-7}, allowing a deep exploration of the capabilities of the simulation algorithms considered and a meaningful comparison between them (Gille 1998, 1999).

1.4.4 Results

In this section, the results of the application of SS and LS for the reliability analysis of the cracked-plate model of Section 1.4.1 are illustrated with reference to case studies 0, 1, 2, and 3 described in Section 1.4.3.

For fair comparison, all methods have been run with a total of $N_T = 50{,}000$ samples in all four cases. The efficiency of the simulation methods under analysis is evaluated in terms of four quantities: the failure probability estimate $\hat{P}(F)$, the sample standard deviation $\hat{\sigma}$ of the failure probability estimate $\hat{P}(F)$, the coefficient of variation (c.o.v.) δ of $\hat{P}(F)$ (defined as the ratio of the sample standard deviation $\hat{\sigma}$ to the estimate $\hat{P}(F)$) and the figure of merit (FOM) of the method (defined as $1/\left(\hat{\sigma}^2 t_{\text{comp}}\right)$, where t_{comp} is the computational time required by the simulation method). The closer the estimate $\hat{P}(F)$ is to the exact (i.e., analytically computed) failure probability $P(F)$, the more accurate the simulation method. The sample standard deviation $\hat{\sigma}$ and the c.o.v. δ of $\hat{P}(F)$ are used to quantify the variability of the failure probability estimator; in particular, the lower the values of $\hat{\sigma}$ and δ, the lower the variability of the corresponding failure probability estimator and thus the higher the efficiency of the simulation method adopted. Finally, the FOM is introduced to take into account the computational time required by the method. The value of the FOM increases as the sample variance $\hat{\sigma}^2$ of the failure probability estimate $\hat{P}(F)$ and the computational time t_{comp} required by the method decrease; thus, in this case the higher the value of the index, the higher the efficiency of the method (Gille 1998, 1999).

The different simulation methods are also compared with respect to two direct performance indicators relative to standard MCS. First, the ratio of the sample standard deviation $\hat{\sigma}_{\text{MC}}$ obtained by standard MCS to that obtained by the simulation method under analysis $\hat{\sigma}_{\text{meth}}$ is computed. This ratio *only* quantifies the improvement in the *precision* of the estimate achieved by using a given simulation method instead of standard MCS. Then, the ratio of the FOM of the simulation method in object, namely FOM_{meth}, to that of standard MCS, namely FOM_{MC}, is considered to quantify the *overall* improvement in *efficiency* achieved by a given simulation method with respect to standard MCS, since it also takes into account the computational time required. Obviously, the higher the values of these two indices for a given method, the higher the efficiency of that method (Gille 1998, 1999).

Table 1.3 reports the values of $\hat{P}(F)$, $\hat{\sigma}$, δ, FOM, $\hat{\sigma}_{\text{MC}}/\hat{\sigma}_{\text{meth}}$, and $\text{FOM}_{\text{meth}}/\text{FOM}_{\text{MC}}$ obtained by standard MCS, SS, and LS in cases 0, 1, 2, and 3 (Section 1.4.3); the *actual* number N_{sys} of system response analyses (i.e., model evaluations) is also reported. Notice that for both SS and LS the actual number N_{sys} of system analyses does not coincide with the total number N_T of random samples drawn (i.e., $N_T = 50{,}000$). In particular, in the SS method, the presence of *repeated* conditional samples in each Markov chain (used to gradually populate the intermediate event regions) allows a *reduction* in the number of model evaluations required: actually, one evaluation is enough for all identical samples (see Appendix 1). In the LS method, instead, the actual number N_{sys} of system analyses is given by $N_{\text{sys}} = N_s + 2 \cdot N_T$: in particular, $N_s = 2000$ analyses are performed to generate the Markov chain used to compute the important unit vector $\boldsymbol{\alpha}$ as the normalized "center of mass" of the failure domain F (Section 1.2.2.2); the $2 \cdot N_T$ analyses are carried out to compute the N_T conditional one-dimensional probability estimates $\left\{\hat{P}^k(F): k = 1, 2, \ldots, N_T\right\}$ by linear interpolation (Equation 1.39 in Appendix 2).

It can be seen that SS performs consistently better than standard MCS and its performance significantly grows as the failure probability to be estimated decreases: for instance, in case 0 (reference), where $P(F) \sim 10^{-3}$, the FOM of SS, namely FOM_{SS}, is only four times larger than that of standard MCS, namely FOM_{MC}; whereas in case 1, where $P(F) \sim 10^{-7}$, the ratio $\text{FOM}_{\text{SS}}/\text{FOM}_{\text{MC}}$ is about 557. On the other hand, LS outperforms SS with respect to both $\hat{\sigma}_{\text{MC}}/\hat{\sigma}_{\text{meth}}$ and $\text{FOM}_{\text{meth}}/\text{FOM}_{\text{MC}}$ in all the cases considered. For instance, in case 2, where the failure probability $P(F)$ to be estimated is very small, i.e., $P(F) = 4.4 \times 10^{-7}$, the ratio $\hat{\sigma}_{\text{MC}}/\hat{\sigma}_{\text{LS}}$ is 155 times larger than the ratio $\hat{\sigma}_{\text{MC}}/\hat{\sigma}_{\text{SS}}$, whereas the ratio $\text{FOM}_{\text{LS}}/\text{FOM}_{\text{MC}}$ is 11,750 times larger than the ratio $\text{FOM}_{\text{SS}}/\text{FOM}_{\text{MC}}$. Notice that for the LS method, even though the determination of the sampling important di-

Table 1.3 Results of the application of standard MCS, SS, and LS to the reliability analysis of cases 0 (reference), 1, 2, and 3 of the cracked-plate model of Section 1.4.2; the values of the performance indicators used to compare the effectiveness of the methods (i.e., $\hat{\sigma}_{MC}/\hat{\sigma}_{meth}$ and FOM_{meth}/FOM_{MC}) are highlighted in bold

Case 0 (Reference)							
	$\hat{P}(F)$	$\hat{\sigma}$	c.o.v., δ	N_{sys}	FOM	$\hat{\sigma}_{MC}/\hat{\sigma}_{meth}$	FOM_{meth}/FOM_{MC}
Standard MCS	1.120×10^{-3}	1.496×10^{-4}	1.336×10^{-1}	50 000	893.65	**1**	**1**
SS	1.274×10^{-3}	7.136×10^{-5}	5.597×10^{-2}	49 929	3936.67	**2.10**	**4.41**
LS	1.169×10^{-3}	5.142×10^{-7}	4.399×10^{-4}	102 000	3.782×10^{7}	**290.92**	**42 318**
Case 1							
	$\hat{P}(F)$	$\hat{\sigma}$	c.o.v., δ	N_{sys}	FOM	$\hat{\sigma}_{MC}/\hat{\sigma}_{meth}$	FOM_{meth}/FOM_{MC}
Standard MCS	4.500×10^{-7}	3.000×10^{-6}	6.667	50 000	2.222×10^{6}	**1**	**1**
SS	4.624×10^{-7}	7.295×10^{-8}	1.578×10^{-1}	49 937	3.762×10^{9}	**41.12**	**1.7×10^{3}**
LS	4.493×10^{-7}	1.791×10^{-10}	3.986×10^{-4}	102 000	3.117×10^{14}	**16 750**	**1.4×10^{8}**
Case 2							
	$\hat{P}(F)$	$\hat{\sigma}$	c.o.v., δ	N_{sys}	FOM	$\hat{\sigma}_{MC}/\hat{\sigma}_{meth}$	FOM_{meth}/FOM_{MC}
Standard MCS	4.400×10^{-7}	3.000×10^{-6}	6.667	50 000	2.222×10^{6}	**1**	**1**
SS	4.679×10^{-7}	6.890×10^{-8}	1.473×10^{-1}	49 888	4.222×10^{9}	**43.54**	**1.9×10^{3}**
LS	4.381×10^{-7}	4.447×10^{-10}	1.015×10^{-3}	102 000	4.959×10^{13}	**6746.7**	**2.2×10^{7}**
Case 3							
	$\hat{P}(F)$	$\hat{\sigma}$	c.o.v., δ	N_{sys}	FOM	$\hat{\sigma}_{MC}/\hat{\sigma}_{meth}$	FOM_{meth}/FOM_{MC}
Standard MCS	3.000×10^{-4}	7.745×10^{-5}	2.582×10^{-1}	50 000	3.334×10^{3}	**1**	**1**
SS	3.183×10^{-4}	2.450×10^{-5}	7.697×10^{-2}	49 907	3.339×10^{4}	**3.16**	**10.01**
LS	3.068×10^{-4}	1.817×10^{-7}	5.923×10^{-4}	102 000	3.028×10^{8}	**426.16**	**9.1×10^{4}**

rection $\boldsymbol{\alpha}$ (Section 1.2.2.2) and the calculations of the conditional one-dimensional failure probability estimates $\left\{\hat{P}^k(F): k = 1, 2, \ldots, N_T\right\}$ (Equation 1.39 in Appendix 2) require much more than N_T system analyses by the model, this is significantly overweighed by the accelerated convergence rate that can be attained by the LS method with respect to SS.

1.4.4.1 Comparison with Other Stochastic Simulation Methods

The results obtained by SS and LS are compared to those obtained by the IS, DR, and OA methods and by a combination of IS and DR (Section 1.3) (Gille 1998, 1999). For DR, the variable x_1 is explicit.

The values of the performance indicators $\hat{\sigma}_{MC}/\hat{\sigma}_{meth}$ and FOM_{meth}/FOM_{MC} obtained by the four methods in cases 0, 1, 2, and 3 are summarized in Table 1.4.

Table 1.4 Values of the performance indicators $\hat{\sigma}_{MC}/\hat{\sigma}_{meth}$ and FOM_{meth}/FOM_{MC} obtained by IS, DR (with variable x_1 specified), OA, and IS + DR when applied for the reliability analysis of cases 0 (reference), 1, 2, and 3 of the cracked-plate model of Section 1.4.2 (Gille 1998, 1999)

Case 0 (reference)	$\hat{\sigma}_{MC}/\hat{\sigma}_{meth}$	FOM_{meth}/FOM_{MC}
IS	17	100
DR (variable $\boldsymbol{x}_1$)	14	14
OA	340	7.7×10^3
IS + DR	194	2.1×10^4
Case 1	$\hat{\sigma}_{MC}/\hat{\sigma}_{meth}$	FOM_{meth}/FOM_{MC}
IS	630	376
DR (variable $\boldsymbol{x}_1$)	856	7.3×10^5
OA	17 255	2.0×10^7
IS + DR	8 300	1.3×10^8
Case 2	$\hat{\sigma}_{MC}/\hat{\sigma}_{meth}$	FOM_{meth}/FOM_{MC}
IS	643	1.5×10^5
DR (variable $\boldsymbol{x}_1$)	242	242
OA	10 852	7.9×10^6
IS + DR	8077	3.6×10^7
Case 3	$\hat{\sigma}_{MC}/\hat{\sigma}_{meth}$	FOM_{meth}/FOM_{MC}
IS	29	289
DR (variable $\boldsymbol{x}_1$)	7	7
OA	4852	4.9×10^5
IS + DR	150	1.2×10^4

Comparing Table 1.3 and Table 1.4, it can be seen that LS performs significantly better than IS and DR in all the case studies considered: in particular, in cases 1 and 2 the values of the performance indicators $\hat{\sigma}_{MC}/\hat{\sigma}_{LS}$ (16,750 and 6746.7) and FOM_{LS}/FOM_{MC} (1.4×10^8 and 2.2×10^7) are more than one order of magnitude larger than those reported in Gille (1998, 1999) for IS (630, 376, and 643, 1.5×10^5 for cases 1 and 2, respectively) and DR (856, 7.3×10^5 and 242, 242 for cases 1 and 2, respectively). Moreover, it is worth noting that in the reference studies by Gille (1998, 1999) a significant number of simulations has been run to properly tune the parameters of the ISDs for the IS method (in particular, 8, 6, 6, and 8 simulations have been performed for cases 0, 1, 2, and 3, respectively), with a significant increase in the associated computational effort.

LS is found to perform slightly worse than OA in all the case studies considered: actually, the values of both $\hat{\sigma}_{MC}/\hat{\sigma}_{LS}$ and FOM_{LS}/FOM_{MC} are slightly lower than those reported in Gille (1998, 1999) for OA. However, it should be considered that in these studies the OA method has been applied to a simplified version of the problem described in Sections 1.4.1 and 1.4.2; actually, only three uncertain variables (i.e., x_1, x_2, and x_4) have been considered by keeping variable x_3 (i.e., F) fixed to its mean value (i.e., 0.99): this certainly reduces the variability of the model output and contributes to the reduction of the variability of the associated failure probability estimator.

Further, LS performs consistently better than the combination of IS and DR in the task of estimating failure probabilities around $10^{-3} \div 10^{-4}$ (for instance, in case 0 $\hat{\sigma}_{MC}/\hat{\sigma}_{IS+DR} = 194$ and $\hat{\sigma}_{MC}/\hat{\sigma}_{LS} = 290$, whereas in case 4 $\hat{\sigma}_{MC}/\hat{\sigma}_{IS+DR} = 150$ and $\hat{\sigma}_{MC}/\hat{\sigma}_{LS} = 426$). In addition, LS performs comparably to the combination of IS and DR in the estimation of failure probabilities around 10^{-7}: actually, in case 1 $\hat{\sigma}_{MC}/\hat{\sigma}_{IS+DR} = 8300$ and $\hat{\sigma}_{MC}/\hat{\sigma}_{LS} = 16{,}750$, whereas in case 2 $\hat{\sigma}_{MC}/\hat{\sigma}_{IS+DR} = 8077$ and $\hat{\sigma}_{MC}/\hat{\sigma}_{LS} = 6746$. However, it has to be noticed again that in the reference studies by Gille (1998, 1999) a significant number of simulations has been run to properly tune the parameters of the ISDs for the IS method (in particular, 4, 8, 8, and 10 simulations have been performed in cases 0, 1, 2, and 3, respectively).

Finally, it is worth noting that in these cases SS performs worse than the other methods proposed.

1.5 Application 2: Thermal-fatigue Crack Growth Model

The thermal-fatigue crack growth model considered in this study is based on the deterministic Paris–Erdogan model which describes the propagation of a manufacturing defect due to thermal fatigue (Paris 1961).

1.5.1 The Mechanical Model

The evolution of the size a of a defect satisfies the following equation:

$$\frac{\mathrm{d}a}{\mathrm{d}N_\mathrm{c}} = C \cdot (f(R) \cdot \Delta K)^m \tag{1.24}$$

where N_c is the number of fatigue cycles, C and m are parameters depending on the properties of the material, $f(R)$ is a correction factor which is a function of the material resistance R, and ΔK is the variation of the intensity factor, defined as

$$\Delta K = \Delta s \cdot Y(a) \cdot \sqrt{\pi a} \tag{1.25}$$

In Equation 1.25, Δs is the variation of the uniform loading (stress) applied to the system and $Y(a)$ is the shape factor of the defect. Let $S_i = \Delta s_i$ be the variation of the uniform normal stress at cycle $i = 1, 2, \ldots, N_\mathrm{c}$. The integration of Equation 1.24 gives

$$\int_{a_0}^{a_{N_\mathrm{c}}} \frac{\mathrm{d}a}{\left(Y(a)\sqrt{\pi a}\right)^m} = C \cdot \sum_{i=1}^{N_\mathrm{c}} (f(R) \cdot S_i)^m \tag{1.26}$$

where a_0 and a_{N_c} are the initial and final size of the defect, respectively. In Equation 1.26 the following approximation can be adopted:

$$\sum_{i=1}^{N_\mathrm{c}} (f(R) \cdot S_i)^m \approx (T - T_0) \cdot N_\mathrm{c} \cdot (f(R) \cdot S)^m \tag{1.27}$$

where T and T_0 are the initial and final times of the thermal-fatigue treatment (of N_c cycles).

The system is considered failed when the size a_{N_c} of the defect at the end of the N_c cycles exceeds a critical dimension a_c, i.e.,

$$a_\mathrm{c} - a_{N_\mathrm{c}} \leqslant 0 \tag{1.28}$$

which in the integral form 1.26 reads

$$\psi\,(a_\mathrm{c}) - \psi\,(a_{N_\mathrm{c}}) \leqslant 0 \tag{1.29}$$

where

$$\psi(a) = \int_{a_0}^{a} \frac{\mathrm{d}a'}{\left(Y(a') \cdot \sqrt{\pi a'}\right)^m} \tag{1.30}$$

Table 1.5 Names, descriptions, and units of measure of the variables of the thermal-fatigue crack growth model

Name	Description	Unit of measure
a_0	Initial size of the defect	[m]
a_c	Critical size of the defect	[m]
T_0	Initial time	[years]
T	Final time	[years]
C	Parameter of the material	–
m	Parameter of the material	–
$f(R)$	Correction factor	–
N_c	Number of cycles per year	–
S	Stress per cycle	[MPa]

Using Equation 1.27, a safety margin $M(T)$ can then be defined as follows:

$$M(T) = \int_{a_0}^{a_c} \frac{\mathrm{d}a}{\left(Y(a) \cdot \sqrt{\pi a}\right)^m} - C \cdot (T - T_0) \cdot N_c \cdot (f(R) \cdot S)^m \qquad (1.31)$$

The failure criterion can then be expressed in terms of the safety margin 1.31:

$$M(T) \leqslant 0 \qquad (1.32)$$

The variables of the model are summarized in Table 1.5.

1.5.2 The Structural Reliability Model

For the purpose of a structural reliability analysis, the thermal-fatigue crack growth model is framed within a probabilistic representation of the uncertainties affecting the nine variables a_0, a_c, T_0, T, C, m, $f(R)$, N_c, and S (hereafter named x_1, x_2, x_3, x_4, x_5, x_6, x_7, x_8, and x_9, respectively).

From Equation 1.32, the probability of system failure $P(F)$ is written as

$$\begin{aligned} P(F) &= P\left[M(T) \leqslant 0\right] \\ &= P\left[\int_{a_0}^{a_c} \frac{\mathrm{d}a}{\left(Y(a) \cdot \sqrt{\pi a}\right)^m} - C \cdot (T - T_0) \cdot N_c \cdot (f(R) \cdot S)^m \leqslant 0\right] \end{aligned} \qquad (1.33)$$

or

$$P(F) = P\left[M(T) \leqslant 0\right]$$
$$= P\left[\int_{x_1}^{x_2} \frac{da}{\left(Y(a)\cdot\sqrt{\pi a}\right)^{x_6}} - x_5\cdot(x_4 - x_3)\cdot x_8\cdot(x_7\cdot x_9)^{x_6} \leqslant 0\right] \quad (1.34)$$

It is worth noting the highly nonlinear nature of Equations 1.33 and 1.34, which increases the complexity of the problem.

1.5.3 Case Studies

Two different case studies, namely case 1 and case 2, are built with reference to the structural reliability model of Section 1.5.2. The characteristics of the PDFs of the uncertain variables of Table 1.5 are summarized in Table 1.6; the values of the exact (i.e., analytically computed) failure probabilities, $P(F)$, for both cases 1 and 2 are also reported in the last row of Table 1.6.

1.5.4 Results

In this section, the results of the application of SS and LS for the reliability analysis of the thermal-fatigue crack growth model of Sections 1.5.1 and 1.5.2 are illustrated with reference to cases 1 and 2 (Table 1.6 of Section 1.5.3).

Table 1.6 Probability distributions and parameters (i.e., means and standard deviations) of the uncertain variables x_1, x_2, ..., x_9 of the thermal-fatigue crack growth model of Section 1.5.2 for cases 1 and 2; the last row reports the values of the corresponding exact (i.e., analytically computed) failure probabilities, $P(F)$ (Gille 1998, 1999). Exp = exponential distribution; LG = Lognormal distribution; N = Normal distribution

	Case 1	Case 2
$x_1(a_0)$	$\text{Exp}(0.61\times10^{-3})$	$\text{Exp}(0.81\times10^{-3})$
$x_2(a_c)$	$N(21.4\times10^{-3}, 0.214\times10^{-3})$	$N(21.4\times10^{-3}, 0.214\times10^{-3})$
$x_3(T_0)$	0	0
$x_4(T)$	40	40
$x_5(C)$	$LG(6.5\times10^{-13}, 5.75\times10^{-13})$	$LG(1.00\times10^{-12}, 5.75\times10^{-13})$
$x_6(m)$	3.4	3.4
$x_7(f(R))$	2	2
$x_8(N_c)$	$N(20, 2)$	$N(20, 2)$
$x_9(S)$	$LG(300, 30)$	$LG(200, 20)$
$P(F)$	3.3380×10^{-4}	1.780×10^{-5}

Table 1.7 Results of the application of standard MCS, SS, and LS to the reliability analysis of cases 1 and 2 of the thermal-fatigue crack growth model of Section 1.5.2; the values of the performance indicators used to compare the effectiveness of the methods (i.e., $\hat{\sigma}_{MC}/\hat{\sigma}_{meth}$ and FOM_{meth}/FOM_{MC}) are highlighted in bold

Case 1	$\hat{P}(F)$	$\hat{\sigma}$	c.o.v., δ	N_{sys}	FOM	$\hat{\sigma}_{MC}/\hat{\sigma}_{meth}$	FOM_{meth}/FOM_{MC}
Standard MCS	2.500×10^{-4}	7.905×10^{-5}	3.162×10^{-1}	40 000	4.001×10^{3}	**1**	**1**
SS	3.006×10^{-4}	3.214×10^{-5}	1.069×10^{-1}	40 019	2.419×10^{4}	**2.46**	**6.05**
LS	3.768×10^{-4}	4.610×10^{-7}	1.223×10^{-3}	82 000	5.737×10^{7}	**171.46**	**1.434×10^{4}**

Case 2	$\hat{P}(F)$	$\hat{\sigma}$	c.o.v., δ	N_{sys}	FOM	$\hat{\sigma}_{MC}/\hat{\sigma}_{meth}$	FOM_{meth}/FOM_{MC}
Standard MCS	1.780×10^{-5}	2.269×10^{-5}	1.102	40 000	4.860×10^{4}	**1**	**1**
SS	1.130×10^{-5}	1.653×10^{-6}	1.462×10^{-1}	39 183	9.341×10^{6}	**13.73**	**192.36**
LS	1.810×10^{-5}	2.945×10^{-8}	1.627×10^{-3}	81 999	1.188×10^{13}	**770.02**	**2.892×10^{5}**

Again for fair comparison all simulation methods have been run with the same total number of samples ($N_T = 40{,}000$) in both cases 1 and 2. The efficiency of the methods has been evaluated in terms of the same indices and performance indicators defined in Section 1.4.4.

Table 1.7 reports the values of $\hat{P}(F)$, $\hat{\sigma}$, δ, FOM, $\hat{\sigma}_{MC}/\hat{\sigma}_{meth}$, and FOM_{meth}/FOM_{MC} obtained by standard MCS, SS, and LS in cases 1 and 2 of Section 1.5.3; the *actual* number N_{sys} of system response analyses (i.e., model evaluations) is also reported.

Also in this application, the LS methodology is found to outperform SS in both cases 1 and 2: for example, in case 2, where the failure probability $P(F)$ to be estimated is around 10^{-5}, the ratio FOM_{LS}/FOM_{MC} is about 1500 times larger than the ratio FOM_{SS}/FOM_{MC}.

1.5.4.1 Comparison with Other Stochastic Simulation Methods

As done for the previous application of Section 1.4, the results obtained by SS and LS have been compared to those obtained by other literature methods, in particular IS and a combination of IS and DR (Section 1.3) which have turned out to give the best results in the case studies considered (Gille 1998, 1999). Notice that the OA method has not been implemented for this application in the reference study (Gille 1998, 1999): this is due to the high dimensionality of the problem which makes the definition of a proper rotation matrix very difficult (step 3 in Section 1.3.3).

The values of the performance indicators $\hat{\sigma}_{MC}/\hat{\sigma}_{meth}$ and FOM_{meth}/FOM_{MC} obtained by IS and IS + DR for cases 1 and 2 of the thermal-fatigue crack growth model of Sections 1.5.1 and 1.5.2 are summarized in Table 1.8.

In this application, LS is found to outperform both IS and the combination of IS and DR: for example, in case 2, the ratio FOM_{LS}/FOM_{MC} is 65 and 35 times larger than FOM_{IS}/FOM_{MC} and FOM_{IS+DR}/FOM_{MC}, respectively. This confirms the capability of the LS method to efficiently probe complex high-dimensional domains of integration.

Table 1.8 Values of the performance indicators $\hat{\sigma}_{MC}/\hat{\sigma}_{meth}$ and FOM_{meth}/FOM_{MC} obtained by IS and IS + DR when applied for the reliability analysis of cases 1 and 2 of the thermal-fatigue crack growth model of Section 1.5.2 (Gille 1998, 1999)

Case 1	$\hat{\sigma}_{MC}/\hat{\sigma}_{meth}$	FOM_{meth}/FOM_{MC}
IS	16.9	424.36
IS + DR	65.4	864.36
Case 2	$\hat{\sigma}_{MC}/\hat{\sigma}_{meth}$	FOM_{meth}/FOM_{MC}
IS	41.1	4.396×10^3
IS + DR	172.4	8.317×10^3

1.6 Summary and Critical Discussion of the Techniques

One of the major obstacles in applying simulation methods for the reliability analysis of engineered systems and structures is the challenge posed by the estimation of small failure probabilities: the simulation of the rare events of failure occurrence implies a significant computational burden (Schueller 2007).

In order to overcome the rare-event problem, the IS method has been introduced (Au and Beck 2003a; Schueller *et al.* 2004). This technique amounts to replacing the original PDF of the uncertain random variables with an ISD chosen so as to generate samples that lead to failure more frequently (Au and Beck 2003a). IS has the capability to considerably reduce the variance compared with standard MCS, provided that the ISD is chosen similar to the theoretical optimal one (Equation 1.11 of Section 1.3.1). However, generally substantial insights on the system stochastic behavior and extensive modeling work is needed to identify a "good" ISD, e.g., by identifying "design points" (Schueller *et al.* 2004), setting up complex kernel density estimators (Au and Beck 2003a) or simply by tuning the parameters of the ISD based on expert judgment and trial-and-error (Gille 1998, 1999; Pagani *et al.* 2005). Overall, this greatly increases the effort associated to the simulation for accurate failure probability estimation. Furthermore, there is always the risk that an inappropriate choice of the ISD may lead to worse estimates compared to standard MCS (Schueller *et al.* 2004).

SS offers a clever way out of this problem by breaking the small failure probability evaluation task into a sequence of estimations of larger conditional probabilities. During the simulation, more frequent samples conditional to intermediate regions are generated from properly designed Markov chains. The method has been proven much more effective than standard MCS in the very high-dimensional spaces characteristic of structural reliability problems in which the failure regions are just tiny bits (Au and Beck 2001).

The strength of SS lies in the generality of its formulation and the straightforward algorithmic scheme. In contrast to some of the alternative methods (e.g., LS and OA), it is not restricted to standard normal spaces and can provide equally good results irrespectively of the joint distribution of the uncertain variables as long as one can draw samples from it. Furthermore, a single run of the SS algorithm leads to the calculation of the probabilities associated with all the conditional events considered: if for example, the probability of exceeding a critical level by a system response statistic of a stochastic system (the mean or a percentile of the displacement, stress, temperature, etc.) is sought, then by appropriate parametrization of the intermediate conditional events, a single run can provide the probabilities of exceedance associated with a wide range of values of the response statistic of interest, irrespective of their magnitude (Au 2005).

On the other hand, a word of caution is in order with respect to the fact that the conditional samples generated during the MCMC simulation are correlated by construction. Since it is demonstrated that a high correlation among conditional samples increases the variance of the SS estimates, a good choice/tuning of the SS parameters (i.e., the conditional probability p_0 and the proposal PDFs for MCMC simulation) is required to avoid it (Au and Beck 2003b). Finally, another drawback of the SS method is the need to express the failure event F in terms of a real-valued parameter crossing a given threshold (i.e., $F = \{Y < y\}$). This parameterization is natural for the cases of practical interest in structural reliability and otherwise specific for other system reliability problems (Zio and Pedroni 2008).

An alternative way to perform robust estimations of small failure probabilities without the extensive modeling effort required by IS is offered by LS. The LS method employs *lines* instead of random *points* in order to probe the high-dimensional failure domain of interest. An "important direction" is optimally determined to point towards the failure domain of interest and a number of conditional, one-dimensional problems are solved along such direction, in place of the original high-dimensional problem (Pradlwarter *et al.* 2005). When the boundaries of the failure domain of interest are not too rough (i.e., approximately linear) and the "important direction" is almost perpendicular to them, only a few simulations suffice to arrive at a failure probability with acceptable confidence. The determination of the important direction requires additional evaluations of the system performance which increases the computational cost (Section 1.2.2.2). Further, for each random sample (i.e., system configuration) drawn, two or three evaluations of the system performance are necessary to estimate the conditional one-dimensional failure probability estimates by linear or quadratic interpolation (Equation 1.39 in Appendix 2). When the "important direction" is not the optimal one, the variance of the estimator will increase. Of particular advantage of LS is its robustness: in the worst possible case where the "important direction" is selected orthogonal to the (ideal) optimal direction, line sampling performs at least as well as standard Monte Carlo simulation (Schueller *et al.* 2004).

Finally, the DR method and the OA method employ simulation concepts similar to those of LS, but with important limitations (Gille 1998, 1999). In the DR method, the failure event of interest is re-expressed in such a way as to highlight *one* (say, x_j) of the input random variables, recognized as more important; then, the failure probability estimate is computed as the expected value of the CDF of x_j conditional on the remaining $(n - 1)$ input variables. By so doing, the zero values contained in the standard MCS estimator (i.e., $I_F(\boldsymbol{x}) = 0$, if $\boldsymbol{x}y \in F$) are removed: this allows one to (1) reach any level of probability (even very small) and (2) reduce the variance of the failure probability estimator (Gille 1998, 1999). Notice that DR can be considered a very special case of LS where the important direction $\boldsymbol{\alpha}$ coincides with the "direction" of the variable x_j, i.e., $\boldsymbol{\alpha} = \left(0, 0, \ldots, x_j, \ldots, 0, 0\right)$. However, such method cannot always be applied: first, the performance function of the system must be analytically known (which is never the case for realistic systems simulated by detailed computer codes); second, the performance function must have the characteristic that one of the variables can be separated from the others (Gille 1998, 1999).

Finally, the OA method performs a sort of importance sampling around the design point of the problem in the standard normal space. Thus, if the design point is actually representative of the most important regions of the failure domain, the OA leads to an impressive reduction in the variance of the failure probability es-

Table 1.9 Synthetic comparison of the stochastic simulation methods considered in this work

Method	Simulation concepts	Decisions	Advantages	Drawbacks
Standard MCS	Repeat random sampling of possible system configurations		Samples the full range of each input variable Consistent performance in spite of complexity and dimension of the problem Accuracy easily assessed No need for simplifying assumptions nor surrogate models No complex elaborations of the original model Identification of nonlinearities, thresholds and discontinuities Simplicity	High computational cost (in presence of long-running models for determining system response and small failure probabilities)
SS	Express a small probability as a product of larger conditional probabilities Generate conditional samples by MCMC simulation	Conditional failure probability p_0 at each simulation level Proposal PDFs for MCMC Simulation	General formulation Straightforward algorithmic scheme No restriction to standard normal space Consistent performance in spite of complex joint PDFs Consistent performance in spite of irregularities in topology and boundary of the failure domain One single run computes probabilities for more than one event Reduced computational effort with respect to other methods	Parametrization of the failure event in terms of intermediate conditional events Correlation among conditional samples: bias in the estimates and possibly increased variance

Table 1.9 (continued)

Method	Simulation concepts	Decisions	Advantages	Drawbacks
LS	Turn a high-dimensional problem in the physical space into one-dimensional problems in the standard normal space Project the problem onto a line $\boldsymbol{\alpha}$ pointing at the important regions of the failure domain Use line $\boldsymbol{\alpha}$ almost perpendicular to the failure domain to reduce the variance of the estimates	One failure point to start the Markov chain for the determination of $\boldsymbol{\alpha}$	No assumptions about regularity of the limit state function (robustness) If limit state function is almost linear, few simulations suffice to achieve acceptable estimation accuracies No necessity to estimate important direction $\boldsymbol{\alpha}$ with excessive accuracy Even in the worst possible case ($\boldsymbol{\alpha}$ orthogonal to optimal direction) the performance is at least comparable to standard MCS	Determination of important direction $\boldsymbol{\alpha}$ requires additional evaluation of system performance (with increase in the computational cost) For each sample drawn, two or three evaluations of system performance are necessary to estimate failure probability (with increase in the computational cost) Essential restriction to standard normal space (Rosenblatt's or Nataf's transformations are required) (Rosenblatt 1952; Nataf 1962)
IS	Repeated random sampling of possible system configurations Sample from ISD to generate more samples in the region of interest (e.g., low probability of occurrence)	Construction/choice of the ISD	If the ISD is similar to optimal one: significant increase in estimation accuracy (or, conversely, reduction in sample size for given accuracy)	Many system behavior insights and and much modeling work needed for identification of good ISD Inappropriate ISD leads to worse estimates compared to Standard MCS

Table 1.9 (continued)

Method	Simulation concepts	Decisions	Advantages	Drawbacks
DR	Express failure event in such a way as to highlight one random variable Estimate failure probability as expected value of the CDF of the chosen variable conditional on the remaining ($n-1$) variables	Random variable to be separated from others	Remove zero values included in the standard MCS estimator (reduced variance) Any probability level can be reached (also the very small ones of rare events)	Analytical expression for the system performance function is required Performance function must have the characteristics that one of the variables can be separated out from the others
OA	Identification of the design point Rotation of system coordinates Solve one-dimensional problems along direction of design point	–	If the design point is representative of the most important regions of the failure domain, then the variance is significantly reduced	Design point frequently not representative of the most important regions of the failure domain (high-dimensional problems) High computational cost associated to design point (nonlinear constrained optimization problem) Rotation matrix difficult to introduce in high-dimensional spaces

timator. However, it is worth noting that the design points and their neighbors do not always represent the most important regions of the failure domain, especially in high-dimensional problems. Moreover, the computational cost associated with the identification of the design points may be quite relevant, which adversely affects the efficiency of the method (Schueller *et al.* 2004). Finally, the implementation of the OA method requires the definition of a rotation matrix in order to modify the coordinate system, which can be very difficult for high-dimensional problems.

A synthetic comparison of the stochastic simulation methods considered in this work is given in Table 1.9 (the "Decisions" column refers to parameters, distributions, and other characteristics of the methods that have to be chosen or determined by the analyst in order to perform the simulation).

Appendix 1. Markov Chain Monte Carlo Simulation

Markov Chain Monte Carlo Simulation

MCMC simulation comprises a number of powerful simulation techniques for generating samples according to any given probability distribution (Metropolis *et al.* 1953).

In the context of the reliability assessment of interest in the present work, MCMC simulation provides an efficient way for generating samples from the multidimensional conditional PDF $q(\boldsymbol{x}|F)$. The distribution of the samples thereby generated tends to the multidimensional conditional PDF $q(\boldsymbol{x}|F)$ as the length of the Markov chain increases. In the particular case of the initial sample $\boldsymbol{x}^1$ being distributed exactly as the multidimensional conditional PDF $q(\boldsymbol{x}|F)$, then so are the subsequent samples and the Markov chain is always stationary (Au and Beck 2001).

In the following it is assumed without loss of generality that the components of $\boldsymbol{x}$ are independent, that is, $q(\boldsymbol{x}) = \prod_{j=1}^{n} q_j(x_j)$, where $q_j(x_j)$ denotes the one-dimensional PDF of x_j (Au and Beck 2001).

To illustrate the MCMC simulation algorithm with reference to a generic failure region F_i, let $\boldsymbol{x}^u = \left\{x_1^u, x_2^u, \ldots, x_j^u, \ldots, x_n^u\right\}$ be the uth Markov chain sample drawn and let $p_j^*(\xi_j|x_j^u)$, $j = 1, 2, \ldots, n$, be a one-dimensional "proposal PDF" for ξ_j, centered at the value x_j^u and satisfying the symmetry property $p_j^*(\xi_j|x_j^u) = p_j^*(x_j^u|\xi_j)$. Such distribution, arbitrarily chosen for each element x_j of $\boldsymbol{x}$, allows generating a "precandidate value" ξ_j based on the current sample value x_j^u. The following algorithm is then applied to generate the next Markov chain sample $\boldsymbol{x}^{u+1} = \left\{x_1^{u+1}, x_2^{u+1}, \ldots, x_j^{u+1}, \ldots, x_n^{u+1}\right\}$, $u = 1, 2, \ldots, N_s - 1$ (Au and Beck 2001):

1. Generation of a candidate sample $\tilde{\boldsymbol{x}}^{u+1} = \left\{\tilde{x}_1^{u+1}, \tilde{x}_2^{u+1}, \ldots, \tilde{x}_j^{u+1}, \ldots, \tilde{x}_n^{u+1}\right\}$: for each parameter x_j, $j = 1, 2, \ldots, n$:
   Sample a precandidate value ξ_j^{u+1} from $p_j^*(\cdot|x_j^u)$;

- Compute the acceptance ratio:

$$r_j^{u+1} = \frac{q_j\left(\xi_j^{u+1}\right)}{q_j\left(x_j^u\right)} \tag{1.35}$$

- Set the new value $\tilde{x}_j^{u+1}$ of the jth element of $\tilde{\boldsymbol{x}}^{u+1}$ as follows:

$$\tilde{x}_j^{u+1} = \begin{cases} \xi_j^{u+1} & \text{with probability } \min\left(1, r_j^{u+1}\right) \\ x_j^u & \text{with probability } 1 - \min\left(1, r_j^{u+1}\right) \end{cases} \tag{1.36}$$

2. Acceptance/rejection of the candidate sample vector $\tilde{\boldsymbol{x}}^{u+1}$: If $\tilde{\boldsymbol{x}}^{u+1} = \boldsymbol{x}^u$ (i.e., no precandidate values have been accepted), set $\boldsymbol{x}^{u+1} = \boldsymbol{x}^u$. Otherwise, check whether $\tilde{\boldsymbol{x}}^{u+1}$ is a system failure configuration, i.e. $\tilde{\boldsymbol{x}}^{u+1} \in F_i$: if it is, then accept the candidate $\tilde{\boldsymbol{x}}^{u+1}$ as the next state, i.e., set $\boldsymbol{x}^{u+1} = \tilde{\boldsymbol{x}}^{u+1}$; otherwise, reject the candidate $\tilde{\boldsymbol{x}}^{u+1}$ and take the current sample as the next one, i.e., set $\boldsymbol{x}^{u+1} = \boldsymbol{x}^u$.

In synthesis, a candidate sample $\tilde{\boldsymbol{x}}^{u+1}$ is generated from the current sample $\boldsymbol{x}^u$ and then either the candidate sample $\tilde{\boldsymbol{x}}^{u+1}$ or the current sample $\boldsymbol{x}^u$ is taken as the next sample $\boldsymbol{x}^{u+1}$, depending on whether the candidate $\tilde{\boldsymbol{x}}^{u+1}$ lies in the failure region F_i or not.

Finally, notice that in this work, the one-dimensional proposal PDF p_j^*, $j = 1, 2, \ldots, n$, is chosen as a symmetrical uniform distribution centered at the current sample x_j, $j = 1, 2, \ldots, n$, with width $2l_j$, where l_j is the maximum step length, i.e., the maximum allowable distance that the next sample can depart from the current one. The choice of l_j is such that the standard deviation of p_j^* is equal to that of q_j, $j = 1, 2, \ldots, n$.

Appendix 2. The Line Sampling Algorithm

The Line Sampling Algorithm

The LS algorithm proceeds as follows (Pradlwarter *et al.* 2005):

1. Determine the *unit* important direction $\boldsymbol{\alpha} = \{\alpha_1, \alpha_2, \ldots, \alpha_j, \ldots, \alpha_n\}$. Any of the methods summarized in Section 1.2.2.2 can be employed to this purpose. Notice that the computation of $\boldsymbol{\alpha}$ implies additional system analyses, which substantially increase the computational cost associated to the simulation method (Section 1.2.2.2).
2. From the original multidimensional joint probability density function $q(\cdot)$: $\Re^n \rightarrow [0, \infty)$, sample N_T vectors $\{\boldsymbol{x}^k : k = 1, 2, \ldots, N_T\}$, with $\boldsymbol{x}^k = \{x_1^k, x_2^k, \ldots, x_j^k, \ldots, x_n^k\}$ by standard MCS.

3. Transform the N_T sample vectors $\{x^k: k = 1, 2, \ldots, N_T\}$ defined in the original (i.e., physical) space of possibly dependent, non-normal random variables (step 2. above) into N_T samples $\{\boldsymbol{\theta}^k: k = 1, 2, \ldots, N_T\}$ defined in the standard normal space where each component of the vector $\boldsymbol{\theta}^k = \{\theta_1^k, \theta_2^k, \ldots, \theta_j^k, \ldots, \theta_n^k\}, k = 1, 2, \ldots, N_T$, is associated with an independent central unit Gaussian standard distribution (Section 1.2.2.2).
4. Estimate N_T conditional "one-dimensional" failure probabilities $\{\hat{P}^k(F): k = 1, 2, \ldots, N_T\}$, corresponding to each one of the standard normal samples $\{\boldsymbol{\theta}^k: k = 1, 2, \ldots, N_T\}$ obtained in step 3 above. In particular, for each random sample $\boldsymbol{\theta}^k$, $k = 1, 2, \ldots, N_T$, perform the following steps (Figure 1.5) (Schueller *et al.* 2004; Pradlwarter *et al.* 2005, 2007):

- Define the sample vector $\tilde{\boldsymbol{\theta}}^k$, $k = 1, 2, \ldots, N_T$, as the sum of a deterministic multiple of $\boldsymbol{\alpha}$ and a vector $\boldsymbol{\theta}^{k,\perp}$, $k = 1, 2, \ldots, N_T$, perpendicular to the direction $\boldsymbol{\alpha}$, i.e.,

$$\tilde{\boldsymbol{\theta}}^k = c^k \boldsymbol{\alpha} + \boldsymbol{\theta}^{k,\perp}, \quad k = 1, 2, \ldots, N_T \tag{1.37}$$

where c^k is a real number in $[-\infty, +\infty]$ and

$$\boldsymbol{\theta}^{k,\perp} = \boldsymbol{\theta}^k - \left\langle \boldsymbol{\alpha}, \boldsymbol{\theta}^k \right\rangle \boldsymbol{\alpha}, \quad k = 1, 2, \ldots, N_T \tag{1.38}$$

In Equation 1.38, $\boldsymbol{\theta}^k$, $k = 1, 2, \ldots, N_T$, denotes a random realization of the input variables in the standard normal space of dimension n and $\langle \boldsymbol{\alpha}, \boldsymbol{\theta}^k \rangle$ is the scalar product between $\boldsymbol{\alpha}$ and $\boldsymbol{\theta}^k$, $k = 1, 2, \ldots, N_T$. Finally, it is worth noting that since the standard Gaussian space is isotropic, both the scalar c^k and the vector $\boldsymbol{\theta}^{k,\perp}$ are also standard normally distributed (Pradlwarter *et al.* 2007).
- Compute the value $\bar{c}^k$ as the intersection between the limit state function $g_\theta(\tilde{\boldsymbol{\theta}}^k) = g_\theta(c^k \boldsymbol{\alpha} + \boldsymbol{\theta}^{k,\perp}) = 0$ and the line $l^k(c^k, \boldsymbol{\alpha})$ passing through $\boldsymbol{\theta}^k$ and parallel to $\boldsymbol{\alpha}$ (Figure 1.5). The value of $\bar{c}^k$ can be approximated by evaluating the performance function $g_\theta(\cdot)$ at two or three different values of c^k (e.g., c_1^k, c_2^k, and c_3^k in Figure 1.5), fitting a first- or second-order polynomial and determining its root (Figure 1.5). Hence, for each standard normal random sample $\boldsymbol{\theta}^k$, $k = 1, 2, \ldots, N_T$, two or three system performance evaluations by the model are required.
- Solve the conditional one-dimensional reliability problem associated to each random sample $\boldsymbol{\theta}^k$, $k = 1, 2, \ldots, N_T$, in which the only (standard normal) random variable is c^k. The associated conditional failure probability $\hat{P}^k(F)$,

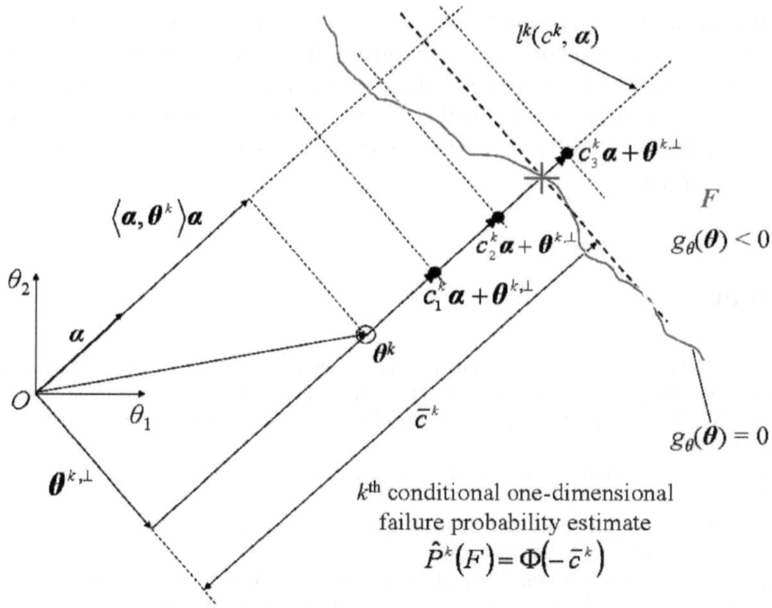

Figure 1.5 The LS procedure (Pradlwarter *et al.* 2005)

$k = 1, 2, \ldots, N_T$, is given by

$$\begin{aligned}\hat{P}^k(F) &= P\left[N(0,\,1) > \bar{c}^k\right] \\ &= 1 - P\left[N(0,\,1) \leqslant \bar{c}^k\right] \\ &= 1 - \Phi\left(\bar{c}^k\right) = \Phi\left(-\bar{c}^k\right)\end{aligned} \tag{1.39}$$

where $\Phi(\cdot)$ denotes the standard normal cumulative distribution function.

5. Using the independent conditional "one-dimensional" failure probability estimates $\left\{\hat{P}^k(F)\colon k = 1,\,2,\,\ldots,\,N_T\right\}$ in Equation 1.39 above, compute the unbiased estimator $\hat{P}(F)$ for the failure probability $P(F)$ as

$$\hat{P}(F) = \frac{1}{N_T}\sum_{k=1}^{N_T}\hat{P}^k(F) \tag{1.40}$$

The variance of the estimator 1.41 is

$$\sigma^2\left(\hat{P}(F)\right) = \frac{1}{N_T(N_T-1)}\sum_{k=1}^{N_T}\left(\hat{P}^k(F) - \hat{P}(F)\right)^2 \tag{1.41}$$

With the described approach the variance of the estimator $\hat{P}(F)$ of the failure probability $P(F)$ is considerably reduced. In general, a relatively low number N_T of simulations has to be carried out to obtain a sufficiently accurate estimate. A single evaluation would suffice for the ideal case in which the limit state function is linear and an LS direction $\boldsymbol{\alpha}$ perpendicular to it has been identified (Koutsourelakis *et al.* 2004).

References

Ahammed M, Malchers ME (2006) Gradient and parameter sensitivity estimation for systems evaluated using Monte Carlo analysis. Reliab Eng Syst Saf 91:594–601.

Ardillon E, Venturini V (1995) Measures de sensibilitè dans les approaches probabilistes. Rapport EDF HP-16/95/018/A.

Au SK (2005) Reliability-based design sensitivity by efficient simulation. Comput Struct 83:1048–1061.

Au SK, Beck JL (2001) Estimation of small failure probabilities in high dimensions by subset simulation. Probab Eng Mech 16(4):263–277.

Au SK, Beck JL (2003a) Importance sampling in high dimensions. Struct Saf 25(2):139–163.

Au SK, Beck JL (2003b) Subset simulation and its application to seismic risk based on dynamic analysis. J Eng Mech 129(8):1–17.

Der Kiureghian A (2000) The geometry of random vibrations and solutions by FORM and SORM. Probab Eng Mech 15(1):81–90.

Fishman GS (1996) Monte Carlo: concepts, algorithms, and applications. New York: Springer.

Freudenthal AM (1956) Safety and the probability of structural failure. ASCE Trans 121:1337–1397.

Fu M (2006) Stochastic gradient estimation. In Henderson SG, Nelson BL (eds) Handbook on operation research and management science: simulation, chap 19. Elsevier.

Hastings WK (1970) Monte Carlo sampling methods using Markov chains and their applications. Biometrika 57:97–109.

Huang B, Du X (2006) A robust design method using variable transformation and Gauss-Hermite integration. Int J Numer Meth Eng 66:1841–1858.

Gille A (1998) Evaluation of failure probabilities in structural reliability with Monte Carlo methods. ESREL '98, Throndheim.

Gille A (1999) Probabilistic numerical methods used in the applications of the structural reliability domain. PhD thesis, Universitè Paris 6.

Koutsourelakis PS, Pradlwarter HJ, Schueller GI (2004) Reliability of structures in high dimensions, Part I: Algorithms and application. Probab Eng Mech (19):409–417.

Metropolis N, Rosenbluth AW, Rosenbluth MN, Taller AH (1953) Equations of state calculations by fast computing machines. J Chem Phys 21(6):1087–1092.

Nataf A (1962) Determination des distribution dont les marges sont donnees. C R Acad Sci 225:42–43.

Nutt WT, Wallis GB (2004) Evaluations of nuclear safety from the outputs of computer codes in the presence of uncertainties. Reliab Eng Syst Saf 83:57–77.

Pagani L, Apostolakis GE, Hejzlar P (2005) The impact of uncertainties on the performance of passive systems. Nucl Technol 149:129–140.

Paris PC (1961) A rational analytic theory of fatigue. Trend Eng Univ Wash 13(1):9.

Patalano G, Apostolakis GE, Hejzlar P (2008) Risk informed design changes in a passive decay heat removal system. Nucl Technol 163:191–208.

Pradlwarter HJ, Pellissetti MF, Schenk CA et al. (2005) Realistic and efficient reliability estimation for aerospace structures. Comput Meth Appl Mech Eng 194:1597–1617.

Pradlwarter HJ, Schueller GI, Koutsourelakis PS, Charmpis DC (2007) Application of line sampling simulation method to reliability benchmark problems. Struct Saf 29:208–221.
Rosenblatt M (1952) Remarks on multivariate transformations. Ann Math Stat 23(3):470–472.
Schueller GI (2007) On the treatment of uncertainties in structural mechanics and analysis. Comput Struct 85:235–243.
Schueller GI, Pradlwarter HJ (2007) Benchmark study on reliability estimation in higher dimension of structural systems – An overview. Struct Saf 29:167–182.
Schueller GI, Stix R (1987) A critical appraisal of methods to determine failure probabilities. Struct Saf 4:293–309.
Schueller GI, Pradlwarter HJ, Koutsourelakis PS (2004) A critical appraisal of reliability estimation procedures for high dimensions. Probab Eng Mech 19:463–474.
Thunnissen DP, Au SK, Tsuyuki GT (2007) Uncertainty quantification in estimating critical spacecraft component temperature. AIAA J Therm Phys Heat Transf (doi: 10.2514/1.23979).
Zio E, Pedroni N (2008) Reliability analysis of discrete multi-state systems by means of subset simulation. Proceedings of the ESREL 2008 Conference, 22–25 September, Valencia, Spain.

Chapter 2
Dynamic Fault Tree Analysis: Simulation Approach

K. Durga Rao, V.V.S. Sanyasi Rao, A.K. Verma, and A. Srividya

Abstract Fault tree analysis (FTA) is extensively used for reliability and safety assessment of complex and critical engineering systems. One of the important limitations of conventional FTA is the inability for one to incorporate complex component interactions such as sequence dependent failures. Dynamic gates are introduced to extend conventional FT to model these complex interactions. This chapter presents various methods available in the literature to solve dynamic fault trees (DFT). Special emphasis on a simulation-based approach is given as analytical methods have some practical limitations.

2.1 Fault Tree Analysis: Static Versus Dynamic

Fault tree analysis has gained widespread acceptance for quantitative reliability and safety analysis. A fault tree is a graphical representation of various combinations of basic failures that lead to the occurrence of undesirable top events. Starting with the top event all possible ways for this event to occur are systematically deduced. The methodology is based on three assumptions: (1) events are binary events, (2) events are statistically independent; and (3) the relationship between events is represented by means of logical Boolean gates (AND; OR; voting). The analysis is carried out in two steps: a qualitative step in which the logical expression of the top event is derived in terms of prime implicants (the minimal cut-sets); a quantitative step in which on the basis of the probabilities assigned to the failure events of the basic components, the probability of occurrence of the top event is calculated.

K. Durga Rao
Paul Scherrer Institut, Villigen PSI, Switzerland

V.V.S. Sanyasi Rao
Bhabha Atomic Research Centre, Mumbai, India

A.K. Verma · A. Srividya
Indian Institute of Technology Bombay, Mumbai, India

The traditional static fault trees with AND, OR, and voting gates cannot capture the behavior of components of complex systems and their interactions such as sequence-dependent events, spares and dynamic redundancy management, and priorities of failure events. In order to overcome this difficulty, the concept of dynamic fault trees (DTFs) is introduced by adding sequential notion to the traditional fault tree approach [1]. System failures can then depend on component failure order as well as combination. This is done by introducing dynamic gates into fault trees. With the help of dynamic gates, system sequence-dependent failure behavior can be specified using DTFs that are compact and easily understood. The modeling power of DTFs has gained the attention of many reliability engineers working on safety critical systems [2].

As an example of sequence dependent failure, consider a power supply system in a nuclear power plant (NPP) where one active system (grid supply) and one standby system (diesel generator (DG) supply) are connected with a switch controller. If the switch controller fails after the grid supply fails, then the system can continue operation with the DG supply. However, if the switch fails before the grid supply fails, then the DG supply cannot be switched into active operation and the power supply fails when the grid supply fails. Thus, the failure criterion depends on the sequence of events also apart from the combination of events.

2.2 Dynamic Fault Tree Gates

The DFT introduces four basic (dynamic) gates: the priority AND (PAND), the sequence enforcing (SEQ), the spare (SPARE), and the functional dependency (FDEP) [1]. They are discussed here briefly.

The *PAND gate* reaches a failure state if all of its input components have failed in a pre-assigned order (from left to right in graphical notation). In Figure 2.1a, a failure occurs if A fails before B, but B may fail before A without producing a failure in G. The truth table for PAND gate is shown in Table 2.1, the occurrence of event (failure) is represented as 1 and its nonoccurrence as 0. In the second case, though, both A and B have failed but due to the undesired order, it is not a failure of the system.

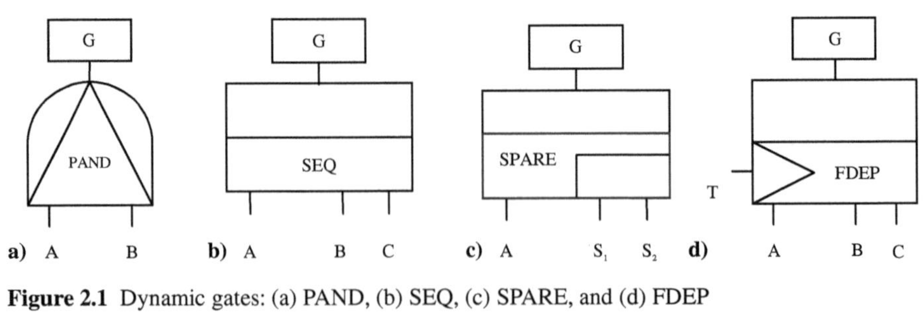

Figure 2.1 Dynamic gates: (a) PAND, (b) SEQ, (c) SPARE, and (d) FDEP

Table 2.1 Truth table for PAND gate with two inputs

A	B	Output
1 (first)	1 (second)	1
1 (second)	1 (first)	0
0	1	0
1	0	0
0	0	0

Example of PAND gate
Fire alarm in a chemical process plant gives signal to fire fighting personnel for further action if it detects any fire. If the fire alarm fails (got burnt in the fire) after giving alarm, then the plant will be in safe state as fire fighting is in place. However, if the alarm fails (failed in standby mode which got undetected) before the fire accident, then the extent of damage would be very high. This can be modeled by PAND gate only as the scenario exactly fits into its definition.

A *SEQ gate* forces its inputs to fail in a particular order: when a SEQ gate is found in a DFT, it never happens that the failure sequence takes place in different orders. While the SEQ gate allows the events to occur only in a pre-assigned order and states that a different failure sequence can never take place, the PAND gate does not force such a strong assumption: it simply detects the failure order and fails just in one case. The truth table for SEQ gate is shown in Table 2.2.

SPARE gates are dynamic gates modeling one or more principal components that can be substituted by one or more backups (spares), with the same functionality (Figure 2.1c). The SPARE gate fails when the number of operational powered spares

Table 2.2 Truth table for SEQ gate with three inputs

A	B	C	Output
0	0	0	0
0	0	1	Impossible
0	1	0	Impossible
0	1	1	Impossible
1	0	0	0
1	0	1	Impossible
1	1	0	0
1	1	1	1

Example of SEQ gate
Considering a scenario where pipe in pumping system fails in different stages. There is a minor welding defect at the joint of the pipe section, which can become a minor leak with time and subsequently it lead to a rupture.

Table 2.3 Truth table for SPARE gate with two inputs

A	B	Output
1	1	1
0	1	0
1	0	0
0	0	0

Example of SPARE gate
Reactor regulation system in NPP consists of dual processor hot standby system. There will be two processors which will be continuously working. Processor 1 will be normally doing the regulation; in case it fails processor 2 will take over.

and/or principal components is less than the minimum required. Spares can fail even while they are dormant, but the failure rate of an unpowered spare is lower than the failure rate of the corresponding powered one. More precisely, λ being the failure rate of a powered spare, the failure rate of the unpowered spare is $\alpha\lambda$, where $0 \leqslant \alpha \leqslant 1$ is the dormancy factor. Spares are more properly called "hot" if $\alpha = 1$ and "cold" if $\alpha = 0$. The truth table for a SPARE gate with two inputs is shown in Table 2.3.

In the *FDEP gate* (Figure 2.1d), there will be one trigger input (either a basic event or the output of another gate in the tree) and one or more dependent events. The dependent events are functionally dependent on the trigger event. When the trigger event occurs, the dependent basic events are forced to occur. In the Markov model of FDEP gate, when a state is generated in which the trigger event is satisfied, all the associated dependent events are marked as having occurred. The separate occurrence of any of the dependent basic events has no effect on the trigger event (see Table 2.4).

Table 2.4 Truth table for FDEP gate with two inputs

Trigger	Output	Dependent event 1	Dependent event 2
1	1	1	1
0	0	0/1	0/1

Example of FDEP gate
In the event of power supply failure, all the dependent systems will be unavailable. The trigger event is the power supply and systems which are drawing power are dependent events.

2.3 Effect of Static Gate Representation in Place of Dynamic Gates

There are two solution strategies to solve DFT, namely, analytical and simulation approaches. They are explained in detail in the following sections. Evaluating dynamic gates and their modeling is resource intensive by both analytical and simulation approaches. It is important to see the benefit achieved while doing such analysis. This is the case especially with probabilistic safety assessment (PSA) of NPP where there are a number of systems with many cut-sets. PAND and SEQ gates are special cases of the static AND gate. Evaluations are shown here with different cases of input parameters to see the sensitivity of the results to the dynamic and static representations of a gate. Consider two inputs for both the gates AND and PAND with their respective failure and repair rates as shown in Table 2.5. Unavailability has been evaluated for both the gates with different cases. It is interesting to note that for all these combinations, the static AND gate yields the result in the same order. However, the PAND gate differs by 2500% with AND gate in Case 1 and Case 3 where $\mu_A \gg \mu_B$. From these results it can be observed that irrespective of values of failure rates, the unavailability is found to be much less in for a dynamic gate when $\mu_A \gg \mu_B$. The difference is marginal in other cases. Nevertheless, the system uncertainty bounds and importance measures can vary with the dynamic modeling in such scenarios. Dynamic reliability modeling reduces any uncertainties that may arise due to the modeling assumptions.

Table 2.5 Comparison with Static AND and PAND

Case	Scenario	Unavailability PAND	AND	% difference
Case 1 $\lambda_A = 4 \times 10^{-2}$; $\lambda_B = 2.3 \times 10^{-3}$ $\mu_A = 1$; $\mu_B = 4.1 \times 10^{-2}$	$\lambda_A \gg \lambda_B$ $\mu_A \gg \mu_B$	8.2×10^{-5}	2.0×10^{-3}	2500%
Case 2 $\lambda_A = 4 \times 10^{-2}$; $\lambda_B = 2.3 \times 10^{-3}$ $\mu_A = 4.1 \times 10^{-2}$; $\mu_B = 1$	$\lambda_A \gg \lambda_B$ $\mu_A \ll \mu_B$	1.9×10^{-3}	2.0×10^{-3}	Negligible
Case 3 $\lambda_A = 2.3 \times 10^{-3}$; $\lambda_B = 4 \times 10^{-2}$ $\mu_A = 1$; $\mu_B = 4.1 \times 10^{-2}$	$\lambda_A \ll \lambda_B$ $\mu_A \gg \mu_B$	4.5×10^{-5}	1.1×10^{-3}	2500%
Case 4 $\lambda_A = 2.3 \times 10^{-3}$; $\lambda_B = 4 \times 10^{-2}$ $\mu_A = 4.1 \times 10^{-2}$; $\mu_B = 1$	$\lambda_A \ll \lambda_B$ $\mu_A \ll \mu_B$	1.9×10^{-3}	2.0×10^{-3}	Negligible

2.4 Solving Dynamic Fault Trees

Several researchers [1–3] proposed methods to solve DFT. Dugan [1, 4, 5], has shown, through a process known as modularization, that it is possible to identify the independent sub-trees with dynamic gates and to use different a Markov model for each of them. It was applied to computer-based fault-tolerant systems successfully. But, with the increase in the number of basic elements, there is problem state-space explosion. To reduce state space and minimize the computational time, an improved decomposition scheme where the dynamic sub-tree can be further modularized (if there exist some independent sub-trees in it) is proposed by Huang [6]. Amari [2] proposed a numerical integration technique for solving dynamic gates. Although this method solves the state-space problem, it cannot be easily applied for repairable systems. Bobbio [3, 7] proposed a Bayesian network-based method to further reduce the problem of solving DTFs with state-space approach. Keeping the importance of sophisticated modeling for engineering systems in dynamic environment, several researchers [8–11] contributed significantly to the development and application of DFT.

However, a state-space approach for solving dynamic gates becomes too large for calculation with Markov models when the number of gate inputs increases. This is the case especially with PSA of NPP where there is a large number of cut-sets. In addition, the Markov model is applicable for exponential failure and repair distributions, and also modeling test and maintenance information on spare components is difficult. Many of the methods to solve DTFs are problem specific and it may be difficult to generalize for all the scenarios. In order to overcome these limitations of the above-mentioned methods, a Monte Carlo simulation approach has been attempted by Karanki *et al.* [11, 12] to implement dynamic gates. Scenarios which may often be difficult to solve with analytical solutions are easily tackled with the Monte Carlo simulation approach. The Monte Carlo simulation-based reliability approach, due to its inherent capability in simulating the actual process and random behavior of the system, can eliminate uncertainty in reliability modeling.

2.5 Modular Solution for Dynamic Fault Trees

Markov models can be used to solve DFTs. The order of occurrence of failure events can be easily modeled with the help of Markov models. Figure 2.2 shows the Markov models for various gates. The shaded state is the failure state in the state-space diagram. In each state, 1 and 0 represent success and failure of the components. However, the solution of a Markov model is much more time and memory consuming than the solution of a standard fault tree model. As the number of components increases in the system, the number of states and transition rates grows exponentially. Development of a state transition diagram can become very cumbersome and a mathematical solution may be infeasible.

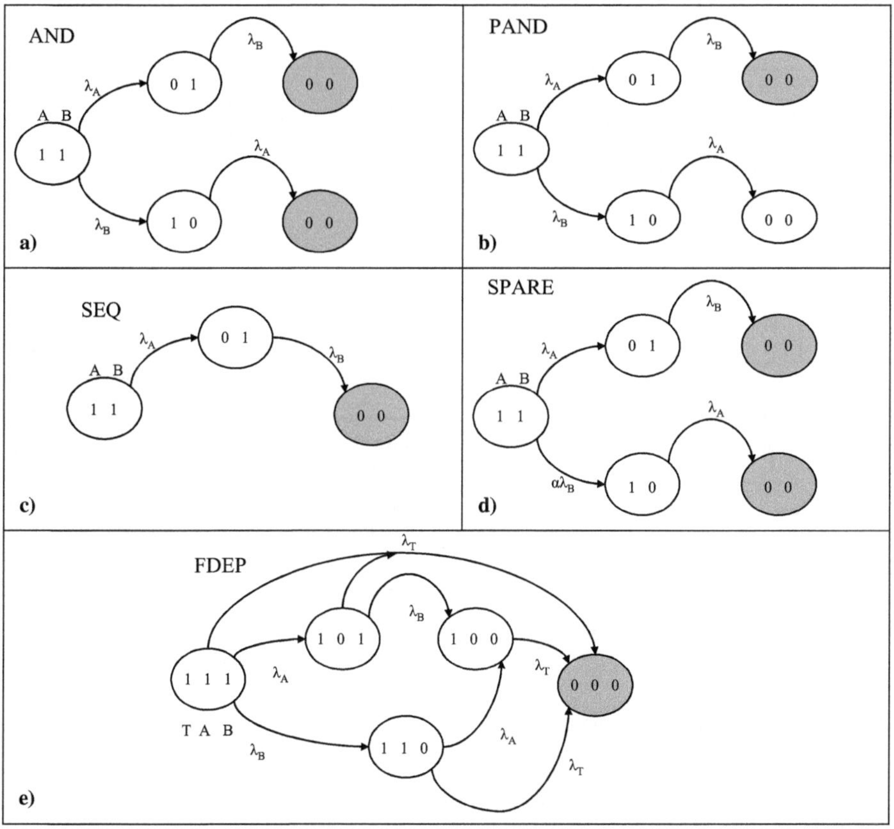

Figure 2.2 Markov models for various gates: (a) AND, (b) PAND, (c) SEQ, (d) SPARE, and (e) FDEP

Dugan [1] proposed a modular approach for solving DFTs. In this approach, the system-level fault tree is divided into independent modules, and the modules are solved separately, then the separate results can be combined to achieve a complete analysis. The dynamic modules are solved with the help of Markov models and the solution of static module is straightforward.

For example, consider the fault tree for dual processor failure; the dynamic module can be identified as shown in Figure 2.3. The remaining module has only static gates. Using a Markov model approach the dynamic module can be solved and plugged into the fault tree for further analysis.

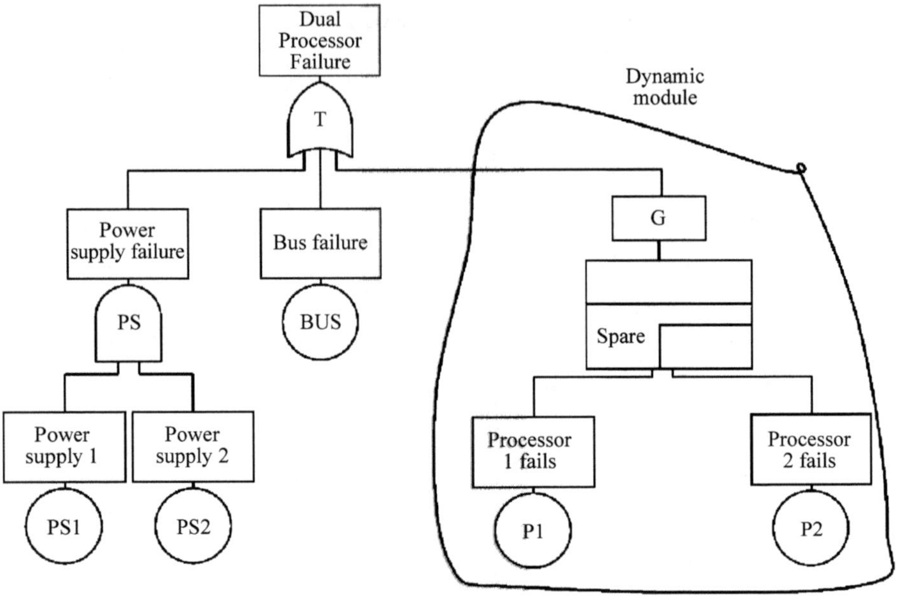

Figure 2.3 Fault tree for dual processor failure

2.6 Numerical Method

Amari [2] proposed a numerical integration technique for solving dynamic gates, which is explained below.

2.6.1 PAND Gate

A PAND gate has two inputs. The output occurs when the two inputs occur in a specified order (left one first and then right one). Let T_1 and T_2 be the random variables of the inputs (sub-trees). Therefore,

$$\begin{aligned} G(t) &= \Pr\{T_1 \leqslant T_2 < t\} \\ &= \int_{x_1=0}^{t} \mathrm{d}G_1(x_1)\left[\int_{x_2=x_1}^{t} \mathrm{d}G_2(x_2)\right] \\ &= \int_{x_1=0}^{t} \mathrm{d}G_1(x_1)\,[G_2(t) - G_2(x_1)] \end{aligned} \tag{2.1}$$

Once we compute $G_1(t)$ and $G_2(t)$, we can easily find $G(t)$ in Equation 2.1 using numerical integration methods. In order to illustrate this computation, a trapezoidal integral is used. Therefore,

$$G(t) = \sum_{i=1}^{m} [G_1(i \times h) - G_1(i-1) \times h] \times [G_2(t) - G_2(i \times h)] \qquad (2.2)$$

where m is the number of time steps/intervals and $h = t/m$ is step size/interval. The number of steps, m, in the above equation is almost equivalent to the number of steps required in solving differential equations corresponding to a Markov chain. Therefore, the gain in these computations can be in the order of n^{3n}. It shows that this method takes much less computational time than the Markov chain solution.

2.6.2 SEQ Gate

A SEQ gate forces events to occur in a particular order. The first input of a SEQ gate can be a basic event or a gate, and all other inputs are basic events.

Considering that the distribution of time to occurrence of input i is G_i, then the probability of occurrence of the SEQ gate can be found by solving the following equation:

$$\begin{aligned} G(t) &= \Pr\{T_1 + T_2 + \cdots + T_m < t\} \\ &= G_1 \times G_2 \times \cdots \times G_m(t) \end{aligned} \qquad (2.3)$$

2.6.3 SPARE Gate

A generic spare (SPARE) gate allows the modeling of heterogeneous spares including cold, hot, and warm spares. The output of the SPARE gate will be true when the number of powered spares/components is less than the minimum number required. The only inputs that are allowed for a SPARE gate are basic events (spare events). Therefore:

1. If all the distributions are exponential, we can get the closed-form solutions for $G(t)$.
2. If the standby failure rate of all spares are constant (not time dependent), then $G(t)$ can be solved using non-homogeneous Markov chains.
3. Otherwise, we need to use conditional probabilities or simulation to solve this part of the fault tree.

Therefore, using the above method, we can calculate the occurrence probability of a dynamic gate without explicitly converting it into a Markov model (except for some cases of the SPARE gate).

2.7 Monte Carlo Simulation Approach for Solving Dynamic Fault Trees

Monte Carlo simulation is a very valuable method which is widely used in the solution of real engineering problems in many fields. Lately the utilization of this method is growing for the assessment of availability of complex systems and the monetary value of plant operation and maintenance [13–16]. The complexity of the modern engineering systems besides the need for realistic considerations when modelling their availability/reliability renders the use of analytical methods very difficult. Analyses that involve repairable systems with multiple additional events and/or other maintainability information are very difficult to solve analytically (DFTs through state-space, numerical integration, Bayesian network approaches). DFTs through simulation approach [12] can incorporate these complexities and can give a wide range of output parameters. Algorithms based on Monte Carlo simulation were also proposed by Juan [17], which can be used to analyze a wide range of time-dependent complex systems, including those presenting multiple states, dependencies among failure/repair times, or non-perfect maintenance policies.

The simulation technique estimates the reliability indices by simulating the actual process and random behavior of the system in a computer model in order to create a realistic lifetime scenario of the system. This method treats the problem as a series of real experiments conducted in a simulated time. It estimates the probability and other indices by counting the number of times an event occurs in simulated time. The required information for the analysis is: probability density functions (PDFs) for time to failure and repair of all basic components with the parameter values; maintenance policies; interval and duration of tests and preventive maintenance.

Components are simulated for a specified mission time for depicting the duration of available (up) and unavailable (down) states. Up and down states will come alternatively; as these states are changing with time they are called state–time diagrams. A down state can be due to unexpected failure and its recovery will depend upon the time taken for repair action. Duration of the state is random for both up and down states. It will depend upon PDF of time to failure and time to repair respectively.

Evaluation of time to failure or time to repair for state–time diagrams. Consider a random variable x that is following an exponential distribution with parameter λ; $f(x)$ and $F(x)$ are given by the following expressions:

$$f(x) = \lambda \exp(-\lambda x) \tag{2.4}$$

$$F(x) = \int_0^x f(x)\mathrm{d}x = 1 - \exp(-\lambda x) \quad \text{Now } x \text{ is derived as a function of } F(x)\,, \tag{2.5}$$

$$x = G(F(x)) = \frac{1}{\lambda} \ln\left(\frac{1}{1 - F(x)}\right) \tag{2.6}$$

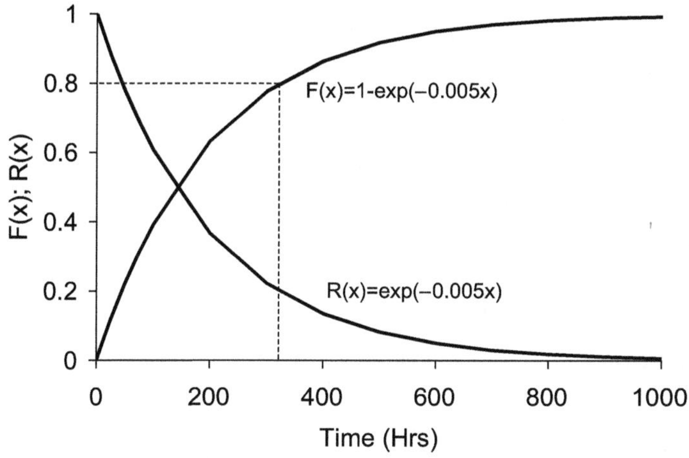

Figure 2.4 Exponential distribution

A uniform random number is generated using any of the standard random number generators. Let us assume 0.8 is generated by a random number generator, then the value of x is calculated by substituting 0.8 in place of $F(x)$ and say $1.8/\text{yr}$ $(5 \times 10^{-3}/\text{h})$ in place of λ in the above equation:

$$x = \frac{1}{5 \times 10^{-3}} \ln\left(\frac{1}{1-0.8}\right) = 321.8\,\text{h}$$

This indicates that the time to failure of the component is 321.8 h (see Figure 2.4). This procedure is applicable similarly for repair time also, and if the shape of the PDF is different, accordingly one has to solve for $G(F(x))$.

The solutions for four basic dynamic gates are explained here through a simulation approach [12].

2.7.1 PAND Gate

Consider a PAND gate having two active components. The active component is the one which is in working condition during normal operation of the system. Active components can be either in success state or failure state. Based on the PDF of failure of components, time to failure is obtained from the procedure mentioned above. The failure is followed by repair whose time depends on the PDF of repair time. This sequence is continued until it reaches the predetermined system mission time. Similarly for the second component, also state–time diagrams are developed.

For generating PAND gate state–time diagrams, both the components' state–time profiles are compared. The PAND gate reaches a failure state if all of its input components have failed in a pre-assigned order (usually from left to right). As shown

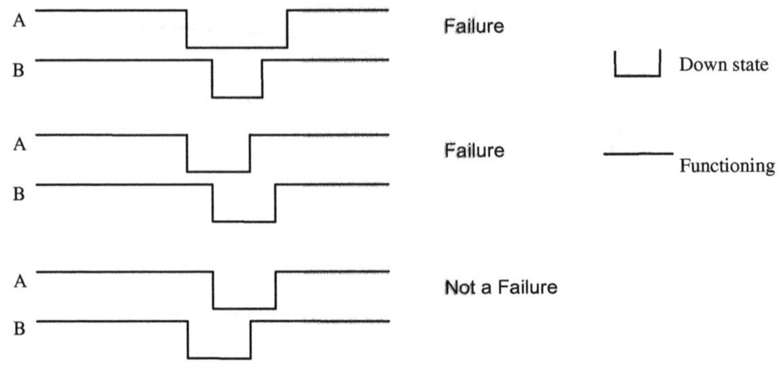

Figure 2.5 PAND gate state–time possibilities

in Figure 2.5 (first and second scenarios), when the first component failed followed by the second component, it is identified as failure and simultaneous down time is taken into account. But, in the third scenario of Figure 2.5, both the components have failed simultaneously but the second component has failed first, hence it is not considered as failure.

2.7.2 SPARE Gate

The SPARE gate will have one active component and remaining spare components. Component state–time diagrams are generated in a sequence starting with the active component followed by spare components in the order left to right. The steps are as follows:

1. *Active components.* Times to failure and times to repair based on their respective PDFs are generated alternatively until they reach mission time.
2. *Spare components.* When there is no demand, it will be in standby state or may be in failed state due to on-shelf failure. It can also be unavailable due to test or maintenance state as per the scheduled activity when there is a demand for it. This makes the component have multiple states and such stochastic behavior needs to be modeled to represent the practical scenario. Down times due to the scheduled test and maintenance policies are first accommodated in the component state–time diagrams. In certain cases test override probability has to be taken into account for its availability during testing. As the failures that occurred during the standby period cannot be revealed until its testing, time from failure until identification has to be taken as down time. It is followed by imposing the standby down times obtained from the standby time to failure PDF and time to repair PDF. Apart from the availability on demand, it is also required to check whether the standby component is successfully meeting its mission. This is incorporated by obtaining the time to failure based on the operating failure PDF and is checked with the mission time, which is the down time of the active

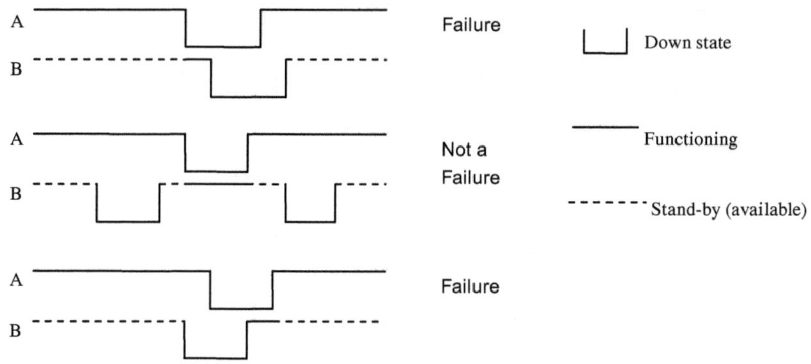

Figure 2.6 SPARE gate state–time possibilities

component. If the first standby component fails before the recovery of the active component, then demand will be passed on to the next spare component.

Various scenarios with the SPARE gate are shown in Figure 2.6. The first scenario shows that demand due to failure of the active component is met by the standby component, but it has failed before the recovery of the active component. In the second scenario, demand is met by the standby component. But the standby failed twice when it is in dormant mode, but it has no effect on success of the system. In the third scenario, the standby component was already in failed mode when the demand came, but it has reduced the overall down time due to its recovery afterwards.

2.7.3 FDEP Gate

The FDEP gate's output is a "dummy" output as it is not taken into account during the calculation of the system's failure probability. When the trigger event occurs, it will lead to the occurrence of the dependent event associated with the gate. Depending upon the PDF of the trigger event, failure time and repair times are generated. During the down time of the trigger event, the dependent events will be virtually in failed state though they are functioning. This scenario is depicted in the Figure 2.7. In the second scenario, the individual occurrences of the dependent events are not affecting the trigger event.

2.7.4 SEQ Gate

It is similar to the priority AND gate but occurrence of events are forced to take place in a particular fashion. Failure of the first component forces the other components to follow. No component can fail prior to the first component. Consider a three-

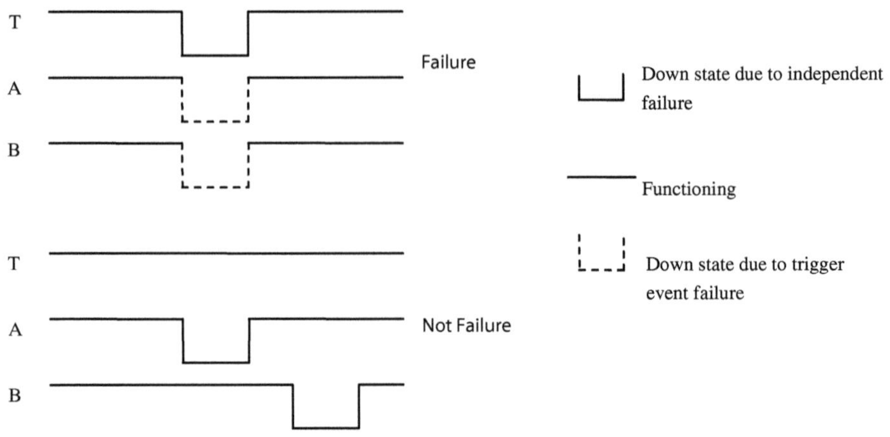

Figure 2.7 FDEP gate state–time possibilities

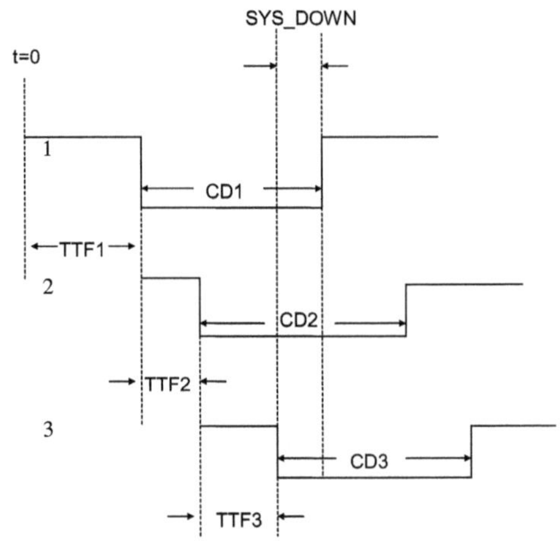

Figure 2.8 SEQ gate state–time possibilities. TTFi = Time to failure for ith component. CDi = Component down time for ith component. SYS_DOWN = System down time

input SEQ gate having repairable components. The following steps are involved with Monte Carlo simulation approach.

1. The component state–time profile is generated for the first component based upon its failure and repair rate. The down time of the first component is the mission time for the second component. Similarly the down time of the second component is the mission time for the third component.
2. When the first component fails, operation of the second component starts. The failure instance of the first component is taken as $t = 0$ for the second component. Time to failure (TTF2) and time to repair/component down time (CD2) is generated for the second component.
3. When the second component fails, operation of the third component starts. The failure instance of the second component is taken as $t = 0$ for the third compo-

nent. Time to failure (TTF3) and time to repair/component down time (CD3) is generated for the third component.
4. The common period in which all the components are down is considered as the down time of the SEQ gate.
5. The process is repeated for all the down states of the first component.

A software tool, DRSIM (Dynamic Reliability with SIMulation) has been developed by the authors to do comprehensive DTF analysis. The following examples have been solved with DRSIM.

2.8 Example 1: Simplified Electrical (AC) Power Supply System of Typical Nuclear Power Plant

Electrical power supply is essential in the operation of the process and safety system of any NPP. The grid supply (off-site-power supply) known as a Class IV supply is the one which feeds all these loads. To ensure high reliability of the power supply, redundancy is provided with the diesel generators known as a Class III supply (also known as on-site emergency supply) in the absence of a Class IV supply to supply the loads. There will be sensing and control circuitry to detect the failure of a Class IV supply which triggers the redundant Class III supply [18]. Loss of the off-site power supply (Class IV) coupled with loss of on-site AC power (Class III) is called station blackout. In many PSA studies [19], severe accident sequences resulting from station blackout conditions have been recognized to be significant contributors to the risk of core damage. For this reason the reliability/availability modelling of AC Power supply system is of special interest in PSA of NPP.

The reliability block diagram is shown in Figure 2.9. Now this system can be modeled with the dynamic gates to calculate the unavailability of overall AC power supply of a NPP.

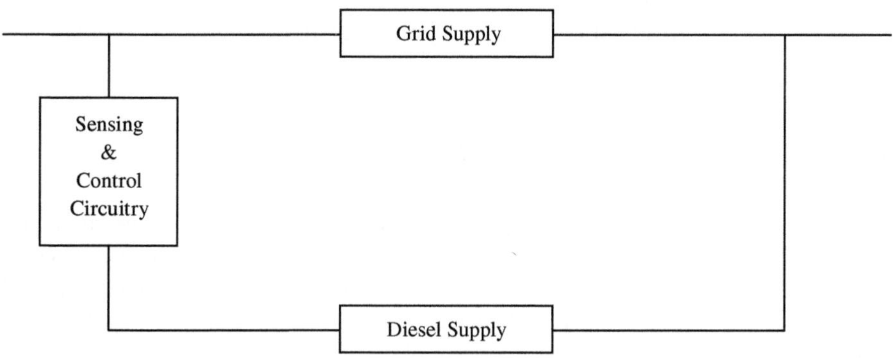

Figure 2.9 Reliability block diagram of electrical power supply system of NPP

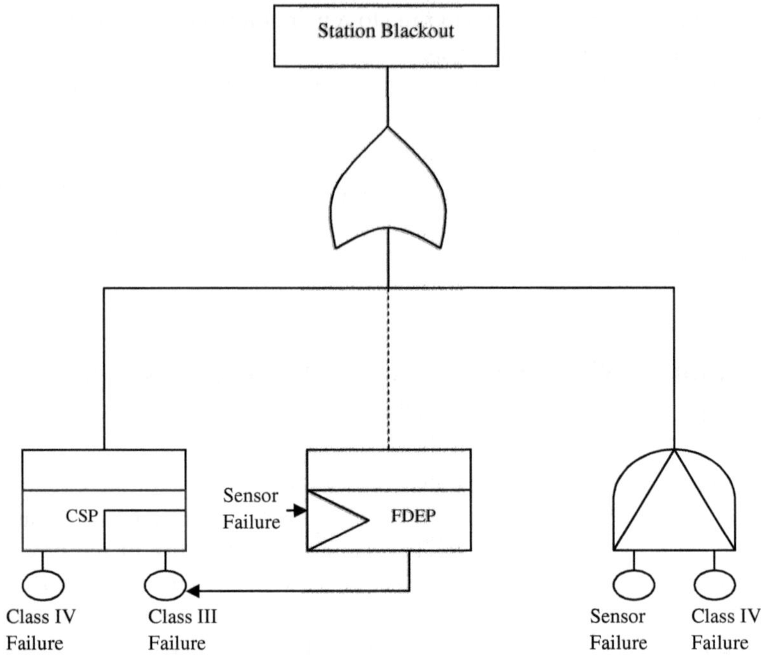

Figure 2.10 Dynamic fault tree for station blackout

The DTF (Figure 2.10) has one PAND gate having two events, namely, sensor and Class IV. If the sensor fails first then it will not be able to trigger the Class III, which will lead to non-availability of power supply. But if it fails after already triggering Class III due to occurrence of Class IV failure first, it will not affect the power supply. As Class III is a standby component to Class IV, it is represented with a spare gate. This indicates their simultaneous unavailability will lead to supply failure. There is a functional dependency gate as the sensor is the trigger signal and Class III is the dependent event.

This system is solved with an analytical approach and Monte Carlo simulation.

2.8.1 Solution with Analytical Approach

Station blackout is the top event of the fault tree. Dynamic gates can be solved by developing state-space diagrams and their solutions give required measures of reliability. However, for subsystems which are tested (surveillance), maintained, and repaired, if any problem is identified during check-up, it cannot be modeled by state-space diagrams. However, there is a school of thought that initial state probabilities can be given as per the maintenance and demand information; this is often debatable. A simplified time-averaged unavailability expression is suggested by IAEA P-4 [20]

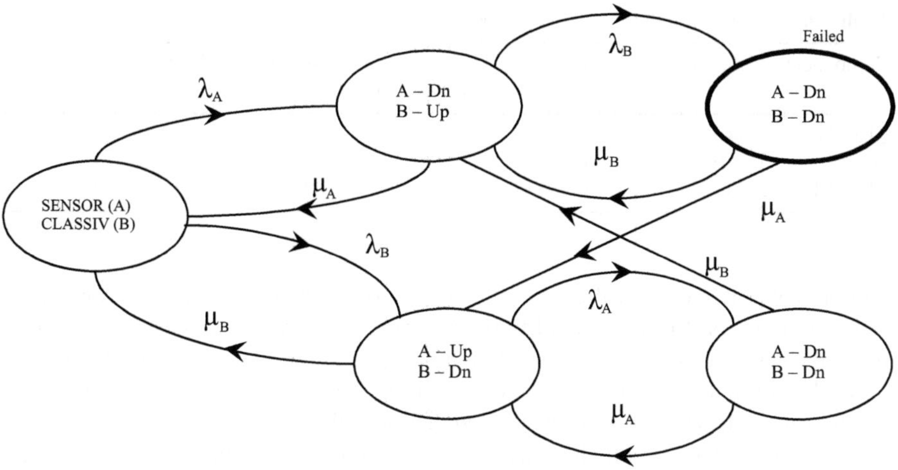

Figure 2.11 Markov (state-space) diagram for PAND gate having sensor and Class IV as inputs

for standby subsystems having exponential failure/repair characteristics. The same is applied here to solve the standby gate. If Q is the unavailability of the standby component, it is expressed by the following equation, where λ is failure rate, T is test interval, τ is test duration, f_m is frequency of preventive maintenance, T_m is duration of maintenance, and T_r is repair time. It is the sum of contributions from failures, test outage, maintenance outage, and repair outage. In order to obtain the unavailability of the standby gate, the unavailability of Class IV is multiplied by the unavailability of the standby component (Q):

$$Q = \left[1 - \frac{1 - e^{-\lambda T}}{\lambda T}\right] + \left[\frac{\tau}{T}\right] + [f_m T_m] + [\lambda T_r] \tag{2.7}$$

The failure of the sensor and Class IV is modeled by a PAND gate in the fault tree. This is solved by a state-space approach by developing a Markov model as shown in Figure 2.11. The bolded state where both the components failed in the required order is the unavailable state and remaining states are all available states. ISOGRAPH software has been used to solve the state-space model. Input parameter values used in the analysis are shown in Table 2.6 [21]. The sum of the both the values (PAND and SPARE) give the unavailability of station blackout scenario which is obtained as 4.847×10^{-6}.

2.8.2 *Solution with Monte Carlo Simulation*

As one can see, the Markov model for a two-component dynamic gate has 5 states with 10 transitions, thus the state space becomes unmanageable as the number of

Table 2.6 Component failure and maintenance information

Component	Failure rate (/h)	Repair rate (/h)	Test period (h)	Test time (h)	Maint. period (h)	Maint. time (h)
Class IV	2.34×10^{-4}	2.59	–	–	–	–
Sensor	1×10^{-4}	0.25	–	–	–	–
Class III	5.33×10^{-4}	0.08695	168	0.0833	2160	8

components increases. In the case of standby components, the time-averaged analytical expression for unavailability is only valid for exponential cases. To address these limitations, Monte Carlo simulation is applied here to solve the problem.

In the simulation approach, random failure/repair times from each component's failure/repair distributions are generated. These failure/repair times are then combined in accordance with the way the components are arranged reliability-wise within the system. As explained in the previous section, the PAND gate and SPARE gate can easily be implemented through the simulation approach. The difference from the normal AND gate to PAND and SPARE gates is that the sequence of failure has to be taken into account and standby behavior including the testing, maintenance, and dormant failures has to be accommodated. The unique advantage with simulation is incorporating non-exponential distributions and eliminating S-independent assumption.

Component state–time diagrams are developed as shown in Figure 2.12 for all the components in the system. For active components which are independent, only two states will be there, one is functioning state (UP – operational state) and second is repair state due to failure (DOWN – repair state). In the present problem, Class IV and sensor are active components, whereas Class III is the standby component. For

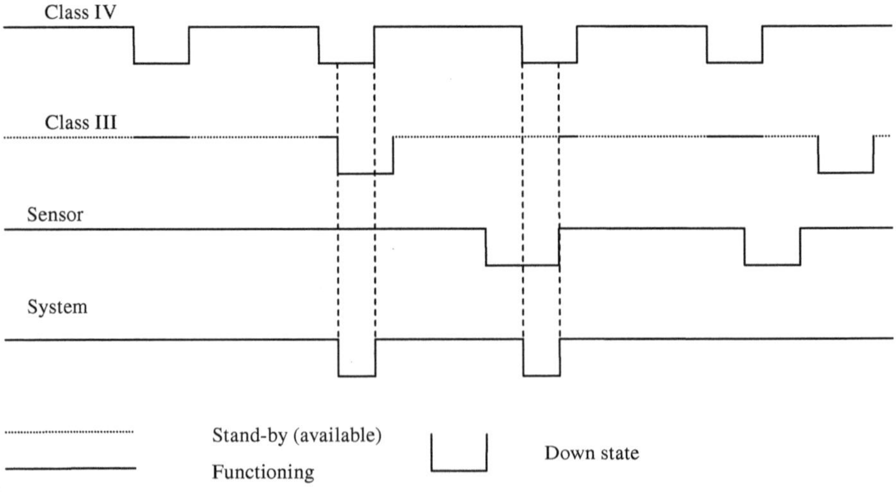

Figure 2.12 State–time diagrams for Class IV, sensor, Class III, and overall system

Class III, generation of state–time diagram involves more calculations than former. It is having six possible states, namely: testing, preventive maintenance, corrective maintenance, standby functioning, standby failure undetected, and normal functioning to meet the demand. As testing and preventive maintenance are scheduled activities, they are deterministic and are initially accommodated in component profile. Standby failure, demand failure and repair are random and according to their PDF the values are generated. The demand functionality of Class III depends on the functioning of sensor and Class IV. Initially after generating the state–time diagrams of sensor and Class IV, the DOWN states of Class IV is identified and sensor availability at the beginning of the DOWN state is checked to trigger the Class III. The reliability of Class III during the DOWN state of Class IV is checked. Monte-Carlo simulation code has been developed for implementing the station blackout studies. The unavailability obtained is 4.8826×10^{-6} for a mission time of 10,000 h with 10^6 simulations which is in agreement with the analytical solution. Failure time, re-

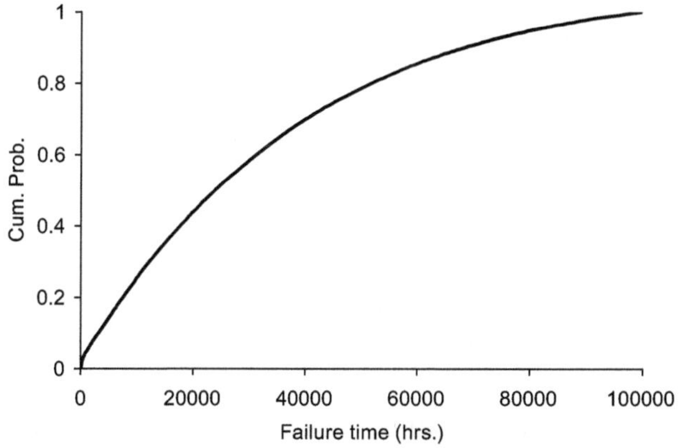

Figure 2.13 Failure time distribution

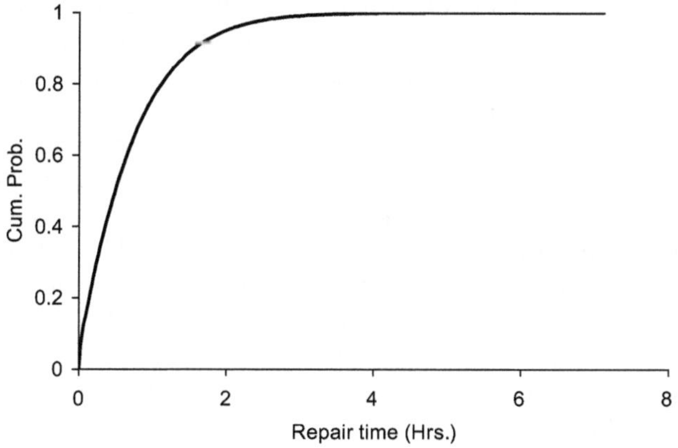

Figure 2.14 Repair time distribution

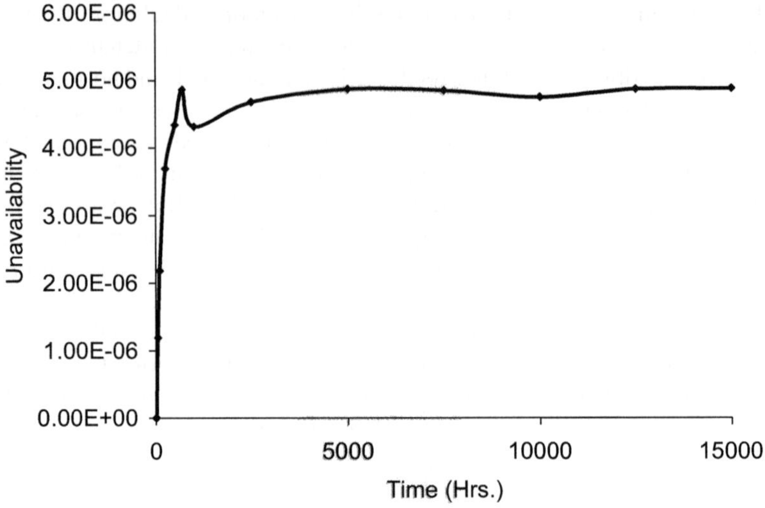

Figure 2.15 Unavailability with time

pair time and unavailability distributions are shown in Figures 2.13, 2.14, and 2.15 respectively.

2.9 Example 2: Reactor Regulation System of a Nuclear Power Plant

The reactor regulation system (RRS) regulates rector power in the NPP. It is a computer-based feedback control system. The regulating system is intended to control the reactor power at a set demand from 10^{-7} FP to 100% FP by a generating control signal for adjusting the position of adjuster rods and adding poison to the moderator in order to supplement the worth of adjuster rods [22–24]. The RRS has a dual-processor hot standby configuration with two systems, namely, system A and system B. All inputs (analog and digital or contact) are fed to system A as well as system B. On failure of system A or B, the control transfer unit (CTU) will automatically change over the control from system A to system B and *vice versa*, if the system to which control is transferred is healthy. Control transfer will also be possible through manual command by an external switch. This command will be ineffective if the system, to which control is desired to be transferred, is declared unhealthy. Transfer logic will be implemented through CTU. To summarize, the above described computer-based system has failures needs to happen in a specific sequence, to be declared as system failure. Dynamic fault tree should be constructed for realistic reliability assessment.

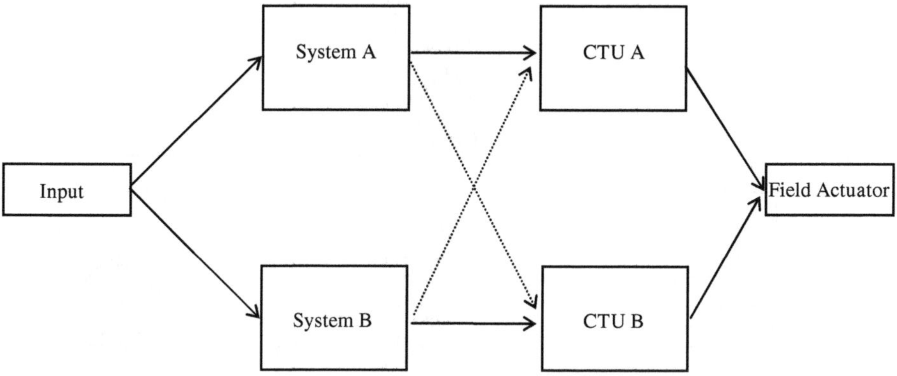

Figure 2.16 Simplified block diagram of reactor regulator system

2.9.1 Dynamic Fault Tree Modeling

The important issue that arises in modeling is the dynamic sequence of actions involved in assessing the system failure. The top event for RRS, "Failure of Reactor Regulation," will have the following sequence of failures to occur:

1. Computer system A or B fails.
2. Transfer of control to hot standby system by automatic mode through relay switching and CTU fails.
3. transfer of control to hot standby system by manual mode through operator intervention and hand switches fails after the failure of auto mode.

PAND and SEQ gates are used, as shown in Figure 2.17, to model these dynamic actions. The PAND gate has two inputs, namely, auto transfer and system A/B failure. Auto transfer failure after the failure of system A/B has no effect as the switching action has already taken place. The sequence gate has two inputs, one from the PAND gate and another from manual action. Chances of manual failure only arise after the failure of AUTO and SYS A/B. Manual action has four events, in which three are hand switch failures and one is OE (operator error). AUTO has only two events, failure of control transfer unit and failure of relay. System A/B has many basic events and failure of any these basic events will lead to the failure, represented by the OR gate.

2.10 Summary

In order to simplify the complex reliability problems, conventional approaches make many assumptions to create a simple mathematical model. Use of the DTF approach eliminates many of the assumptions that are inevitable with conventional approaches to model the complex interactions. It is found that in certain scenarios, assuming

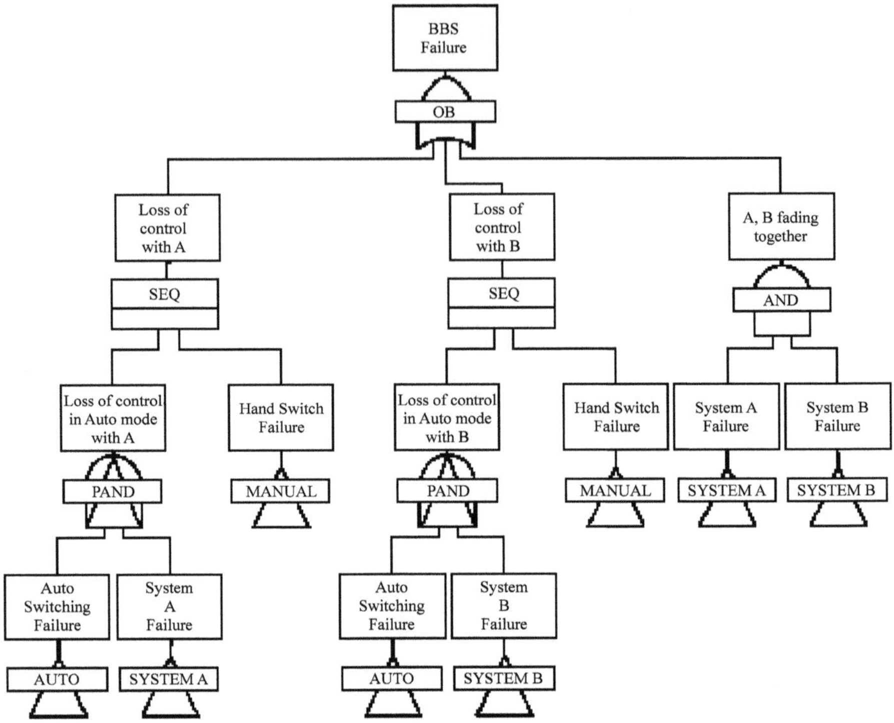

Figure 2.17 Dynamic fault Tree of DPHS-RRS

static AND in place of PAND can give erroneous results by several orders. This is explained in Section 2.3 with an example (PAND/AND with two inputs). The difference in the results is significant where the repair rate of first component is larger than the second component (repair time of first component is smaller than the second), irrespective of their failure rates.

The solution for dynamic gates through analytical approaches such as Markov models, Bayesian belief methods and numerical integration method have limitations in terms of number of basic events, non-exponential failure or repair distributions, incorporating test and maintenance policies and in a situation where the output of one dynamic gate being input to another dynamic gate. The Monte Carlo simulation-based DTF approach, due to its inherent capability in simulating the actual process and random behavior of the system, can eliminate these limitations in reliability modeling. Although computational time is the constraint, the incredible development in the computer technology for data processing at unprecedented speed levels is further emphasizing the use of a simulation approach to solve dynamic reliability problems. In Section 2.7 all the basic dynamic gates (PAND, SEQ, SPARE, and FDEP) are explained with Monte Carlo simulation approach. Examples demonstrate application of DTF in practical problems.

Acknowledgements The authors are grateful to Shri H.S. Kushwaha, Dr. A.K. Ghosh, Dr. G. Vinod, Mr. Vipin Saklani, and Mr. M. Pavan Kumar for their invaluable support provided during the studies on DFT.

References

1. Dugan JB, Bavuso SJ, Boyd MA (1992) Dynamic fault-tree for fault-tolerant computer systems. IEEE Trans Reliab 41(3):363–376
2. Amari S, Dill G, Howald E (2003) A new approach to solve dynamic fault trees. In: Annual IEEE reliability and maintainability symposium. Institute of Electrical and Electronics Engineers, New York, pp 374–379
3. Bobbio A, Portinale L, Minichino M, Ciancamerla E (2001) Improving the analysis of dependable systems by mapping fault trees into Bayesian networks. Reliab Eng Syst Saf 71:249–260
4. Dugan JB, Sullivan KJ, Coppit D (2000) Developing a low cost high-quality software tool for dynamic fault-tree analysis. IEEE Trans Reliab 49:49–59
5. Meshkat L, Dugan JB, Andrews JD (2002) Dependability analysis of systems with on-demand and active failure modes using dynamic fault trees. IEEE Trans Reliab 51(3):240–251
6. Huang CY, Chang YR (2007) An improved decomposition scheme for assessing the reliability of embedded systems by using dynamic fault trees. Reliability Eng Syst Saf 92(10):1403–1412
7. Bobbio A, Daniele CR (2004) Parametric fault trees with dynamic gates and repair boxes. In: Proceedings annual IEEE reliability and maintainability symposium. Institute of Electrical and Electronics Engineers, New York, pp 459–465
8. Manian R, Coppit DW, Sullivan KJ, Dugan JB (1999) Bridging the gap between systems and dynamic fault tree models. In: Proceedings Annual IEEE reliability and maintainability symposium. Institute of Electrical and Electronics Engineers, New York, pp 105–111
9. Cepin M, Mavko B (2002) A dynamic fault tree. Reliab Eng Syst Saf 75:83–91
10. Marseguerra M, Zio E, Devooght J, Labeau PE (1998) A concept paper on dynamic reliability via Monte Carlo simulation. Math Comput Simul 47:371–382
11. Karanki DR, Rao VVSS, Kushwaha HS, Verma AK, Srividya A (2007) Dynamic fault tree analysis using Monte Carlo simulation. In: 3rd International conference on reliability and safety engineering, IIT Kharagpur, Udaipur, India, pp 145–153
12. Karanki DR, Vinod G., Rao VVSS, Kushwaha HS, Verma AK, Ajit S (2009) Dynamic fault tree analysis using Monte Carlo simulation in probabilistic safety assessment. Reliab Eng Syst Saf 94:872–883
13. Zio E, Podofillinia L, Zille V (2006) A combination of Monte Carlo simulation and cellular automata for computing the availability of complex network systems. Reliab Eng Syst Saf 91:181–190
14. Marquez AC, Heguedas AS, Iung B (2005) Monte Carlo-based assessment of system availability. Reliab Eng Syst Saf 88:273–289
15. Zio E, Marella M, Podollini L (2007) A Monte Carlo simulation approach to the availability assessment of multi-state systems with operational dependencies. Reliab Eng Syst Saf 92:871–882
16. Zio, E. Podofillinia, L. Levitin, G (2004) Estimation of the importance measures of multi-state elements by Monte Carlo simulation. Reliab Eng Syst Saf 86:191–204
17. Juan A, Faulin J, Serrat C, Bargueño V (2008) Improving availability of time-dependent complex systems by using the SAEDES simulation algorithms. Reliab Eng Syst Saf 93(11):1761–1771
18. Saraf RK, Babar AK, Rao VVSS (1997) Reliability Analysis of Electrical Power Supply System of Indian Pressurized Heavy Water Reactors. Bhabha Atomic Research Centre, Mumbai, BARC/1997/E/001
19. IAEA-TECDOC-593 (1991) Case study on the use of PSA methods: Station blackout risk at Millstone unit 3. International Atomic Energy Agency, Vienna

20. IAEA (1992) Procedure for conducting probabilistic safety assessment of nuclear power plants (level 1). Safety series No. 50-P-4. International Atomic Energy Agency, Vienna
21. IAEA TECDOC 478 (1988) Component reliability data for use in probabilistic safety assessment. International Atomic Energy Agency, Vienna
22. Dual processor hot standby reactor regulating system (1995) Specification No. PPE-14484. http://www.sciencedirect.com/science?_0b=ArticleURL&_udi=B6V4T-4TN82FN-1&_user=971705&_coverDate=04%2F30%2F2009&_rdoc=1&_fmt=high&_orig=search&_sort=d&_docanchor=&view=c&_searchStrId=1202071465&_rerunOrigin=google&_acct=C000049641&_version=1&_urlVersion=0&_userid=971705&md5=c499df740691959e0d0b59f20d497316
23. Gopika V, Santosh TV, Saraf RK, Ghosh AK (2008) Integrating safety critical software system in probabilistic safety assessment. Nucl Eng Des 238(9):2392–2399
24. Khobare SK, Shrikhande SV, Chandra U, Govindarajan G (1998) Reliability analysis of microcomputer circuit modules and computer-based control systems important to safety of nuclear power plants. Reliab Eng Syst Saf 59:253–258

Chapter 3
Analysis and Improvements of Path-based Methods for Monte Carlo Reliability Evaluation of Static Models

Héctor Cancela, Pierre L'Ecuyer, Matías Lee, Gerardo Rubino, and Bruno Tuffin

Abstract Many dependability analyses are performed using static models, that is, models where time is not an explicit variable. In these models, the system and its components are considered at a fixed point in time, and the word "static" means that the past or future behavior is not relevant for the analysis. Examples of such models are reliability diagrams, or fault trees. The main difficulty when evaluating the dependability of these systems is the combinatorial explosion associated with exact solution techniques. For large and complex models, one may turn to Monte Carlo methods, but these methods have to be modified or adapted in the presence of rare important events, which are commonplace in reliability and dependability systems. This chapter examines a recently proposed method designed to deal with the problem of estimating reliability metrics for highly dependable systems where the failure of the whole system is a rare event. We focus on the robustness properties of estimators. We also propose improvements to the original technique, including its combination with randomized quasi-Monte Carlo, for which we prove that the variance converges at a faster rate (asymptotically) than for standard Monte Carlo.

Héctor Cancela
Universidad de la República, Uruguay

Pierre L'Ecuyer
Université de Montréal, Canada

Matías Lee
Universidad Nacional de Córdoba, Argentina

Gerardo Rubino · Bruno Tuffin
INRIA, France

P. Faulin, A. Juan, S. Martorell, and J.E. Ramírez-Márquez (eds), *Simulation Methods for Reliability and Availability of Complex Systems*. © Springer 2010

3.1 Introduction

Dependability analysis of complex systems is sometimes performed using dynamic stochastic models. The system is represented by some type of stochastic process such as a Markov or a semi-Markov one, and different dependability metrics (reliability, point availability, interval availability, etc.) are evaluated as functions of the process at a fixed point in time (e.g., reliability or point availability), over a finite interval (e.g., interval availability), or in equilibrium (e.g., asymptotic availability). But in many cases, the system is considered in a context where the time variable plays no specific role. These models are called *static*, and are widely used in engineering, taking the form of specific mathematical objects such as reliability networks, reliability diagrams, fault trees, etc.

The basic scheme is the following. The system is composed of M components that typically are subsystems of the original one, and are considered as atoms in the modeling effort. Each component and the whole system can be in two different states, either operational or failed. The set of states of the M components is a *configuration* or *state-vector* of the system (hence, there are at most 2^M such configurations, since not all configurations are necessarily possible in the model of a specific system). We assume that the probability of each configuration is known. The main system dependability metric is the *reliability* R of the system, the probability that the whole system is operational, or equivalently, its *unreliability* $U = 1 - R$, the probability that the whole system fails. The reliability is the sum of the probabilities of all the configurations leading to an operational state for the whole system, and the unreliability is the corresponding sum of the probabilities of all the configurations leading to a failed system. In such a static context, R is sometimes also called the availability of the system. The function Φ mapping the configurations into one of the two possible system states is called the structure function of the system. It provides the information about the way the M components are organized from the dependability point of view, that is, the way the combination of operational and failed components leads to an operational or failed system. The different modeling frameworks (reliability networks, fault trees, etc.) can be seen as different languages that allow for a compact representation of structure functions.

Suppose that the components behave independently, and that for each component we know the probability that it is in the operational state. We number the components from 1 to M, and r_i is the probability that component i is working. Coding by 1 the operational state (of a component, or of the whole system) and by 0 the failed state, we have that

$$R = \sum_{x:\Phi(x)=1} p(x) ,$$

where x denotes a configuration and $p(x)$ its probability, $x = (x_1, \ldots, x_M)$, and x_i is the state of component i in configuration x. The independence assumption on the states of the components means that for any configuration x we have

$$p(x) = \prod_{i:x_i=1} r_i \prod_{j:x_j=0} (1 - r_j) \, .$$

We are interested in the case where $R \approx 1$, or equivalently, $U \approx 0$, the usual situation in many areas, typically in the analysis of critical systems. These are systems where a failure may produce losses in human lives (transporting facilities, nuclear plants, etc.) or huge losses in monetary terms (information systems, telecommunication networks, etc.), so that system design is extremely conservative, ensuring very low failure probability. This is a rare event context, the rare event being the system failure. If X is a random configuration (randomness coming from the fact that the component state is assumed to be a random variable), then, the rare event is "$\Phi(X) = 0$," and $\Pr(\Phi(X) = 0) = 1 - R = U$. Since these are binary random variables, $R = \mathrm{E}(\Phi(X))$ and $U = \mathrm{E}(1-\Phi(X))$, where E( ) denotes the expectation operator.

In this chapter, we address the problem of estimating U (or R) using Monte Carlo, where the structure function is given by means of a graph. Think of a communication network represented by an undirected graph $G = (V, E)$, where V is the set of nodes and E is the set of edges, also referred to as links in this context. The graph is supposed to be connected and without loops. The components are, for instance, the edges, and recall they are assumed to operate independently. Associated with edge i we have its (elementary) reliability r_i (or equivalently, its unreliability u_i); if X_i is the binary random variable "state of component i," we have $r_i = \Pr(x_i = 1)$, and $u_i = \Pr(x_i = 0)$, $r_i + u_i = 1$. A configuration is a vector $x = (x_1, \ldots, x_M)$, where x_i is the state of component (here, edge) i. We denote by $\mathrm{Op}(x)$ the subset of operational edges in configuration x, that is, $\mathrm{Op}(x) = \{i \in E : x_i = 1\}$, and by $G(x)$ the graph $G(x) = (V, \mathrm{Op}(x))$. It remains to specify when the whole system works, that is, to define the structure function. For this purpose, two nodes are selected in V, denoted by the letters s (as source) and t (as terminal). The system (the network) works under the configuration x if nodes s and t belong to the same connected component of the random graph $G(X)$. That is, $\Phi(x) = 1$ iff in $G(x)$, nodes s and t are connected. This model is the basic one in the network reliability area, and it corresponds to the typical model in reliability block diagrams. Computing $R = \Pr(\Phi(X) = 1)$ is an NP-complete problem, even in very restricted classes of graphs. More specifically, the use of the many combinatorial approaches for computing R or U cannot deal with models of moderate size (around, say, 100 components) and simulation is the only available evaluation tool.

For general presentations about the computation of the reliability or the unreliability in these static contexts, or about bounding them, as well as complexity issues, see [1–4]. In these references the reader can also find material about Monte Carlo estimation of these basic dependability metrics.

3.2 Standard Monte Carlo Reliability Evaluation

The standard estimation procedure for U (or R) simply consists in building a sequence $X^{(1)}, X^{(2)}, \cdots, X^{(n)}$ of independent copies of the random configuration X, and checking in each graph of the corresponding sequence $G(X^{(1)}), \ldots, G(X^{(n)})$ if s and t are connected. The ratio between the number of times s and t are not connected and n, is then an unbiased estimator $\hat{U}$ of U. Formally,

$$\hat{U} = \frac{1}{n}\sum_{i=1}^{n} 1\left(\Phi\left(X^{(i)}\right) = 0\right),$$

where $1(A)$ is the indicator function of event A. The variance of $\hat{U}$ being $\mathrm{Var}(\hat{U}) = \sigma_n^2 = U(1-U)/n$, a confidence interval for U, with level $\alpha \in [0,1]$, is obtained from the central limit theorem:

$$U \in \left(\hat{U} - z_{1-\alpha/2}\sqrt{U\,(1-U)\,/n},\ \hat{U} + z_{1-\alpha/2}\sqrt{U\,(1-U)\,/n}\right),$$

with probability $1-\alpha$, where $z_{1-\alpha/2}$ is the $1-\alpha/2$ quantile of the normal law with mean 0 and variance 1. In many interesting and important systems, the reliability of the components is close to one and the path redundancy in the graph makes that the probability of the existence of at least a path between the two selected nodes is extremely high. Both factors make the unreliability of the whole network very small. This precludes the use of the standard estimation approach, since we have to wait for a long time (on average) before observing a system failure. In other words, the cost in time of the standard procedure is very high.

To formalize this situation, assume that the unreliability of link i is $u_i = a_i\varepsilon^{b_i}$ with $a_i, b_i > 0$ and $0 < \varepsilon \ll 1$. Recall that a cut in the graph (with respect to nodes s and t) is a set of edges such that if we delete them from the graph, s and t become unconnected. A *mincut* is a cut that does not contain strictly another cut. Nodes s and t are unconnected if and only if for at least one (min)cut in the graph, all the edges that compose it are down. If γ is a mincut, we can denote by C_γ the event "all the edges in γ are down," and write

$$U = \Pr\left(\bigcup_{\text{all mincuts } \gamma} C_\gamma\right).$$

Observing that, due to the independence of the components' states, $\Pr\left(C_\gamma\right) = \prod_{i\in\gamma} u_i$, we see that U is a polynomial in ε and that $U = \Theta(\varepsilon^c)$ for some $c > 0$ (recall that the graph is connected, so, there is at least one cut separating nodes s and t).

The real number ε is a way to parameterize rarity: as ε goes to zero, the system failure event becomes increasingly rare. The relative error [5,6] when estimating U using $\hat{U}$, defined as the ratio between the square root of the variance of the estimator and its mean, i.e. $\sqrt{U(1-U)/n}/U$ (also called *relative variance*, or *coefficient of variation*) is $\approx (nU)^{-1/2}$ when ε is small, and increases as ε decreases. We want this relative error to be small, but not at any price! This means that we would like to avoid using an important computing effort in order to obtain specific error levels. That is, the CPU time required to compute the estimator from a sample of size n must also be taken into account. For this purpose, we consider the work-normalized relative variance (WNRV) of the estimator $\hat{U}$, defined by

$$\text{WNRV}\left(\hat{U}\right) = \frac{\sigma_n^2 \tau_n}{U^2},$$

where τ_n is the mean time needed to compute $\hat{U}$ using a sample of size n. Here, this time is essentially linear in n. What we want now is that this ratio remains bounded when $\varepsilon \to 0$. In other words, no matter how rare the system failure is, we would like to be able to estimate it accurately with "reasonable" effort. This property is called bounded WNRV (BWNRV), and it does not hold for $\hat{U}$, because WNRV($\hat{U}$) is proportional to $1/U$, and $1/U \to \infty$ when $\varepsilon \to 0$.

In this work we discuss efficient Monte Carlo methods for the estimation of the unreliability U of the network, by combining two approaches. First, as in other works, we use easy-to-get knowledge about the network, namely its path structure, to follow a conditional approach allowing us to bound the target metrics (this is based on ideas presented in [7]). We show, in particular, how to derive methods having BWNRV in the homogeneous-components case. We also exhibit a counter-example in the heterogeneous case, that is, a case of unbounded WNRV. Secondly, we explore the randomized quasi-Monte Carlo (RQMC) technique in this context, in order to further reduce the variance of the estimators. These methods are usually effective mostly to estimate the integrals of smooth functions over the unit hypercube, when the function depends only or mostly on a few coordinates. They often perform poorly for discontinuous integrands. However, in our case, RQMC performs very nicely both theoretically (with a provably faster convergence rate) and empirically. Numerical results illustrate and compare the effectiveness of the different techniques considered, as well as their combination.

For general material about Monte Carlo approaches in this area, in addition to some general references [2–4] given earlier, the reader can see [8] where many different procedures are described. In the same book [9], completely devoted to rare-event estimation using Monte Carlo techniques, other chapters contain related material focused on other aspects of the problems and the methods available to solve them.

3.3 A Path-based Approach

In [7] a technique for facing the problem of rarity is proposed. The idea is to start by building a set $\boldsymbol{P} = \{P_1, P_2, \ldots, P_H\}$ of elementary paths (no node appears more than once in the path) connecting nodes s and t, such that any pair of paths does not share any link (that is, $\boldsymbol{P}$ is a set of edge-disjoint paths between source and terminal). As we will recall later, this is not a computationally expensive task (compared to the cost of Monte Carlo procedures) that can be performed in polynomial time.

Let $p_h = \prod_{i \in P_h} r_i$ denote the probability that all links of path P_h work. Assume $X^{(1)}, X^{(2)}, \ldots$ is a sequence of independent copies of the random configuration X, and that $G(X^{(1)}), G(X^{(2)}), \ldots$ is the associated sequence of random partial graphs of G. The main idea of the method is to consider the random variable F equal to the index of the first graph in this list where every path in $\boldsymbol{P}$ has at least one link that does not work. Clearly, F is geometrically distributed with parameter $q = \prod_{h=1}^{H} (1 - p_H)$: that is, $\Pr(F > f) = (1-q)^f$, $f \geqslant 1$. In particular, $\mathrm{E}(F) = 1/q$.

Let us write $P_h = (i_{h,1}, \ldots, i_{h,M_h})$ for $P_h \in \boldsymbol{P}$, and let $b_h = \min_{1 \le m \le M_h} b_{i_{h,m}}$ be the order (in ε) of the most reliable edge of P_h. We then have $1 - p_h = \Theta\left(\varepsilon^{b_h}\right)$ and $q = \Theta\left(\varepsilon^{b}\right)$ where $b = \sum_{h=1}^{H} b_h > 0$. Observe that $q \to 0$ as $\varepsilon \to 0$. The fact that $\mathrm{E}(F) = 1/q$ means that, on the average, we have to wait for $1/q$ samples to find a graph where at least a link is failed in each path of $\boldsymbol{P}$. This suggests sampling first from F. If the value f is obtained for F, then we assume that in a "virtual" sequence of copies of $G(X)$, in the first $f - 1$ elements nodes s and t are always connected. It remains to deal with the fth copy. Let Y be a binary random variable defined as follows: if C is the event "every path in $\boldsymbol{P}$ has at least one link that does not work", then $\Pr(Y = 1) = \Pr(\Phi(x) = 1 | C)$. According to this "interpretation" of the sampling of F, the state of the network in the fth graph is modeled by Y.

We need now a sampling procedure for Y. Consider a path $P_h = (i_{h,1}, i_{h,2}, \ldots, i_{h,M_h})$ belonging to $\boldsymbol{P}$. Let W_h be the random variable giving the index of the first failed edge of P_h in the order of the links in the path, $W_h \in \{1, 2, \ldots, M_h\}$. For each path P_h in $\boldsymbol{P}$, we have [7]

$$\Pr(W_h = w) = \frac{r_{i_{h,1}} r_{i_{h,2}} \cdots r_{i_{h,w-1}} \left(1 - r_{i_{h,w}}\right)}{1 - r_{i_{h,1}} r_{i_{h,2}} \cdots r_{i_{h,M_h}}},$$

which simply translates the definition of W_h into a formula. Sampling Y consists in first sampling the state of every link in the model, and then checking by a standard procedure, typically a depth-first search or a breadth-first search method, if s and t are unconnected or not. Since we are assuming that in every path of $\boldsymbol{P}$, at least one link is failing, we first sample the states of the components of P_h for $h = 1, 2, \ldots, H$, then the states of the remaining edges in the graph. To sample the states of the links in P_h, we first sample from the distribution of W_h. Assume we get value w. We set the states of edges $i_{h,1}, i_{h,2}, \ldots, i_{h,w-1}$ (that is, random variables $X_{i_{h,1}}, \ldots, X_{i_{h,w-1}}$) to 1 and that of edge $i_{h,w}$ to 0. The states of the remaining edges in P_h, if any, are sampled from their *a priori* original Bernoulli distributions, and the same for the edges not belonging to any path in $\boldsymbol{P}$. Then, we sample from Y, obtaining either 1 or 0 according to the fact that nodes s and t are respectively not connected or connected, and we interpret this as a sample of the state of a network where we know that in every path in $\boldsymbol{P}$ at least one link is failed.

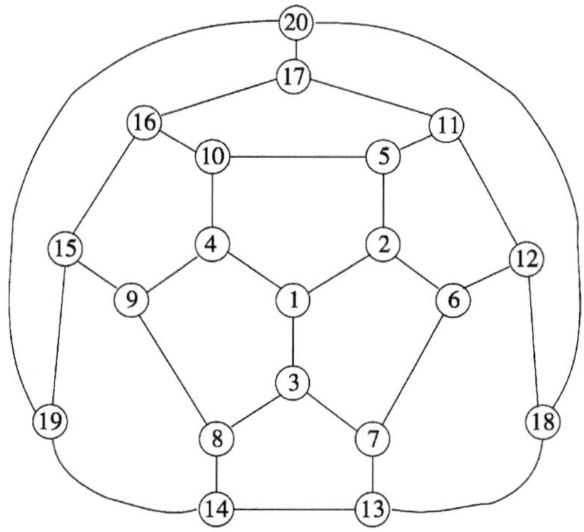

Figure 3.1 A "dodecahedron" (20 nodes, 30 links). All links have reliability $1 - \varepsilon$

Resuming, we will build, say, K independent copies $F_1, \ldots, F_K$ of F together with K independent copies $Y_1, \ldots, Y_K$ of Y, and will use as an estimator of U the number

$$\tilde{U} = \frac{\sum_{k=1}^{K} Y_k}{\sum_{k=1}^{K} F_k} .$$

To illustrate the gain obtained with this algorithm, let us consider the "dodecahedron" shown in Figure 3.1, a structure often used as a benchmark for network reliability evaluation techniques. We consider the homogeneous case, where all links have the same unreliability ε. The source and the terminal are nodes 1 and 20.

The gain in efficiency with respect to the standard procedure is captured by the ratio between the WNRV values for $\hat{U}$ and U. We call *relative efficiency* of U with respect to $\hat{U}$ the ratio $\sigma_n^2 \tau_n / \left(\tilde{\sigma}_n^2 \tilde{\tau}_n^2\right)$ with $\tilde{\sigma}_n^2$ and $\tilde{\tau}_n^2$ the variance and the mean computation time of U for a sample of size n. We estimated the system unreliability for $n = 10^7$ replications, for three cases: $\varepsilon = 0.1$, 0.01, and 0.001. The estimated relative efficiency was, respectively, of 18.9, 188.3, and 3800.2. This illustrates the power of the approach.

3.4 Robustness Analysis of the Algorithm

In [7], it is pointed out that we can still use a fixed number of samples n, by calling F a random number W of times, where $W = \max\left\{K \geqslant 1 : \sum_{k=1}^{K} F_k \leqslant n\right\}$, and using the unbiased estimator

$$U^* = \frac{1}{n} \sum_{k=1}^{W} Y_k \ .$$

In other words, we are "wasting" some results (the last ones) of the virtual sampling process associated with $\hat{U}$. The variance of U^* is then $\text{Var}(U^*) = \sigma_n^2 = U(1-U)/n$, because this is simply an efficient way of implementing the standard estimator.

The point is that while we have not modified the variance with respect to the standard estimator, we did obtain an important gain in time. Let us denote by τ_n^* the average cost in time for the sampling process (that is, sampling W times from the geometric distribution and sampling W times random variable Y). The WNRV of this procedure is $\sigma_n^2 \tau_n^* / U^2$. Here, τ_n^* is proportional to $\text{E}(W)$, that is, to nq, leading to

$$\text{WNRV}\left(U^*\right) = \Theta\left(\varepsilon^{b-c}\right) ,$$

where we recall that $U \approx a\varepsilon^c$ for some constant $a > 0$, and that $b = b_1 + b_2 + \ldots + b_H$ where the most reliable edge in path P_h has unreliability $\approx d\varepsilon^{bh}$ for some constant $d > 0$.

Recall that the desirable property (BWNRV) is to have $\text{WNRV}(U^*)$ bounded when ε gets small. This means that the estimation remains "efficient" for a given computational time budget, no matter how small ε is. We see that the estimator U^* does not always have this property, and that a sufficient condition for BWNRV is then $b \geqslant c$, as pointed out in [10].

In Figure 3.2 we see a trivial example where a three-node model is analyzed using the U^* estimator. We assume homogeneous edges, i.e., edges with reliabilities of the same order of magnitude. In this case, the BWNRV property holds. Indeed, the reader can check that $U = 2\varepsilon^2 - \varepsilon^3 \approx 2\varepsilon^2$ (we are setting all the unreliabilities to the same value ε) and that the variance for a single crude estimation is $U(1-U) = 2\varepsilon^2 - \varepsilon^3 - 4\varepsilon^4 + 4\varepsilon^5 - \varepsilon^6 \approx 2\varepsilon^2$. Letting P_1 be the path (s,t) and P_2 the path (s,u,t), the probabilities that all links of P_1 and P_2 work are $p_1 = 1-\varepsilon$ and $p_2 = (1-\varepsilon)^2$ respectively. Thus $q = (1-p_1)(1-p_2)$, which here is exactly equal to the target, the system unreliability U, and then $q \approx 2\varepsilon^2$. As a consequence, the BWNRV property is verified. We see that $c = 2$ and that $b = 2$ as well, so that the given sufficient condition is satisfied.

Consider now the "bridge" in Figure 3.3, where the links are no longer homogeneous with respect to their reliabilities (or unreliabilities). In the picture, the unreliabilities of the links are indicated.

The unreliability of the system is

$$U = \varepsilon^4(2 + \varepsilon^4 - 2\varepsilon^5 - 2\varepsilon^6 + 2\varepsilon^7) = 2\varepsilon^4 + o(\varepsilon^4) \ .$$

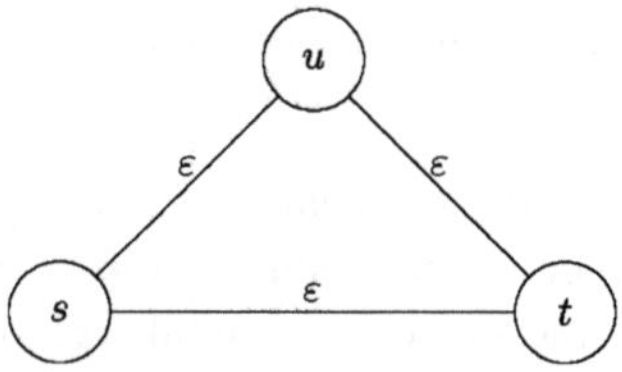

Figure 3.2 A simple "triangle" illustrating the path-based method leading to bounded relative efficiency. The unreliabilities are all equal to ε. There are two paths between s and t, path $P_1 = (s,t)$ and path $P_2 = (s,u,t)$. The probability p_1 that all links in P_1 work is $1-\varepsilon$ and, for P_2, we have $p_2 = (1-\varepsilon)^2$, leading to $q = (1-p_1)(1-p_2) \approx 2\varepsilon^2$. We have $U = 2\varepsilon^2 - \varepsilon^3 \approx 2\varepsilon^2$. Finally, WNRV $= \sigma^2\tau/U \approx 1$, thus bounded

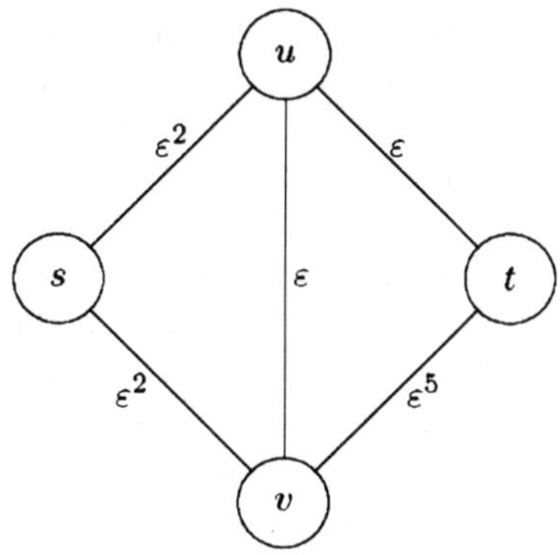

Figure 3.3 A "bridge" illustrating the path-based method leading to unbounded WNRV

The computations are longer here, but we can check that whatever the set of disjoint paths between s and t, we always have $b < 4$. So, in this case, the path-based method has not the BWNRV property. For the details, there are three possible sets of disjoints minpaths: $\boldsymbol{P}_1 = \{(s,u,v,t)\}$, $\{\boldsymbol{P}_2 = \{(s,v,u,t)\}$ and $\boldsymbol{P}_3 = \{(s,u,t),(s,v,t)\}$. For each set $\boldsymbol{P}_i$, let us denote by q_i the corresponding probability that at least one link in each path is not working. We have

$$\begin{aligned}
q_1 &= 1-(1-\varepsilon^2)(1-\varepsilon)(1-\varepsilon) = 2\varepsilon - 2\varepsilon^3 + \varepsilon^4 \approx 2\varepsilon \\
q_2 &= 1-(1-\varepsilon^2)(1-\varepsilon)(1-\varepsilon^5) \approx \varepsilon \\
q_3 &= (1-(1-\varepsilon^2)(1-\varepsilon^5))(1-(1-\varepsilon^2)(1-\varepsilon)) \approx \varepsilon^2\varepsilon = \varepsilon^3 \,.
\end{aligned}$$

Then, for the three cases, BWNRV is not verified because we respectively have for the three cases WNRV $= \Theta(\varepsilon^{-3})$ for $\boldsymbol{P}_1$, WNRV $= \Theta(\varepsilon^{-3})$ for $\boldsymbol{P}_2$ and WNRV $= \Theta(\varepsilon^{-1})$ for $\boldsymbol{P}_3$.

Coming back to the homogeneous case, illustrated by the elementary example of Figure 3.2, let us show that it is always possible to find a set of paths $\boldsymbol{P}$ leading to the BWNRV property of the corresponding estimator U^*. This has been briefly stated in [10]. We provide a more detailed proof here.

Theorem 3.1. *Assume that the unreliabilities of the links are homogeneous in ε, that is, that for any link i in the graph, we have $u_i = a_i\varepsilon$. Then, it is always possible to* *find a set of minpaths* $\boldsymbol{P}$ *such that the corresponding estimator* U^* *has the BWNRV property.*

Proof: First, observe that it is useless to put the same exponent, say β, to the factor ε in the link unreliabilities, since we can then rename ε^β as the new ε in the analysis.

The breadth of a graph is the size of a minimal size mincut. Let K be the number of mincuts in the graph, which we arbitrary order and number from 1 to K. Let C_k be the event "all links in the kth mincut are failed." Writing

$$U = \Pr(C_1 \cup \cdots \cup C_K) ,$$

and using Poincare's formula for expanding this expression, we see that the term with the lowest power in ε is of the form $a\varepsilon^c$ where c is precisely the breadth of the graph. For this, just observe that for each mincut C_k of minimal size c, $\Pr(C_k) = \Theta(\varepsilon^c)$, and that for any other $\Pr(C_j)$ and for all terms of the form $\Pr(C_i \cap C_j \cap \ldots)$ we obtain $\Theta(\varepsilon^d)$, $d > c$.

The second observation comes from the theory of flows in graphs, where a basic result states that if c is the breadth, then there exist c disjoint paths from s to t. For an effective way to find them, they come for instance directly as a byproduct of the marking process in Ford–Fulkerson algorithm (for finding a maximal flow from s to t), which runs in time polynomial in the size of the graph [11]. Then, we just see that with the previous notation, for each of the $H = c$ minpaths, $b_h = 1$ and thus $b = c$, which is sufficient for having the BWNRV property. ■

3.5 Improvement

The estimator $\hat{U}$ does not have the same variance as U^* and is more difficult to analyze; it actually has a (slightly) smaller variance and the same computational cost. The goal in [7] is to point out that the standard estimator can still be very useful when dealing with rare events if an efficient implementation is possible. That means, in particular, to keep F as a geometric random variable.

Looking now for efficiency improvements, we can replace the random variable F by its mean (instead of sampling it). Let us look at what happens in this case. If F is replaced by its expected value, then exactly one in $1/q$ independent graphs will have at least one failed link on each path of $\boldsymbol{P}$. Recall that Y is a Bernoulli random variable that is 1 if the graph is failed and 0 otherwise, conditioned on the fact that at least one link on each selected path is failed. The random variable $Z = qY$ is then an (unbiased) estimator of U over such a block. This is known as a *conditional Monte Carlo* estimator [12]: the usual estimator has been replaced by its conditional expectation given Y. A confidence interval for U is obtained by considering independent copies of Z and applying standard procedures.

Define p as the probability that $Y = 1$. Obviously, $U = qp$, and $\mathrm{Var}(Z) = q^2\mathrm{Var}(Y) = q^2 p(1-p)$. If we look at the ratio of the WNRV of Z (considering the expected value of F) over the WNRV of the estimator U^* (obtained by employing

Table 3.1 Result of the estimation, the variance of the estimator, and the relative efficiency with respect to the original method for three cases, where the system failure event becomes rarer (ε going from 0.1 to 0.001). The model is the "bridge" described in Figure 3.3

u_i, for all link i	Estimation	Variance	Rel. efficiency (relation 3.1)
0.1	2.1×10^{-2}	3.1×10^{-11}	2.4
0.01	2.0×10^{-4}	3.9×10^{-14}	2.0
0.001	2.0×10^{-6}	4.0×10^{-19}	2.0

Table 3.2 Evaluation of the graph given in Figure 3.4 when the elementary unreliability of all links is equal to 0.1, 0.01, and 0.001

u_i, for all link i	Estimation	Variance	Rel. efficiency
0.1	1.9×10^{-2}	8.9×10^{-11}	1.4
0.01	2.0×10^{-5}	3.7×10^{-16}	1.3
0.001	2.0×10^{-8}	2.0×10^{-22}	1.2

the geometric distribution) and if we neglect the time to generate the geometric random variable, we get the following relative efficiency:

$$\frac{\text{WNRV}\,(Z)}{\text{WNRV}\,(U^*)} = \frac{qU\,(1-U)}{q^2 p\,(1-p)} = \frac{1-qp}{1-p} > 1\,. \tag{3.1}$$

This shows that the conditional Monte Carlo estimator always yields an efficiency improvement that we are able to characterize, by reducing the WNRV. The cost (in CPU time) is also reduced because there is no longer a need for sampling from a geometric law. Note that in general, conditional Monte Carlo *always* reduces the variance.

Let us illustrate this improvement on a few examples. Consider first the bridge shown in Figure 3.3, but with all its links identical. For the path-based method, we use the two symmetric paths between s and t of size 2: $\boldsymbol{P} = \{(s,u,t),(s,v,t)\}$. In Table 3.1 we see that the improvement roughly doubles the efficiency of the original approach.

Now, we evaluate the unreliability in the case of the topology given in Figure 3.4 with homogeneous links, where $s = 1$ and $t = 14$. The breadth of the graph is $c = 3$, so, to use an estimation procedure having the BWNRV property, we need three disjoint elementary paths between s and t. The three paths chosen are $P_1 = (1,2,6,8,9,13,14)$, $P_2 = (1,3,7,10,14)$, and $P_3 = (1,4,7,11,12,14)$.

In Table 3.2 we show the relative efficiency of the proposed improvement for this "reducible" architecture. As we can see, the efficiency improvement is still significant, while less than in the previously presented small bridge example.

Finally, we consider in Table 3.3 the more challenging dodecahedron structure given in Figure 3.1. We performed the same experiments as with previous examples, in order to show that in this case there is no improvement over the original method (relative efficiency close to 1). The reason is that given the density of the graph, the

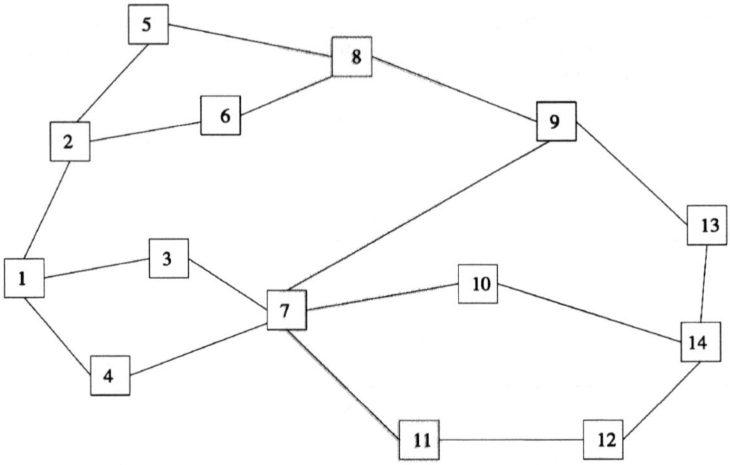

Figure 3.4 We call this example a "reducible" topology, because there are many series-parallel simplifications possible here, when $s = 1$ and $t = 14$. After those reductions, the result is a bridge (see [4] for instance). In the homogeneous case, we can easily see, after some algebra, that when every link has the same unreliability $u_i = \varepsilon$, the system unreliability is $U = 24\varepsilon^3 + o(\varepsilon^3)$. The model is the "reducible" architecture

Table 3.3 Evaluation of the graph given in Figure 3.1 when the elementary unreliability of each links is equal to 0.1, 0.01, 0.001

u_i, for all link i	Estimation	Variance	Rel. efficiency
0.1	2.9×10^{-3}	2.9×10^{-12}	1.02
0.01	2.0×10^{-6}	4.1×10^{-18}	1.01
0.001	2.0×10^{-9}	4.3×10^{-25}	1.01

probability $p = \Pr(Y = 1)$ is small, leading to a relative efficiency of $(1-qp)/(1-p) \approx 1$.

In the next section, we show that the efficiency can be improved further by using RQMC on top of the method proposed earlier.

3.6 Acceleration by Randomized Quasi-Monte Carlo

The previous sections make use of Monte Carlo methods. Very roughly, the basic idea is to choose sample points randomly and independently according to a given distribution. This random choice of points ensures that asymptotically, the empirical distribution of the estimator converges to the theoretical one at a speed of $O(n^{-1/2})$ for a sample size n. This rate can be improved thanks to better spreading of points (which are then no longer independent). This is the basic principle of *quasi-Monte Carlo* (QMC) methods [13]. In practice, randomized versions called RQMC are often used in order to obtain an unbiased estimator and allow error estimation. We will now explain briefly the QMC and RQMC methods before applying them to our static reliability problem.

Note that RQMC is not an appropriate method to handle the problem of rare events, but once that problem is handled (in our case via a path-based conditional Monte Carlo approach), RQMC can improve the efficiency by an additional order of magnitude.

3.6.1 Quasi-Monte Carlo Methods

In most simulation studies by computer (including ours), a single (random) realization of the model is defined as a function of a uniform random variable over $(0, 1)^M$, or equivalently from M independent unidimensional uniform random variables over (0, 1), where M is possibly unbounded; those uniform random variates are actually replaced in practice by the output of a pseudorandom generator in Monte Carlo methods. To describe QMC and RQMC techniques, we will therefore use (without loss of generality) the framework of an estimation over the hypercube $(0, 1)^M$.

Suppose we want to estimate

$$E\left[f\left(U\right)\right] = \int\limits_{[0,1]^M} f(u)\mathrm{d}u \ ,$$

where U is uniformly distributed over $[0, 1]^M$. While Monte Carlo methods use a sample $\{U_i, 1 \leqslant I \leqslant n\}$ of n independent random variables with the same distribution than U to get $1/n \sum_{i=1}^{n} f\left(U_i\right)$ as the estimator, QMC methods [13, 14] replace the independent U_is by a sequence of *deterministic* points $\Xi = \{\xi_n, n \geqslant 1\}$ in $[0, 1]^M$. A basic requirement is that the sequence is asymptotically uniformly distributed, in the sense that the proportion of points among the first n in the sequence Ξ falling in any (multivariate) interval B, namely $A_n(B, \Xi) = \#\{\xi_i, 1 \leqslant I \leqslant n\colon \xi_i \in B\}/n$, converges to $\lambda(B)$ as $n \to \infty$, where $\lambda(B)$ is the Lebesgue measure of B. There exist several different measures of the *discrepancy* between the empirical distribution of the first n points of the sequence and the uniform distribution. One of them is the *star discrepancy*, defined as

$$D_n^*\left(\Xi\right) = \sup_{[0,x)\subset[0,1)^M} \left| \frac{A_n\left([0,x), \Xi\right)}{n} - \lambda\left([0,x)\right) \right| ,$$

which takes the sup over all intervals with one corner at the origin. A sequence Ξ is actually asymptotically uniformly distributed if and only if $D_n^*(\Xi) \to 0$ as $n \to \infty$.

Discrepancy measures are helpful to bound the error in the estimation of the integral $\int_{[0,1]^M} f\left(u\right) \mathbf{d}_u$. Using the star discrepancy, the Koksma–Hlawka bound [13] is

$$\left| \frac{1}{n} \sum_{k=1}^{n} f\left(\xi^{(k)}\right) - \int_{[0,1]^M} f(u)\mathrm{d}u \right| \leqslant V(f) D_n^*(\Xi) ,$$

where $V(f)$ is the variation of the function f in the sense of Hardy and Krause [13]. For the best-known sequences Ξ, we have $D_n^*(\Xi) = \mathrm{O}(n^{-1}(\log n)^M)$ [13]; these are named *low-discrepancy sequences*. In this chapter we use one class of low-discrepancy sequences called the *Sobol' sequences* [15]. Those sequences are instances of (t, M)-sequences in base 2, which means that for a certain integer $t \geqslant 0$, $\forall m \geqslant t$, if we consider a set of 2^m successive points of the form $\{\xi^{(j)} : k2^m \leqslant k < (j+1)2^m\}$, for any $k \geqslant 0$ and $m > t$, and any dyadic interval

$$E = \prod_{i=1}^{M} \left[a_i 2^{-2d_i}, (a_i + 1)\, 2^{-2d_i}\right] , \quad \text{where } a_i, b_i \in N, d_i \geqslant 0, a_i \in \{0, 1\} , \tag{3.2}$$

of size $\lambda(E) = 2^{t-m}$ with $m \geqslant t$, then the number of points in E is exactly 2^t. This means that for any function f which is constant in each dyadic interval of size 2^{t-m}, the integration error by a set of 2^m successive points of the above form is always zero.

QMC methods therefore asymptotically outperform MC, but from the practical side, evaluating the error is a very difficult task in general. The worst-case error bounds such as the Koksma–Hlawka bound are too hard to compute in practice and are often much too large to be useful anyway. Even if the bound converges asymptotically at rate $\mathrm{O}(n^{-1}(\log n)^M)$, it often takes an astronomically large value of n before this bound becomes meaningful, as soon as the dimension M exceeds 10 or so [16]. Nevertheless, QMC methods are typically more effective than what the bounds tell us. RQMC methods permit one to estimate the error without relying on these bounds.

3.6.2 Randomized Quasi-Monte Carlo Methods

RQMC methods randomly perturb a low-discrepancy sequence *without losing its good distribution over* $[0, 1]^M$. A simple illustration of this is when all the points are shifted by the same uniform vector U. That is, Ξ is replaced by its randomly shifted version $\{V_k := (\xi_k + U) \bmod 1,\ k \geqslant 1\}$, where "mod 1" means that we retain only the fractional part of each coordinate. Thus, the whole sequence is somehow just translated over the interval. Other types of randomization exist [14]; some of them are adapted to the structure of the low-discrepancy sequence. For the Sobol' sequence, a *random digital shift* generates a uniform point in $[0, 1]^M$, expands each of its coordinates in base 2, and adds the digits modulo 2 to the corresponding digits of each point of the sequence. This randomization preserves the (t, M)-sequence property. With this particular sequence and randomization, if we assume that f has bounded variation, the variance of $(1/n)\sum_{k=1}^{n} f(V_k)$ is $O(n^{-2}(\log n)^{2M})$, which converges faster than the Monte Carlo rate of $O(1/n)$. The convergence speed can be even faster for specific classes of smooth functions (with square-integrable high-order partial derivatives, for example) and adapted randomized sequences [14, 17].

To estimate the error, one can make m of independent replicates of the randomization, and estimate the variance in a classic way by the sample variance of these m replicates. The central limit theorem applies when $m \to \infty$. In practice, confidence intervals are often computed by assuming (heuristically) that the average is approximately normally distributed even when m is small.

QMC/RQMC error bounds degrade rapidly when the dimension M increases, because the $(\log n)^M$ term becomes more important and a much larger value of n is required before this term is dominated by the $1/n$ term. As a general rule of thumb, QMC/RQMC is more effective when the dimension M is small, but sometimes it also works well in practice even when M is large [14]. This happens when the integrand f depends mostly on just a few coordinates, or can be decomposed (approximately) as a sum of terms where each term depends only on a small number of coordinates [18]. We then say that the integrand has *low effective dimension*.

3.6.3 Application to Our Static Reliability Problem

We now examine how to apply RQMC to our static reliability problem, starting with a crude implementation. We need to sample the status of M links. The state of the jth link in the ith replicate is sampled from the jth coordinate of the ith point of the low-discrepancy sequence: if this coordinate is less than r_j, then the state is 1, otherwise it is 0. Let ψ be the indicator function mapping each point $y = (y_1, \ldots, y_M) \in [0, 1]^M$ to a vector state $x = (x_1, \ldots, x_M)$ in $\{0, 1\}^M$, defined by $x_j = 1$ if $y_j < r_j$, and $x_j = 0$ otherwise. This mapping partitions the unit hypercube $[0, 1]^M$ into 2^M rectangular boxes, each one sharing one corner with the hypercube. The indicator function $\Phi \circ \psi$, where "$\circ$" denotes the composition operator, takes a constant value over each of those boxes. It is equal to 0 for states in which the system is failed, and 1 for the other states. The reliability is therefore $R = \int_{[0,1]^M} \Phi \circ \psi(y)\,dy$ and the unreliability $U = \int_{[0,1]^M} (1 - \Phi) \circ \psi(y)\,dy$. We let a *minimal state vector* be any vector $z \in \{0, 1\}^M$ such that $\Phi(z) = 1$, and $\Phi(x) = 0$ for all $x < z$. Let N_p be the number of minimal state vectors (they correspond to elementary paths in the graph). We similarly define a *maximal state vector* as any vector $z \in \{0, 1\}^M$ such that $(1 - \Phi)(z) = 1$, and $(1 - \Phi)(x) = 0$ for all $x > z$. Let N_c be the number of maximal state vectors (corresponding to minimal cuts in the graph). Observe that the estimation error is the same when estimating the reliability or the unreliability, i.e.,

$$\left|\frac{1}{n}\sum_{i=1}^{n}\Phi\circ\psi\left(y_i\right)-\int_{[0,1]^M}\Phi\circ\psi(y)\mathrm{d}y\right|$$
$$=\left|\frac{1}{n}\sum_{i=1}^{n}(1-\Phi)\circ\psi\left(y_i\right)-\int_{[0,1]^M}(1-\Phi)\circ\psi\left(y\right)\mathrm{d}y\right|. \tag{3.3}$$

Theorem 3.2. *We have the worst-case error bound*

$$\left|\frac{1}{n}\sum_{i=1}^{n}\Phi\circ\psi\left(y_i\right)-\int_{[0,1]^M}\Phi\circ\psi\left(y\right)\mathrm{d}y\right|\leqslant\left(2^{\min(N_p,N_c)}-1\right)D_n^*\left(\Xi\right).$$

Proof: Let $\left\{\pi_1,\cdots,\pi_{N_p}\right\}$ be the set of minimal state vectors. For each π_ℓ, we define the corresponding sub-interval P_ℓ of $[0,1]^M$ by

$$P_\ell=\prod_{i=1}^{M}[0,\alpha_i)\quad\text{where}\quad\begin{cases}\alpha_i=r_i & \text{if } \ell\text{th coordinate of } \pi_\ell \text{ is } 1,\\ \alpha_i=1 & \text{otherwise.}\end{cases}$$

Note that these P_ℓs are not disjoint. The subset of $[0,1]^M$ on which $\Phi\circ\psi(y)=1$ is $B=\bigcup_{\ell=1}^{N_P}P_\ell$. Furthermore,

$$\left|\frac{1}{n}\sum_{k=1}^{n}\Phi\circ\psi\left(\xi^{(k)}\right)-\int_{[0,1]^M}\Phi\circ\psi\left(y\right)\mathrm{d}y\right|\leqslant\left|\frac{1}{n}\sum_{k=1}^{n}1_B\left(\xi^{(k)}\right)-\lambda\left(B\right)\right|.$$

Applying the Poincaré formula and the triangular inequality,

$$\left|\frac{1}{n}\sum_{k=1}^{n}\Phi\circ\psi\left(\xi^{(k)}\right)-\int_{[0,1]^M}\Phi\circ\psi\left(y\right)\mathrm{d}y\right|$$
$$=\left|\sum_{\ell=1}^{N_\mathrm{p}}(-1)^{\ell-1}\sum_{1\leqslant h_1<\cdots<h_\ell\leqslant N_\mathrm{p}}\left(\frac{1}{n}\sum_{\xi^{(k)}\in\bigcap_{j=1}^{\ell}P_{h_j}}1_{\bigcap_{j=1}^{\ell}P_{h_j}}\left(\xi^{(k)}\right)\right.\right.$$
$$\left.\left.-\lambda\left(\bigcap_{j=1}^{\ell}P_{h_j}\right)\right)\right|$$

$$
\begin{aligned}
&\leqslant \sum_{\ell=1}^{N_p} \sum_{1\leqslant h_1<\cdots<h_\ell\leqslant N_p} \left| \frac{1}{n} \sum_{\xi^{(k)}\in\bigcap_{j=1}^{\ell} P_{h_j}} 1_{\bigcap_{j=1}^{\ell} P_{h_j}} \left(\xi^{(k)}\right) - \lambda\left(\bigcap_{j=1}^{\ell} P_{h_j}\right) \right| \\
&\leqslant \sum_{\ell=1}^{N_p} \sum_{1\leqslant h_1<\cdots<h_\ell\leqslant N_p} D_n^*(\Xi) \\
&= \left(2^{N_p}-1\right) D_n^*(\Xi).
\end{aligned}
$$

Proceeding in exactly the same way for computing the error when estimating the unreliability from the set of maximal states instead of minimal ones, we get

$$
\left| \frac{1}{n} \sum_{k=1}^{n} (1-\Phi)\circ\psi\left(\xi^{(k)}\right) - \int_{[0,1]^M} (1-\Phi)\circ\psi(y)\,\mathrm{d}y \right| \leqslant \left(2^{N_c}-1\right) D_n^*(\Xi).
$$

From Equation 3.3 and combining the two above inequalities, we obtain the theorem. ■

This result provides a worst-case error bound that converges asymptotically as $\mathrm{O}\left(n^{-1}(\log n)^{N_p}\right)$. The corresponding RQMC variance is $\mathrm{O}\left(n^{-2}(\log n)^{2N_p}\right)$. We may nevertheless need a very large n before this RQMC approach beats MC when N_p is large.

To apply RQMC with our path-based technique based on conditional Monte Carlo, the random variable Y for the ith replicate is sampled by first generating the first non-working link on each path from the initial coordinates of the point ξ_i, and then sampling all the other links (whose state is not yet known) from the remaining coordinates of ξ_i. The overall dimension of the integrand is again M, because in the worst case we may need to sample all links, if the first link on each path is failed. Nevertheless, the number of required coordinates (or uniform random numbers) is often smaller than M, and the first few coordinates are more important. As a result, the RQMC method tends to be more effective. A worst-case error bound in terms of the discrepancy $D_n^*(\Xi)$ can also be obtained, as for the crude implementation of RQMC discussed earlier.

3.6.4 Numerical Results

We made an experiment to compare MC and RQMC for our three typical examples, the bridge, the dodecahedron, and the reducible topology, in each case with three values of the links reliability ε: 0.9, 0.99, and 0.999. For RQMC, we use the first n points of a Sobol' sequence with a random digital shift and we perform $m = 500$ independent randomizations. For MC, we make nm independent replications (same

Table 3.4 Confidence interval half-widths for MC and for RQMC using the same total computing budget, and their ratio. The RQMC estimates are based on 500 independent replicates with n points. All edges in the network have reliability ε

Topology	ε	n	Half-width MC	Half-width RQMC	Ratio
Bridge	0.9	2^{10}	9.70×10^{-5}	1.61×10^{-5}	0.166
Bridge	0.9	2^{14}	2.43×10^{-5}	1.55×10^{-6}	6.41×10^{-2}
Bridge	0.9	2^{20}	3.03×10^{-6}	4.21×10^{-8}	1.39×10^{-2}
Bridge	0.99	2^{10}	1.08×10^{-6}	1.05×10^{-7}	9.68×10^{-2}
Bridge	0.99	2^{14}	2.71×10^{-7}	8.15×10^{-9}	3.01×10^{-2}
Bridge	0.99	2^{20}	3.39×10^{-8}	2.20×10^{-10}	6.48×10^{-3}
Bridge	0.999	2^{10}	1.09×10^{-8}	7.25×10^{-10}	6.62×10^{-2}
Bridge	0.999	2^{14}	2.74×10^{-9}	3.17×10^{-11}	1.16×10^{-2}
Bridge	0.999	2^{20}	3.42×10^{-10}	1.19×10^{-12}	3.47×10^{-3}
Dodecahedron	0.9	2^{10}	9.30×10^{-5}	6.89×10^{-5}	0.741
Dodecahedron	0.9	2^{14}	2.33×10^{-5}	1.29×10^{-5}	0.556
Dodecahedron	0.9	2^{18}	5.81×10^{-6}	2.58×10^{-6}	0.444
Dodecahedron	0.99	2^{10}	1.10×10^{-7}	5.28×10^{-8}	0.479
Dodecahedron	0.99	2^{14}	2.77×10^{-8}	7.62×10^{-9}	0.275
Dodecahedron	0.99	2^{18}	6.93×10^{-9}	1.37×10^{-9}	0.197
Dodecahedron	0.999	2^{10}	1.13×10^{-10}	4.84×10^{-11}	0.430
Dodecahedron	0.999	2^{14}	2.83×10^{-11}	5.45×10^{-12}	0.193
Dodecahedron	0.999	2^{18}	7.07×10^{-12}	7.92×10^{-13}	0.112
Reducible	0.9	2^{10}	1.64×10^{-4}	8.18×10^{-5}	0.499
Reducible	0.9	2^{14}	4.09×10^{-5}	1.58×10^{-5}	0.386
Reducible	0.9	2^{18}	1.02×10^{-5}	2.49×10^{-6}	0.244
Reducible	0.99	2^{10}	2.36×10^{-7}	5.57×10^{-8}	0.236
Reducible	0.99	2^{14}	5.91×10^{-8}	9.96×10^{-9}	0.168
Reducible	0.99	2^{18}	1.48×10^{-8}	1.63×10^{-9}	0.111
Reducible	0.999	2^{10}	2.44×10^{-10}	3.70×10^{-11}	0.152
Reducible	0.999	2^{14}	6.10×10^{-11}	5.07×10^{-12}	8.31×10^{-2}
Reducible	0.999	2^{18}	1.53×10^{-11}	7.38×10^{-13}	4.83×10^{-2}

total sample size). In both cases, we compute the half-width of a 95% confidence interval on the unreliability, using the path-based technique with conditional Monte Carlo. We then compute the ratio of the confidence interval half-width of MC over that of RQMC. The results are in Table 3.4, where "half-width MC" is the half-width for MC, "half-width RQMC" is that for RQMC, "ratio" is the ratio between the two.

We see that RQMC brings a significant variance reduction in all cases, even on reasonable-size topologies such as the dodecahedron. Also, the larger the cardinality n of the RQMC point set, the more the variance is reduced.

The fact that the improvements are smaller as the model size increases is due to the sensitivity of QMC methods with respect to the dimension of the problem.

Basically, when the dimension is higher, the low-discrepancy sequence needs more time to "distribute" its points well [14].

3.7 Conclusions

We have proposed and examined simulation techniques for static rare-event models. Our discussion emphasizes the importance of an efficiency measure that account for both the accuracy of Monte Carlo methods and the cost (in CPU time) of the estimation procedures. A key concept that captures these ideas in the context of rare-event simulation is the notion of bounded work-normalized relative variance (BWNRV). The application that we considered is the analysis of a reliability metric in a static model. Our analysis was completed by proposals designed to improve efficiency in the considered estimation algorithms.

A last technical remark now on the BWNRV property: the computing time used in the definition may have unbounded relative variance itself, which may lead to a noisy work-normalized variance [5,6]. In that case, we cannot assert that the probability that the estimator is within a value δ of its mean for a given computational budget c, goes to 0 uniformly in ε when c increases. Our definition only looks at the first moment of the computational time, which is less stringent. Considering also the second moment is a subject of further research.

References

1. Colbourn CJ (1987) The combinatorics of network reliability. Oxford University Press, New York
2. Gertbakh IB (1989) Statistical reliability theory. Marcel Dekker, New York
3. Ball MO, Colbourn CJ, Provan JS (1995) Network reliability. In: Handbook of operations research: network models. Elsevier North-Holland, Amsterdam, The Netherlands, pp 673–762
4. Rubino G (1998) Network reliability evaluation. In: Bagchi K, Walrand J (eds) State-of-the-art in performance modeling and simulation, Chap 11. Gordon and Breach, London
5. El Khadiri M, Rubino G (2000) A time reduction technique for network reliability analysis. In: MCQMC'00: 4th international conference on Monte Carlo and quasi-Monte Carlo methods in scientific computing. MCQMC, Hong Kong
6. Cancela H, Rubino G, Tuffin B (2005) New measures of robustness in rare event simulation. In: Kuhl ME, Steiger NM, Armstrong FB, Joines JA (eds) Proceedings of the 2005 winter simulation conference, Orlando, FL, pp 519–527
7. Rubino G, Tuffin B (eds) (2009) Rare event simulation. John Wiley, Chichester, West Sussex, UK
8. Cancela H, El Khadiri M, Rubino G (2009) Rare event analysis by Monte Carlo techniques in static models. In: Rubino G, Tuffin B (eds) Rare event simulation. John Wiley, Chichester, West Sussex, UK
9. Glynn PW, Rubino G, Tuffin B (2009) Robustness properties and confidence interval reliability issues. In: Rubino G, Tuffin B (eds) Rare event simulation. John Wiley, Chichester, West Sussex, UK

10. L'Ecuyer P (2009) Quasi-Monte Carlo methods with applications in finance. Finance Stoch 13(3):307–349
11. L'Ecuyer P, Blanchet JH, Tuffin B, Glynn PW (2010) Asymptotic robustness of estimators in rare-event simulation. ACM Trans Model Comput Simul 20(1):91–99
12. Bratley P, Fox BL, Schrage LE (1987) A guide to simulation, 2nd edn. Springer, New York
13. Niederreiter H (1992) Random number generation and quasi-Monte Carlo methods. CBMS-NSF, SIAM, Philadelphia
14. Owen AB (1997) Scrambled net variance for integrals of smooth functions. Ann Stat 25(4):1541–1562
15. Owen AB (1998) Latin supercube sampling for very high-dimensional simulations. ACM Trans Model Comput Simul 8(1):71–102
16. Sedgewick R (2001) Algorithms in C, Part 5: Graph algorithms, 3rd edn. Addison-Wesley Professional, Indianapolis, IN
17. Sobol' IM (1967) The distribution of points in a cube and the approximate evaluation of integrals. USSR Comput Math Math Phys 7:86–112
18. Tuffin B (1997) Variance reductions applied to product-form multi-class queuing network. ACM Trans Model Comput Simul 7(4):478–500

Chapter 4
Variate Generation in Reliability

Lawrence M. Leemis

Abstract This chapter considers (1) the generation of random lifetimes via density-based and hazard-based methods, (2) the generation of certain stochastic processes that are useful in reliability and availability analysis, and (3) the generation of random lifetimes for the accelerated life and proportional hazards models. The accurate modeling of failure time distributions is critical for the development of valid Monte Carlo and discrete-event simulation models for applications in reliability and survival analysis. Once an accurate model has been established, it is oftentimes the case that the complexity of the model requires an analysis by simulation. The associated variate generation algorithms for common stochastic models are introduced here. Although the generation of random lifetimes is typically applied to reliability and survival analysis in a simulation setting, their use is widespread in other disciplines as well. The more diverse wider literature on generating random objects includes generating random combinatorial objects, generating random matrices, generating random polynomials, generating random colors, generating random geometric objects, and generating random spawning trees.

4.1 Generating Random Lifetimes

This section concerns algorithms for generating continuous, positive random variables, referred to generically here as "lifetimes." Although the main two applications areas are reliability (e.g., a machine or product lifetime, see, for example, [31]) and survival analysis (e.g., patient survival time after an organ transplant, see, for example, [25]), their use is widespread in other disciplines (e.g., sociological applications as in [1]). The discussion here is limited to generating continuous lifetimes, as opposed to discrete or mixed lifetimes, due to their pervasiveness in the reliability and survival analysis literature.

Department of Mathematics, The College of William & Mary, Williamsburg, VA, USA

P. Faulin, A. Juan, S. Martorell, and J.E. Ramírez-Márquez (eds), *Simulation Methods for Reliability and Availability of Complex Systems*. © Springer 2010

There is a subtle but important distinction between a random variable and a random variate. A random variable is a rule that assigns a real number to an outcome of an experiment. A random variate is a realization, or instantiation of a random variable, which is typically generated by a computer. Devroye [13] and Hörmann *et al.* [21] provide comprehensive treatments of random variate generation.

In all of the variate generation algorithms considered here, we assume that the continuous random lifetime T has positive support. We generically refer to T as a "lifetime." The four functions described below each completely characterize the distribution of T: the survival function, the probability density function (pdf), the hazard function, and the cumulative hazard function (chf).

The survival function, also known as the reliability function and complementary cumulative distribution function (cdf), is defined by

$$S(t) = P(T \geqslant t) \quad t \geqslant 0 \tag{4.1}$$

and is a nonincreasing function of t satisfying $S(0) = 1$ and $\lim_{t \to \infty} S(t) = 0$. The survival function is important in the study of systems of components since it is the appropriate argument in the structure function to determine system reliability [31]. Notice that $S(t)$ is the fraction of the population that survives to time t as well as the probability that a single item survives to time t. For continuous random variables, $S(t) = 1 - F(t)$, where $F(t) = P(T \leqslant t)$ is the cdf.

When the survival function is differentiable,

$$f(t) = -S'(t) \quad t \geqslant 0 \tag{4.2}$$

is the associated pdf. For any interval (a, b), where $a < b$,

$$P(a < T < b) = \int_a^b f(t)\mathrm{d}t \tag{4.3}$$

The hazard function, also known as the rate function, failure rate, and force of mortality, can be defined by

$$h(t) = \frac{f(t)}{S(t)} \quad t \geqslant 0 \tag{4.4}$$

The hazard function is popular in reliability because it has the intuitive interpretation as the amount of risk associated with an item that has survived to time t. The hazard function is mathematically equivalent to the intensity function for a nonhomogeneous Poisson process (NHPP), and the failure time corresponds to the first event time in the process. Competing risks models are naturally formulated in terms of $h(t)$, as shown subsequently.

The chf can be defined by

$$H(t) = \int_0^t h(\tau)\, d\tau \quad t \geqslant 0 \tag{4.5}$$

Any one of the four functions that describes the distribution of T can be used to determine the others, e.g. $H(t) = -\log S(t)$. This ends the discussion of the four functions that define the distribution of the random lifetime T. We now begin the discussion of generating the associated random variates.

We assume that there is a reliable source of pseudo-random numbers available, and will use U, with or without subscripts, to denote one instance of such a $U(0, 1)$ random variable. Any of the standard discrete-event simulation textbooks (e.g., [24] or [2]) will have a discussion of random number generation techniques. The purpose of this chapter is to present an overview of several random variate generation techniques that convert random numbers to random variates which are useful in the analysis of reliability and availability problems. The algorithms are broken into density-based algorithms and hazard-based algorithms, and there are analogies between the two sets of algorithms. A classification of algorithms known as "special properties" (e.g., the sum of independent and identically distributed exponential random variables has the Erlang distribution) consists of neither density-based nor hazard-based algorithms.

There are a number of properties of variate generation algorithms that are important in evaluation of the algorithms presented here. These include synchronization (one random number produces one random variate), monotonicity (a monotone relationship between the random numbers and the associated random variates), robustness with respect to all values of parameters, number of lines of code, expected marginal time to generate a variate, set-up time, susceptibility to computer round-off, memory requirements, portability, etc. The choice between the various algorithms presented here must be made based on these criteria. Some of these properties are necessary for implementing "variance reduction techniques" [24].

4.1.1 Density-based Methods

There are three density-based algorithms for generating lifetimes: (1) the inverse cdf technique, (2) composition, which assumes that the pdf can be written as a convex combination, and (3) acceptance–rejection, a majorization technique. These three algorithms are introduced in the subsections that follow. Proofs of the results that provide a basis for these algorithms are given in [13].

4.1.1.1 Inverse Cumulative Distribution Function Technique

The inverse-cdf technique is the algorithm of choice for generating a continuous random lifetime T. It is typically the fastest of the algorithms in terms of marginal execution time, produces a random variate from a single random number, and is monotone. The inverse-cdf technique is based on the probability integral transformation, which states that $F_T(T) \sim U(0,1)$, where $F_T(t)$ is the cdf of the random variable T. This results in the following algorithm for generating a random variable T:

generate $U \sim U(0,1)$
$T \leftarrow F^{-1}(U)$
return T

The inverse-cdf technique works well on distributions that have a closed-form expression for $F^{-1}(u)$, e.g., the exponential, Weibull, and log logistic distributions. Distributions that lack a closed-form expression for $F^{-1}(u)$, e.g., the gamma and beta distributions, can still use the technique by numerically integrating the pdf, although this tends to be slow. This difficulty leads to the development of other techniques for generating random variates.

4.1.1.2 Composition

When the cdf of the random variable T can be written as the convex combination of n cdfs, i.e.,

$$F_T(t) = \sum_{i=1}^{n} p_i F_i(t) \quad t \geqslant 0 \tag{4.6}$$

where $\sum_{i=1}^{n} p_i = 1$ and $p_i > 0$ for $i = 1, 2, \ldots, n$, then the composition algorithm can be used to generate a random lifetime T. Distributions that can be written in this form are also known as finite mixture distributions and there is a significant literature in this area [16–30]. The algorithm generates a random component distribution index I, then generates a lifetime from the chosen distribution using the inverse-cdf (or any other) technique.

generate I with probability p_i
generate $U \sim U(0,1)$
$T \leftarrow F_I^{-1}(U)$
return T

This algorithm is not synchronized because two random numbers (one to generate the index I and another to generate T) are required. This difficulty can be overcome with a small alteration of the algorithm. In either case, however, the algorithm is not monotone. This creates difficulties in applying various variance reduction techniques to a simulation.

4.1.1.3 Acceptance–Rejection

The third and final algorithm is the acceptance–rejection technique, which requires finding a majorizing function $f^*(t)$ that satisfies

$$f^*(t) \geqslant f(t) \quad t \geqslant 0 \tag{4.7}$$

where $f(t)$ is the pdf of the random lifetime T. To improve execution time, it is best to find a majorizing function with minimum area. A scaled majorizing function $g(t)$ is

$$g(t) = \frac{f^*(t)}{\int_0^\infty f^*(\tau)\mathrm{d}\tau} \quad t \geqslant 0 \tag{4.8}$$

which is a legitimate pdf with associated cdf $G(t)$. The acceptance–rejection algorithm proceeds as described below. Here, and throughout the chapter, indentation is used in the algorithms to indicate nesting.

repeat
    generate $U \sim U(0, 1)$
    $T \leftarrow G^{-1}(U)$
    generate $S \sim U(0, f^*(T))$
until $(S \leqslant f(T))$
return T

The acceptance–rejection algorithm uses a geometrically distributed number of $U(0, 1)$'s to generate the random variate T. For this reason the acceptance–rejection algorithm is not synchronized.

4.1.2 Hazard-based Methods

There are three hazard-based algorithms for generating lifetimes: (1) the inverse chf technique, an inversion technique that parallels the inverse-cdf technique, (2) competing risks, a linear combination technique that parallels composition, and (3) thinning, a majorization technique that parallels acceptance–rejection. These three algorithms are introduced in the subsections that follow. Random numbers are again denoted by U and the associated random lifetimes are denoted by T.

4.1.2.1 Inverse Cumulative Hazard Function Technique

If T is a random lifetime with chf H, then $H(T)$ is an exponential random variable with a mean of one. This result, which is an extension of the probability integral transformation, is the basis for the inverse-chf technique. Therefore,

generate $U \sim U(0, 1)$
$T \leftarrow H^{-1}(-\log(1 - U))$
return T

generates a single random lifetime T. This algorithm is easiest to implement when H can be inverted in closed form. This algorithm is monotone and synchronized. Although the sense of the monotonicity is reversed, $1 - U$ can be replaced with U in order to save a subtraction. For identical values of U the inverse-cdf technique and the inverse-chf technique generate the same random variate T.

4.1.2.2 Competing Risks

Competing risks [10, 11] is a linear combination technique that is analogous to the density-based composition method. The competing risks technique applies when the hazard function can be written as the sum of hazard functions, each corresponding to a "cause" of failure

$$h(t) = \sum_{j=1}^{k} h_j(t) \quad t \geqslant 0 \tag{4.9}$$

where $h_j(t)$ is the hazard function associated with cause j of failure acting in a population. The minimum of the lifetimes from each of these risks corresponds to the system lifetime. Competing risks is most commonly used to analyze a series system of k components, but can also be used in actuarial applications with k causes of failure. The competing risks model is also used for modeling competing failure modes for components that have multiple failure modes. The algorithm to generate a lifetime T is

for j **from** 1 **to** k
    generate $T_j \sim h_j(t)$
$T \leftarrow \min\{T_1, T_2, \ldots, T_k\}$
return T

The $T_1, T_2, \ldots, T_k$ values can be generated by any of the standard random variate generation algorithms.

4.1.2.3 Thinning

The thinning algorithm, which was originally suggested by Lewis and Shedler [28] for generating the event times in an NHPP, can be adapted to produce a single lifetime by returning only the first event time generated. The random variable T has hazard function $h(t)$. A majorizing hazard function $h^*(t)$ must be found that satisfies $h^*(t) \geqslant h(t)$ for all $t \geqslant 0$. The algorithm is

$T \leftarrow 0$
repeat
    generate Y from $h^*(t)$ **given** $Y > T$
    $T \leftarrow T + Y$
    generate $S \sim U(0, h^*(T))$
until $S \leqslant h(T)$
return T

Generating Y in the repeat-until loop can be performed by inversion or any other method. The name thinning comes from the fact that T can make several steps, each of length Y, that are thinned out before the repeat-until loop terminal condition is satisfied.

4.2 Generating Stochastic Processes

Most discrete-event simulation models have stochastic elements that mimic the probabilistic nature of the system under consideration. The focus in this section is on the generation of a sample realization of a small group of stochastic point processes. In a reliability setting, these stochastic point processes typically represent the failure times of a repairable system. The models are much more general, however, and are used by probabilists to model the arrival times of customers to a queue, the arrival times of demands in an inventory system, or the times of the births of babies. These stochastic models generalize to model events that occur over time or space. A close match between the failure time model and the true underlying probabilistic mechanism associated with failure times of interest is required for successful simulation modeling. The general question considered here is how to generate a sequence of failures in a repairable system (where repair time is considered negligible) when the underlying stochastic process is known. It is typically the case that a data set of failure times has been collected on the system of interest. We begin by introducing probabilistic models for sequences of failure times, which are special cases of what are known as "point processes," where "events" occur at points in time. A special case of a point process is a "counting process," where event occurrences increment a counter.

4.2.1 Counting Processes

A continuous-time, discrete-state stochastic process is often characterized by the counting function $\{N(t), t = 0\}$ which represents the total number of "events" (failures in a reliability setting) that occur by time t [42]. A counting process satisfies the following properties:

1. $N(t) \geqslant 0$;
2. $N(t)$ is integer-valued;
3. if $s < t$, then $N(s) \leqslant N(t)$;
4. for $s < t$, $N(t) - N(s)$ is the number of events in $(s, t\,]$.

Two important properties associated with some counting processes are independent increments and stationarity. A counting process has independent increments if the number of events that occur in mutually exclusive time intervals are independent. A counting process is stationary if the distribution of the number of events that occur in any time interval depends only on the length of the interval. Thus, the stationarity property should only apply to counting processes with a constant rate of occurrence of events.

Counting processes can be used in modeling events as diverse as earthquake occurrences [43, 44], storm occurrences in the Arctic Sea [26], customer arrival times to an electronics store [45], and failure times of a repairable system [37].

We establish some additional notation at this point which will be used in some results and process generation algorithms that follow. Let $X_1, X_2, \ldots$ represent the times between events in a counting process. Let $T_n = X_1 + X_2 + \cdots + X_n$ be the time of the nth event. With these basic definitions in place, we now define the Poisson process, which is the most fundamental of the counting processes.

4.2.2 Poisson Processes

A Poisson process is a special type of counting process that is a fundamental base case for defining many other types of counting processes.

Definition. [42] The counting process $\{N(t), t \geqslant 0\}$ is said to be a Poisson process with rate λ, $\lambda > 0$, if

1. $N(0) = 0$;
2. the process has independent increments;
3. the number of events in any interval of length t is Poisson distributed with mean λt.

The single parameter λ controls the rate at which events occur over time. Since λ is a constant, a Poisson process is often referred to as a homogeneous Poisson process. The third condition is equivalent to

$$P\left(N(t+s) - N(s) = n\right) = \frac{(\lambda t)^n \, \mathrm{e}^{-\lambda t}}{n!} \qquad n = 0, 1, 2, \ldots \tag{4.10}$$

and the stationarity property follows from it.

Although there are many results that follow from the definition of a Poisson process, three are detailed in this paragraph that have applications in discrete-event simulation. Proofs are given in any introductory stochastic processes textbook. First, given that n events occur in a given time interval $(s,t]$, the event times have the same distribution as the order statistics associated with $n$independent observations drawn from a uniform distribution on $(s,t]$, see, for example, [41]. Second, the times between events in a Poisson process are independent and identically distributed exponential random variables with pdf $f(x) = \lambda \exp\{-\lambda x\}$, for $x > 0$. Since the mode of the exponential distribution is 0, a realization of a Poisson process typically exhibits significant clustering of events. Since the sum of n independent and identically distributed exponential (λ) random variables is Erlang (λ, n), T_n has a cdf that can be expressed as a summation:

$$F_{T_n}(t) = 1 - \sum_{k=0}^{n-1} \frac{(\lambda t)^k}{k!} \mathrm{e}^{-\lambda t} \qquad t > 0 \tag{4.11}$$

Third, analogous to the central limit theorem, which shows that the sum of arbitrarily distributed random variables is asymptotically normal, the superposition of renewal processes converges asymptotically to a Poisson process [46].

The mathematical tractability associated with the Poisson process makes it a popular model. It is the base case for queueing theory (e.g., the M/M/1 queue as defined in [20]) and reliability theory (e.g., the models for repairable systems described in [31]). Its rather restrictive assumptions, however, limit its applicability. For this reason, we consider the following variants of the Poisson process that can be useful for modeling more complex failure time processes: the renewal process, the alternating renewal process, and the NHPP. These variants are typically formulated by generalizing an assumption or a property of the Poisson process. Details associated with these models can be found, for example, in [40] or [35].

4.2.3 Renewal Processes

A renewal process is a generalization of a Poisson process. Recall that in a Poisson process, the inter-event times $X_1, X_2, \ldots$ are independent and identically distributed exponential (λ) random variables. In a renewal process, the inter-event times are independent and identically distributed random variables from any distribution with positive support. One useful classification of renewal processes [8] concerns the coefficient of variation σ/μ of the distribution of the times between failures. This classification divides renewal processes into underdispersed and overdispersed processes. A renewal process is underdispersed (overdispersed) if the coefficient of variation of the distribution of the times between failures is less than (greater than) 1. An extreme case of an underdispersed process is when the coefficient of variation is 0 (i.e., deterministic inter-event times), which yields a deterministic renewal process. The underdispersed process is much more regular in its event times. In the case of a repairable system with underdispersed failure times, for example, it is easier to determine when it is appropriate to replace an item in order to avoid experiencing a potentially catastrophic failure. There is extreme clustering of events, on the other hand, in the case of an overdispersed renewal process, and replacement policies are less effective.

4.2.4 Alternating Renewal Processes

An alternating renewal process is a generalization of a renewal process that is often used to model the failure and repair times of a repairable item. Unlike the other models presented here, repair is explicitly modeled by an alternating renewal process. Let $X_1, X_2, \ldots$ be independent and identically distributed random variables with positive support and cdf $F_X(x)$ that represent the times to failure of a repairable item. Let $R_1, R_2, \ldots$ be independent and identically distributed random variables with positive support and cdf $F_R(r)$ that represent the times to repair of a repairable item. Care must be taken to assure that $X_1, X_2, \ldots$ are indeed identically distributed, i.e., the item is neither improving nor deteriorating. Assuming that the alternating renewal process begins at time 0 with the item functioning, then:

- X_1 is the time of the first failure;
- $X_1 + R_1$ is the time of the first repair;
- $X_1 + R_1 + X_2$ is the time of the second failure;
- $X_1 + R_1 + X_2 + R_2$ is the time of the second repair, etc.

Thus the times between events for an alternating renewal process alternate between two distributions, each with positive support.

4.2.5 Nonhomogeneous Poisson Processes

An NHPP is another generalization of a Poisson process which allows for an failure rate $\lambda(t)$ (known as the intensity function) that can vary with time.

Definition. [42] The counting process $\{N(t), t \geqslant 0\}$ is said to be an NHPP with intensity function $\lambda(t)$, $t \geqslant 0$, if

1. $N(0) = 0$;
2. the process has independent increments;
3. $P(N(t+h) - N(t) \geqslant 2) = o(h)$;
4. $P(N(t+h) - N(t) = 1) = \lambda(t)h + o(h)$,

where a function $f(\cdot)$ is said to be $o(h)$ if $\lim_{h \to 0} f(h)/h = 0$.

An NHPP is often appropriate for modeling a series of events that occur over time in a nonstationary fashion. Two common application areas are the modeling of arrivals to a waiting line (queueing theory) and the failure times of a repairable system (reliability theory) with negligible repair times. The cumulative intensity function

$$\Lambda(t) = \int_0^t \lambda(\tau)\,\mathrm{d}\tau \quad t > 0 \tag{4.12}$$

gives the expected number of events by time t, i.e., $\Lambda(t) = E[N(t)]$. As stated in [6], the probability of exactly n events occurring in the interval $(a, b]$ is given by

$$\frac{\left[\int_a^b \lambda(t)\mathrm{d}t\right]^n \mathrm{e}^{-\int_a^b \lambda(t)\mathrm{d}t}}{n!} \quad \text{for } n = 0, 1, \ldots \tag{4.13}$$

4.2.6 Markov Models

Markov models are characterized by exponential transition times between discrete states. We present one such Markov model here, which is known as a continuous-time Markov chain (CTMC). These models are characterized by the following properties:

- At any time $t > 0$, the state of the process $X(t)$ assumes a discrete value.
- The times between transitions from one state to another state are exponentially distributed.

These models are the continuous analog of discrete-time Markov chain models, where both state and time are discrete. The set of all possible discrete states that the CTMC can assume is denoted by M. The transition rates from state to state are typically organized in an infinitesimal generator matrix G that satisfies the following properties:

- An off-diagonal element of G, denoted by g_{ij}, is the rate of transition from state i to state j. If the transition from state i to state j is impossible, then $g_{ij} = 0$.
- A diagonal element of G is the opposite of the sum of the other elements in row i by convention. This implies that the row sums of G are zero. It also implies that the opposite of the diagonal elements of G denote the rate associated with the holding time in a particular state.

A comprehensive treatment of Markov processes is given in [7].

4.2.7 Other Variants

Other variants of a Poisson process have been proposed. For brevity, we outline three such variants. Details are given in [40]. Mixed Poisson processes can be formulated in terms of an NHPP with cumulative intensity function $\Lambda(t)$ and a random variable L with positive support. The associated counting process $N(L\Lambda(t))$ is a mixed Poisson process. Transforming the time scale with the random variable L results in a process that does not, in general, have independent increments. Ross [42] provides an illustration from the insurance industry where L models the claim rate (which varies from one policyholder to the next) and $\Lambda(t)$ is linear. Doubly stochastic Poisson processes generalize the notion of transforming the time scale by embedding a stochastic process within another stochastic process. The random variable L from a mixed Poisson process is replaced with a stochastic process with non-decreasing paths. Markov-modulated Poisson processes are also a special case of doubly stochastic processes. Compound Poisson processes are formulated with a homogeneous or nonhomogeneous Poisson process and a sequence of independent and identically distributed random variables $D_1, D_2, \ldots$. The function

$$C(t) = \begin{cases} \sum_{i=1}^{N(t)} D_i & \text{if } N(t) > 0 \\ 0 & \text{otherwise} \end{cases} \tag{4.14}$$

defines a process that increases by $D_1, D_2, \ldots$ at each event time. This would be an appropriate model for an automobile insurance company whose claims occur according to a Poisson process with claim values $D_1, D_2, \ldots$ and $C(t)$ models the total claim amounts that have occurred by time t. Similarly, if $D_1, D_2, \ldots$ are independent and identically distributed random variables with support on the non-negative integers, then a compound Poisson process can be used to model batch failures.

4.2.8 Random Process Generation

The algorithms presented in this section generate a sequence of random event times (in our setting they are failure times, or possibly repair times for the stochastic models described in the previous sections) on the time interval $(0, S]$, where S is a real, fixed constant. If the next-event approach is taken for placing events onto the calendar in a discrete-event simulation model, then these algorithms should be modified so that they take the current event time as an argument and return the next event time. All processes are assumed to begin at time 0. The random event times that are generated by the counting process are denoted by T_1, T_2, and random numbers (i.e., $U(0, 1)$ random variables) are denoted by U or $U_1, U_2, \ldots$. If just $T_0 = 0$ is returned, then no events were observed on $(0, S]$.

4.2.8.1 Poisson Processes

Since the times between events in a Poisson process are independent and identically distributed exponential (λ) random variables, the following algorithm generates the event times of a Poisson process on $(0, S]$:

$T_0 \leftarrow 0$
$i \leftarrow 0$
while $T_i \leqslant S$
  $i \leftarrow i + 1$
  generate $U_i \sim U(0, 1)$
  $T_i \leftarrow T_{i-1} - \log(1 - U_i)/\lambda$
return $T_1, T_2, \ldots, T_{i-1}$

4.2.8.2 Renewal Processes

Event times in a renewal process are generated in a similar fashion to a Poisson process. Let $F_X(x)$ denote the cdf of the inter-event times X_1, X_2, ... in a renewal process. The following algorithm generates the event times on $(0, S]$:

$T_0 \leftarrow 0$
$i \leftarrow 0$
while $T_i \leqslant S$
  $i \leftarrow i + 1$
  generate $U_i \sim U(0, 1)$
  $T_i \leftarrow T_{i-1} + F_X^{-1}(U_i)$
return $T_1, T_2, \ldots, T_{i-1}$

4.2.8.3 Alternating Renewal Processes

Event times in an alternating renewal process are generated in a similar fashion to a renewal process, but the inter-event time must alternate between $F_X(x)$, the cdf of the times to failure X_1, X_2, ... and $F_R(r)$, the cdf of the times to repair R_1, R_2, ... using the binary toggle variable j. The following algorithm generates the event times on $(0, S]$:

$T_0 \leftarrow 0$
$i \leftarrow 0$
$j \leftarrow 0$
while $T_i \leqslant S$
  $i \leftarrow i + 1$
  generate $U_i \sim U(0, 1)$
  if $j = 0$
    $T_i \leftarrow T_{i-1} + F_X^{-1}(U_i)$
    $j \leftarrow j + 1$

else
$T_i \leftarrow T_{i-1} + F_R^{-1}(U_i)$
$j \leftarrow j - 1$
return $T_1, T_2, \ldots, T_{i-1}$

4.2.8.4 Nonhomogeneous Poisson Processes

Event times can be generated for use in discrete-event simulation as $\Lambda^{-1}(E_1)$, $\Lambda^{-1}(E_2)$, ..., where E_1, E_2, ... are the event times in a unit Poisson process [6]. This technique is often referred to as "inversion," and is implemented below:

$T_0 \leftarrow 0$
$E_0 \leftarrow 0$
$i \leftarrow 0$
while $T_i \leqslant S$
$i \leftarrow i + 1$
generate $U_i \sim U(0, 1)$
$E_{i-} \leftarrow E_{i-1} - \log(1 - U_i)$
$T_i \leftarrow \Lambda^{-1}(E_i)$
return $T_1, T_2, \ldots, T_{i-1}$

The inversion algorithm is ideal when $\Lambda(t)$ can be inverted analytically, although it also applies when $\Lambda(t)$ needs to be inverted numerically. There may be occasions when the numerical inversion of $\Lambda(t)$ is so onerous that the thinning algorithm devised by Lewis and Shedler [28] might be preferable. This algorithm assumes that the modeler has determined a majorizing value λ^* that satisfies $\lambda^* \geqslant \lambda(t)$ for all $t > 0$:

$T_0 \leftarrow 0$
$i \leftarrow 0$
while $T_i \leqslant S$
$t \leftarrow T_i$
repeat
generate $U \sim U(0, 1)$
$t \leftarrow t - \log(1 - U)/\lambda^*$
generate $U \sim U(0, \lambda^*)$
until $U \leqslant \lambda(t)$
$i \leftarrow i + 1$
$T_i \leftarrow t$
return $T_1, T_2, \ldots, T_{i-1}$

The majorizing value λ^* can be generalized to a majorizing function $\lambda^*(t)$ to decrease the CPU time by minimizing the probability of "rejection" in the **repeat-until** loop.

4.2.8.5 Continuous-time Markov Chains

Event times T_1, T_2, ... and associated states X_0, X_1, X_2, ... can be generated via inversion on the time interval $(0, S]$. Prior to implementing the algorithm, one needs:

- the initial state distribution p_0 which is defined on the finite state space M or some subset of M;
- the infinitesimal generator matrix G.

The following algorithm generates the event times on $(0, S]$:

$T_0 \leftarrow 0$
generate $X_0 \sim p_0$
while $T_i \leqslant S$
    $i \leftarrow i + 1$
    generate $U \sim U(0, 1)$
    $T_i \leftarrow T_{i-1} + \log(1 - U)/g_{X_{i-1}X_{i-1}}$
    generate X_i from the probability vector
        $-g_{X_{i-1}X_i}/g_{X_{i-1}X_{i-1}}$ for $X_i \neq X_{i-1}$
return $T_1, T_2, \ldots, T_{i-1}$; $X_0, X_1, \ldots, X_{i-1}$

All of the stochastic process models given in this section are elementary. More complex models are considered in [36].

4.3 Survival Models Involving Covariates

The accelerated life and proportional hazards lifetime models can account for the effects of covariates on a random lifetime [9]. Variate generation for these models is a straightforward extension of the basic methods for generating random lifetimes when the covariates do not depend on time. Variate generation algorithms for Monte Carlo simulation of NHPPs are a simple extension of the inverse-chf technique.

The effect of covariates (explanatory variables) on survival often complicates the analysis of a set of lifetime data. In a medical setting, these covariates are usually patient characteristics such as age, gender, or blood pressure. In reliability, these covariates are exogenous variables such as the turning speed of a machine tool or the stress applied to a component that affect the lifetime of an item. We use the generic term item here to refer to a manufactured product or organism whose survival time is of interest. Two common models to incorporate the effect of the covariates on lifetimes are the accelerated life and Cox proportional hazards models. The roots of the accelerated life model are in reliability and the roots of the proportional hazards model are in biostatistics. Bender *et al.* [3, 4] indicate an increased interest in the use of random variate generation in medical models.

The $q \times 1$ vector $\boldsymbol{z}$ contains covariates associated with a particular item. The covariates are linked to the lifetime by the function $\psi(\boldsymbol{z})$, which satisfies $\psi(0) = 1$ and $\psi(\boldsymbol{z}) \geqslant 0$ for all z. A popular choice is the log linear form $\psi(\boldsymbol{z}) = e^{\boldsymbol{\beta}'\boldsymbol{z}}$ where $\boldsymbol{\beta}$ is a $q \times 1$ vector of regression coefficients.

4.3.1 Accelerated Life Model

The chf for T in the accelerated life model is

$$H(t) = H_0\left(t\psi(z)\right), \quad t \geqslant 0 \tag{4.15}$$

where H_0 is a baseline chf. When $z = 0$, $H(t) = H_0(t)$. In this model, the covariates accelerate $[\psi(z) > 1]$ or decelerate $[\psi(z) < 1]$ the rate that the item moves through time.

4.3.2 Proportional Hazards Model

The chf for T in the proportional hazards model is

$$H(t) = \psi(z)H_0(t) \quad t \geqslant 0 \tag{4.16}$$

In this model, the covariates increase $[\psi(z) > 1]$ or decrease $[\psi(z) < 1]$ the hazard function associated with the lifetime of the item by the factor $\psi(z)$ for all values of t. This model is known in medicine as the "Cox model" and is a standard model for evaluating the effect of covariates on survival. We do not explicitly consider the estimation of the regression coefficients $\boldsymbol{\beta}$ here since the focus is on random lifetime generation. Cox and Oakes [9], O'Quigley [39], and others give the details associated with estimation of $\boldsymbol{\beta}$ and most modern statistical packages estimate these coefficients using built-in numerical methods.

4.3.3 Random Lifetime Generation

All of the algorithms for variate generation for these models are based on the fact that $H(T)$ is exponentially distributed with a mean of one. Therefore, equating the chf to $-\log(1-U)$, where $U \sim U(0,1)$, and solving for t yields the appropriate generation technique [27].

In the accelerated life model, since time is being expanded or contracted by a factor $\psi(z)$, variates are generated by

$$T \leftarrow \frac{H_0^{-1}\left(-\log\left(1-U\right)\right)}{\psi(z)} \tag{4.17}$$

In the proportional hazards model, equating $-\log(1-U)$ to $H(T)$ yields the random variate generation formula

$$T \leftarrow H_0^{-1}\left(\frac{-\log\left(1-U\right)}{\psi(z)}\right) \tag{4.18}$$

In addition to generating individual lifetimes, these random variate generation techniques can be applied to point process models that include covariates. A renewal process, for example, with time between events having a chf $H(t)$ can be simulated by using the appropriate generation formula for the two cases shown above. These random variate generation formulas must be modified, however, to generate random variates from an NHPP.

In an NHPP, the hazard function, $h(t)$, is analogous to the intensity function, $\lambda(t)$, which governs the rate at which events occur. To determine the appropriate method for generating random variates from an NHPP model which involves covariates, assume that the last event in a point process has occurred at time a. The chf for the time of the next event, conditioned on survival to time a, is

$$H_{T|T>a}(t) = H(t) - H(a) \quad t \geqslant a \tag{4.19}$$

In the accelerated life model, where $H(t) = H_0\left(t\psi(z)\right)$, the time of the next event is generated by

$$T \leftarrow \frac{H_0^{-1}\left(H_0\left(a\psi\left(z\right)\right) - \log\left(1 - U\right)\right)}{\psi(z)} \tag{4.20}$$

Equating the conditional chf to $-\log(1 - U)$, the time of the next event in the proportional hazards case is generated by

$$T \leftarrow H_0^{-1}\left(H_0\left(a\right) - \frac{\log\left(1 - U\right)}{\psi(z)}\right) \tag{4.21}$$

Table 4.1 summarizes the random variate generation algorithms for the accelerated life and proportional hazards models (the last event occurred at time a).

The $1 - U$ could be replaced with U in this table to save a subtraction, although the sense of the monotonicity would be reversed, i.e., small random numbers are mapped to large variates. The renewal and NHPP algorithms are equivalent when $a = 0$ (since a renewal process is equivalent to an NHPP restarted at zero after each event), the accelerated life and proportional hazards models are equivalent when $\psi(z) = 1$, and all four cases are equivalent when $H_0(t) = \lambda t$ (the exponential baseline case) because of the memoryless property associated with the exponential distribution.

Table 4.1 Lifetime generation in regression survival models

	Renewal	NHPP
Accelerated life	$T \leftarrow a + \frac{H_0^{-1}(-\log(1-U))}{\psi(z)}$	$T \leftarrow \frac{H_0^{-1}(H_0(a\psi(z))-\log(1-U))}{\psi(z)}$
Proportional hazards	$T \leftarrow a + H_0^{-1}\left(\frac{-\log(1-U)}{\psi(z)}\right)$	$T \leftarrow H_0^{-1}\left(H_0(a) - \frac{\log(1-U)}{\psi(z)}\right)$

4.4 Conclusions and Further Reading

The discussion here has been limited to the generation of random lifetimes (with and without covariates) and random stochastic processes because of the emphasis in this volume. There are many other quantities that can be generated that might be of use in reliability and availability analysis. These range from generating combinatorial objects [13,38,47], to generating random matrices [5,12,15,18,29,32], to generating random polynomials [14], to shuffling playing cards [34], to generating random spawning trees [17] to generating random sequences [23], to generating Markov chains [19,22,33,42].

References

1. Allison PD (1984) Event history analysis: regression for longitudinal event data. Sage Publications, Newbury Park, CA, USA
2. Banks J, Carson JS, Nelson BL, Nicol DM (2005) Discrete-event system simulation, 4th edn. Prentice-Hall, Upper Saddle River, NJ, USA
3. Bender R, Augustin T, Blettner M (2005) Generating survival times to simulate Cox proportional hazards models. Stat Med 24:1713–1723
4. Bender R, Augustin T, Blettner M (2006) Letter to the editor. Stat Med 25:1978–1979
5. Carmeli M (1983) Statistical theory and random matrices. Marcel Dekker, New York, NY, USA
6. Çinlar E (1975) Introduction to stochastic processes. Prentice-Hall, Upper Saddle River, NJ, USA
7. Clarke AB, Disney RL (1985) Probability and random processes: a first course with applications, 2nd edn. John Wiley, New York, NY, USA
8. Cox DR, Isham V (1980) Point processes. Chapman and Hall, Boca Raton, FL, USA
9. Cox DR, Oakes D (1984) Analysis of survival data. Chapman and Hall, Boca Raton, FL, USA
10. Crowder, MJ (2001) Classical competing risks. Chapman and Hall/CRC Press, Boca Raton, FL, USA
11. David HA, Moeschberger ML (1978) The theory of competing risks. Macmillan, New York, NY, USA
12. Deift P (2000) Orthogonal polynomials and random matrices: A Riemann-Hilbert approach. American Mathematical Society, Providence, RI, USA
13. Devroye L (1986) Non-uniform random variate generation. Springer, New York, NY, USA
14. Edelman A, Kostlan E (1995) How many zeros of a random polynomial are real? Bulletin of the Am Math Soc 32(1):1–37
15. Edelman A, Kostlan E, Shub M (1994) How many eigenvalues of a random matrix are real? J Am Math Soc 7:247–267
16. Everitt BS, Hand DJ (1981) Finite mixture distributions. Chapman and Hall, Boca Raton, FL, USA
17. Fishman GS (1996) Monte Carlo: Concepts, algorithms, and applications. Springer, New York, NY, USA
18. Ghosh S, Henderson SG (2003) Behavior of the NORTA method for correlated random vector generation as the dimension increases. ACM Trans Model Comput Simul 13:276–294
19. Gilks WR, Richardson S, Spiegelhalter DJ (1996) Markov chain Monte Carlo in practice. Chapman and Hall/CRC, Boca Raton, FL, USA
20. Gross D, Harris CM (1998) Fundamentals of queueing theory, 3rd edn. John Wiley, Hoboken, NJ, USA

21. Hörmann W, Leydold J, Derflinger G (2004) Nonuniform automatic random variate generation. Springer, New York, NY, USA
22. Hastings WK (1970) Monte Carlo sampling methods using Markov chains and their applications. Biometrika 57:97–109
23. Knuth DE (1998) The art of computer programming, Vol 2: Seminumerical algorithms, 3rd edn. Addison-Wesley, Reading, MA, USA
24. Law AM (2007) Simulation modeling and analysis, 4th edn. McGraw-Hill, New York, NY, USA
25. Lawless JF (2003) Statistical models and methods for lifetime data, 2nd edn. John Wiley, New York, NY, USA
26. Lee S, Wilson JR, Crawford MM (1991) Modeling and simulation of a nonhomogeneous Poisson process having cyclic behavior. Commun Stat Simul Comput 20(2&3):777–809
27. Leemis LM (1987) Variate generation for the accelerated life and proportional hazards models. Oper Res 35(6):892–894
28. Lewis PAW, Shedler GS (1979) Simulation of nonhomogeneous poisson processes by thinning. Naval Res Logist Quart 26(3)403–413
29. Marsaglia G, Olkin I (1984) Generating correlation matrices. SIAM J Sci Stat Comput 5(2):470–475
30. McLachlan G, Peel D (2000) Finite mixture models. John Wiley, New York, NY, USA
31. Meeker WQ, Escobar LA. (1998) Statistical methods for reliability data. John Wiley, New York, NY, USA
32. Mehta ML (2004) Random matrices, 3rd edn. Elsevier, Amsterdam, The Netherlands
33. Metropolis N, Rosenbluth AW, Rosenbluth MN, Teller AH, Teller E (1953) Equations of state calculations by fast computing machine. J Chem Phys 21:1087–1091
34. Morris SB (1998) Magic tricks, card shuffling and dynamic computer memories. The Mathematical Association of America, Washington D.C., USA
35. Nelson BL (2002) Stochastic modeling: analysis and simulation. Dover Publications, Mineola, NY, USA
36. Nelson BL, Ware P, Cario MC, Harris CA, Jamison SA, Miller JO, Steinbugl J, Yang J (1995) Input modeling when simple models fail. In: Alexopoulos C, Kang K, Lilegdon WR, Goldsman D (eds) Proceedings of the 1995 winter simulation conference, IEEE Computer Society, Washington DC, USA, pp 3–100
37. Nelson WB (2003) Recurrent events data analysis for product repairs, disease recurrences and other applications. ASA/SIAM, Philadelphia, PA, USA
38. Nijenhuis A, Wilf HS (1978) Combinatorial algorithms for computers and calculators, 2nd edn. Academic Press, Orlando, FL, USA
39. O'Quigley J (2008) Proportional hazards regression. Springer, New York, NY, USA
40. Resnick SI (1992) Adventures in stochastic processes. Birkhäuser, New York, NY, USA
41. Rigdon SE, Basu AP (2000) Statistical methods for the reliability of repairable systems. John Wiley, New York, NY, USA
42. Ross SL (2003) Introduction to probability models, 8th edn. Academic Press, Orlando, FL, USA
43. Schoenberg FP (2003) Multidimensional residual analysis of point process models for earthquake occurrences. J Am Stat Assoc 98:464
44. Schoenberg FP (2003) Multidimensional residual analysis of point process models for earthquake occurrences. J Am Stat Assoc 98:789–795
45. White KP (1999) Simulating a nonstationary Poisson process using bivariate thinning: The case of "typical weekday" arrivals at a consumer electronics store. In: Farrington P, Nembhard PA, Sturrock HB, Evans GW (eds) Proceedings of the 1999 winter simulation conference, ACM, New York, NY, USA, pp 458–461
46. Whitt W (2002) Stochastic-process limits: an introduction to stochastic-process limits and their application to queues. Springer, New York, NY, USA
47. Wilf HS (1989) Combinatorial algorithms: an update. SIAM, Philadelphia, PA, USA

Part II
Simulation Applications in Reliability

Chapter 5
Simulation-based Methods for Studying Reliability and Preventive Maintenance of Public Infrastructure

Abhijit Gosavi and Susan Murray

Abstract In recent times, simulation has made significant progress as a tool for improving the performance of complex stochastic systems that arise in various domains in the industrial and service sectors. In particular, what is remarkable is that simulation is being increasingly used in diverse domains, e.g., devising strategies needed for emergency response to terrorist threats in homeland security systems and civil engineering of bridge structures for motor vehicle transport. In this chapter, we will focus on (1) describing some of the key decision-making problems underlying (a) response to emergency bomb-threat scenarios in a public building, and (b) prevention of catastrophic failures of bridges used for motor-vehicle transport; (2) providing an overview of simulation-based technologies that can be adopted for solving the associated problems. Our discussion will highlight some *performance measures* applicable to emergency response and prevention that can be estimated and improved upon via discrete-event simulation. We will describe two problem domains in which measurement of these metrics is critical for optimal decision-making. We believe that there is a great deal of interest currently, within both the academic world and the government sector, in enhancing our homeland security systems. Simulation already plays a vital role in this endeavor. The nature of the problems in this chapter is unconventional and quite unlike that seen commonly in classical simulation-based domains of manufacturing and service industries.

5.1 Introduction

Simulation is known to be a powerful tool for modeling systems that are too complex for analytical models. However, analytical models are often preferred in decision-making because they generate exact or near-exact closed forms of the objective function (performance measure). Hence not only are they more amenable to optimization but they are also capable of providing structural insights. Through the 1970s

---

Missouri University of Science and Technology in Rolla, Missouri, USA

P. Faulin, A. Juan, S. Martorell, and J.E. Ramírez-Márquez (eds), *Simulation Methods for Reliability and Availability of Complex Systems*. © Springer 2010

and 1980s, analysts involved in devising emergency response strategies related to public infrastructure systems were partial to analytical models (Walker *et al.* 1979; Larson 1975; Kolesar and Swersey 1985). Simulation models were *perceived* as black-boxes and as tools lacking in the ability of providing the much sought-after theoretical structural insights needed in decision-making (Kleijnen 2008). This perception has changed to a great extent in recent years, and increasing attention is now being paid to simulation-based models because of (1) the enhanced power of computers that has dramatically increased our ability to simulate a very large number of scenarios via simulation – thereby making simulation a viable, practical tool for decision-making (Kleijnen 2008) and (2) path-breaking advances in solving certain classes of optimization problems within simulators (Bertsekas and Tsitsiklis 1996; Sutton and Barto 1998; Gosavi 2003). It is also worth emphasizing that although the structural insights that simulation can offer are usually of an empirical nature, they are not as limited by the restrictive assumptions that must be made in analytical closed-form-type models.

In this chapter, we will describe in detail two problem domains where simulation can be used for performance improvement. These domains are (1) a terrorist attack (a combined bomb and chemical threat) in a public building and (2) the development of effective maintenance strategies for a public bridge susceptible to natural disasters. The first case study is from the field of *consequence management*, which is the response to terrorist activities or natural disasters by the local and/or federal government (i.e., police departments, Federal Emergency Management Agency, the military). The second case study is drawn from civil and structural engineering, where simulation is gradually playing an important role in developing strategies for structure management (Melchers 1999; Juan *et al.* 2008b). We will define some of the core performance metrics that are of interest in both the design and operation stages. Our interest is in the metrics for which simulation is the most effective technique for measurement and performance improvement. We will also discuss why and how simulation is an effective approach for measuring these metrics. While both of the above-described domains are not as well-known in the literature as manufacturing and service organizations, such as factories and banks, we believe that the novel nature of these studies illustrate emerging areas for applying simulation.

The rest of this chapter is organized as follows. In Section 5.2, we will outline the need for using simulation. Section 5.3.1 is devoted to discussing how simulation can be used for decision-making in a public building where a bomb threat or a terror attack has been made. In Section 5.3.2, we briefly describe the problem of reliability measurement of a bridge used by motor vehicles. We conclude with a summary of our discussion in Section 5.4.

5.2 The Power of Simulation

The greatest advantage of simulation is that it can avoid the simplifying assumptions that are often necessary in analytical models to derive performance metrics.

These assumptions can render the models less useful in the practical world (Henderson and Mason 2004). Also, simulation can be easily explained to managers and personnel involved in the operational activities, thereby increasing the probability of its application. As discussed above, recent advances in the theory of numerical optimization have allowed us to combine simulation with optimization techniques – thus increasing its power for decision-making. Simulation has been used in the past for ambulance planning (Fujiwara *et al.* 1987; Harewood 2002; Ingolfsson *et al.* 2003; Henderson and Mason 2004) and emergency evacuation (Pidd *et al.* 1996). Simulation allows us to make decisions regarding a number of important issues. The following represents only a small sample in the context of the domains we study in this chapter.

- What tactics work best in improving the ability of humans to respond quickly in an emergency?
- How many first responders are required, and where should they be stationed?
- When during its lifetime and how often should a bridge receive preventive maintenance?
- What is the reliability of a bridge at any given point of time in its life?

In case studies that follow, we will discuss in detail how simulation of both Monte Carlo and discrete-event type can be used fruitfully for decision-making. *Monte Carlo* simulation can be used to determine the probability of failure of a bridge. This failure probability is essential to generate a strategy for optimal preventive maintenance of a bridge. When the distributions of the random variables underlying the failure mechanisms follow *arbitrary* distributions, which is often the case with complex bridges, simulation is the only tool that can be used to determine the failure probability. Similarly, in modeling emergency response, *discrete-event* simulation can be used to model the dynamics of a terrorist attack inside a building. The complex dynamics of such an event are governed by so many random variables that, usually, simulation is the only modeling tool that can accurately capture the behavior of the entire system.

Simulation programs can be written in simulation-specific software such as ARENA or CSIM, or in more basic languages such as C, C++, Java, or even in generic software such as MATLAB that permit easy combination of simulation with optimization modules. Such software or compilers for basic languages are now becoming increasingly available (even free in some cases). This has led to a dramatic increase in the use of simulation in the industry and in academic settings; see Hlupic (2000) for a survey of numerous commercial software and their users.

5.3 Case Studies

This section is devoted to presenting the two case studies. Section 5.3.1 presents an overview of the emergency response case study in which discrete-event simulation is used as a tool to model the dynamics of an incident of bomb threat in a public building. We describe the case study, the simulation model, and how it is useful for decision-making purposes. Section 5.3.2 presents an overview of the bridge-maintenance case study. In this case study, Monte Carlo simulation plays a critical role in determining the failure probability of structures in the bridge. We also describe briefly how the subsequent analysis that requires simulation as an input is performed to devise preventive maintenance strategies.

5.3.1 Emergency Response

Simulation has become a very useful tool for modeling human performance in an emergency situation such as a terrorist attack or a natural disaster. Usually, in an emergency situation, the police, fire, and health administration officials are responsible for responding to the crisis. Response to the crisis requires sending the right number of personnel and ensuring that the problem is solved with minimum casualties. The events that take place in this emergency can be mathematically modeled via stochastic processes. Mathematical stochastic-process tools, e.g., Markov chains, Brownian motion, or renewal processes, which can lead to closed form formulations of the associated performance metrics and objective functions, are generally insufficient to model the event dynamics of the *entire system* within one model. Discrete-event simulation is the most powerful tool to model the underlying stochastic processes in this situation.

The US Army has developed a simulation tool that can be used to improve the performance of emergency response personnel in responding to a bomb threat in a public building. The tool has also been used for training personnel in addressing such situations. In what follows, we will present the main concepts underlying a simulation tool called IMPRINT (see `http://www.arl.army.mil/ARL`) that has been developed for human performance modeling. We will begin with an overview of the problem studied.

5.3.1.1 A Bomb Threat

A university campus in the USA was subject to a bomb and anthrax threat inside one of its buildings (Martin 2007). A graduate student arrived at an engineering building and claimed to have a document which detailed a plan for destroying numerous buildings on the campus. A call was made to the police informing them that a student was making threats and that the student carried a knife, a firearm, and a powder of some kind. About seven agencies responded to the threat, and they included, in addition to the local police, the fire department, a weapons of mass destruction (WMD) team, the FBI, and a local unit of Homeland Security. The situation was defused by the police by the use of tasers. The student had to be tasered three times. It was discovered that the "bomb" that the student was carrying was in fact soil and the powdery substance was powdered sugar. The activities of all the personnel in this incident can be accurately modeled within a discrete-event simulator. In the next subsection, we will describe the main events that have to be accounted for in the simulation tool, IMPRINT.

5.3.1.2 IMPRINT

The model developed in IMPRINT starts with the generation of a suspicious call. This triggers the dispatch of two officers to the building, which requires a random amount of time. The officers upon arriving at the scene make an initial assessment of the situation. Then, they call for backup from their force, approach the suspect, and then successfully capture the suspect. The task of capturing the suspect is performed in three stages of tasering. Immediately after the suspect is captured, a number of other people who could possibly have been exposed to the suspicious powder are quarantined. The powder is sent for inspection to the WMD department; people remain quarantined until the results of the analysis are available. When the results come back negative, the quarantined people are released. The buildings are shut down for a finite amount of time. As soon as the police arrive at the scene and find that the call is not a hoax, they call for backup, which also alerts a number of other agencies, such as WMD and the fire department. This in turn triggers a sequence of parallel events, e.g., closing down of a number of buildings where the suspect could have possibly spread the powder prior to making threats. As is clear, this incident involves a number of chains of simultaneously occurring events, which are inter-dependent, and the duration of most activities associated with the events (e.g., approaching subject, fire personnel arriving to the scene) are random variables. Together, these factors make simulation the only viable tool for performance analysis.

It needs to be pointed out that the study of the performance of the personnel involved in which various events are linked together via communication amongst the entities is also called *consequence management* (Menk and Mills 1999). In particular, in consequence management, an important task is to study the impact of communication between chains of events that occur in parallel but are inter-dependent due to inter-communication. Simulation is especially suitable for modeling these scenarios, and is hence a popular tool in this field.

IMPRINT was designed by the military, keeping in mind that external and internal stressors affect the humans (response personnel) involved. The software provides decision-makers with insight into time delays, success rates, and the interaction between the first responders in a serious threat scenario. IMPRINT is like any other simulation package such as ARENA and PROMODEL. It uses a graphic user-friendly interface to develop the model, and at the back-end uses the standard time-advance mechanisms of simulation packages to generate a large number of samples (Law and Kelton 2000) of performance measures from which values of long-run measures can be estimated accurately.

The goal of the study in Murray and Ghosh (2008) (see also Gosakan 2008) was to evaluate the capability of the personnel to operate effectively under environmental stressors. They used IMPRINT to simulate the system, and measured numerous

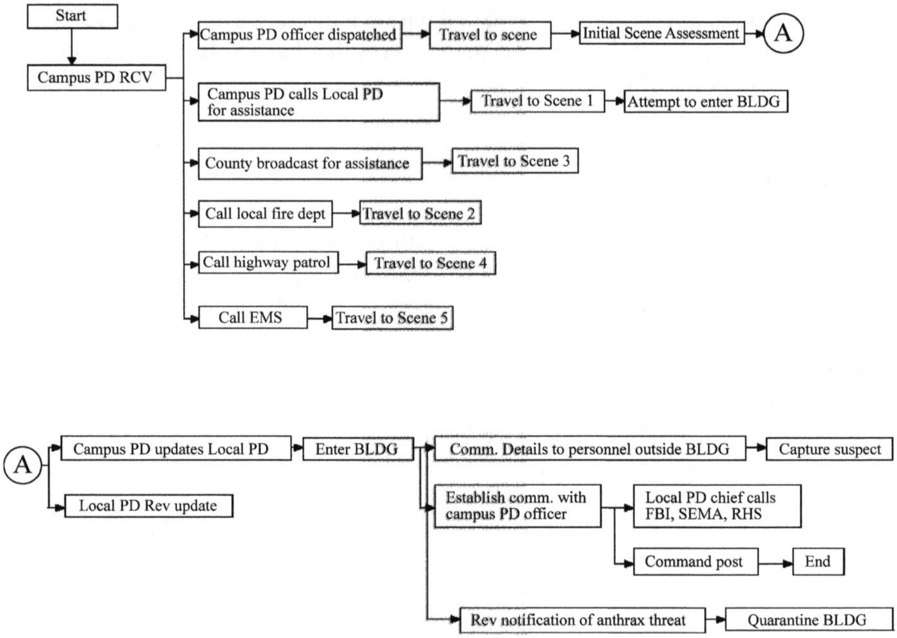

Figure 5.1 Task network model for the university bomb threat

performance measures. See Figure 5.1 for a schematic of their model. The performance metrics of interest in their study were: availability of the system as a whole, mission performance time (mean and standard deviation) of the emergency personnel (see Figure 5.2), accuracy of their performance (frequency of failures), and the workload profiles of the responders. See also Table 5.1, which shows that IMPRINT can be used to provide details of which action occurred at a given point of time. In addition, we would like to provide some additional numerical results that will be of interest to the reader. The IMPRINT model shows that the mission time takes an average of 11 : 13 : 29 (read the units as hours : minutes : seconds . milliseconds). The minimum is 9 : 28 : 37 . 28 and maximum is 12 : 27 : 07 . 75. The powder that is gathered has to be tested for whether it is actually anthrax. A what-if scenario can be analyzed here: if that test has a failure of 40% (in case of failure, one must re-test), the mission time has an average of 11 : 43 : 14 and the maximum is 14 : 51 : 39; the minimum does not change. Also, the statistics for the time to secure the building and making sure there is no other terrorist in the building are: average of 10 : 56 : 15 with a maximum of 13 : 06 : 08 and a minimum of 8 : 47 : 65. Such information about the range of time values and the impact of various what-ifs is useful for those planning and evaluating the procedures used by first responders to emergencies.

The mathematical formulas underlying the measurement of these performance metrics are similar to standard metrics in discrete-event simulation, e.g., mean wait in a single-server, single-channel queue. Since these ideas are elementary in simulation, we do not present the formulas (the beginner to simulation studies is directed

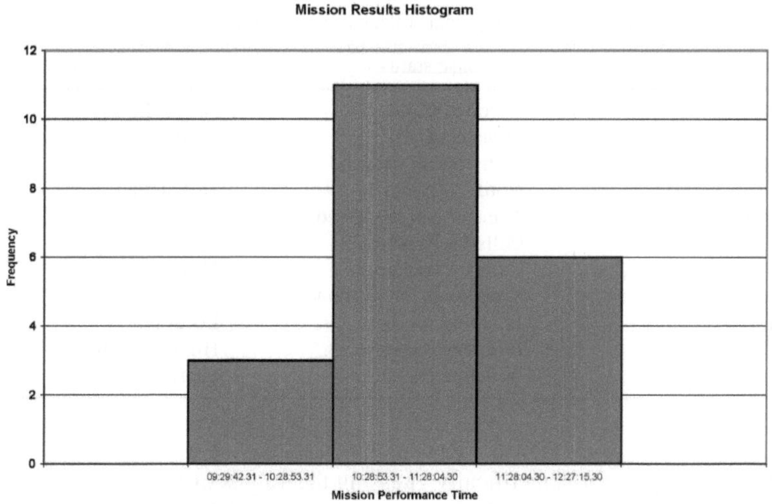

Figure 5.2 Frequency distribution bar chart of the mission time

to Chapter 1 of Law and Kelton 2000). Instead we focus on the on some of the questions that can be answered by IMPRINT in helping improve the quality of the first response in the future in a similar situation:

- How would a delay in the arrival of the bomb-detection unit impact the system as a whole and some key performance metrics, such as the number of casualties?
- How would the performance measures change had the event occurred in the vicinity of the university's nuclear reactor? What additional measures could reduce the risk of endangering the entire human community around the university?
- If the university put into place an emergency mass notification system that employed cell phones, what impact would that have on the evacuation processes?
- How effective would an additional decontamination team be in terms of the performance of the responders?
- What resource requirements would be necessary if there were multiple attackers in different parts of the university at the same time?

It is to be noted that because of the numerous random variables and the interdependencies of the activities, these systems are extremely complex, making mathematical models intractable and possibly making simulation the only viable tool for answering the questions posed above. Furthermore, answering these questions accurately is *imperative* for designing systems that can effectively deal with terrorist or criminal threats in the future.

Determining the values of the input variable to the IMPRINT model require the collection of data from other events in which emergency personnel had to respond. On occasions such data is available, and when it is not, e.g., under a scenario in which the personnel are under stress factors of different kinds, one can use re-

Table 5.1 Portion of the event tracking data collected in a simulation

Clock	Response status	Responding name
45.38585175	Receives Call 1	Dispatcher
45.38585175	Transfers Call 1	Dispatcher
45.38585175	Transfer call received	Sgt
60	Notifies Officer 1	Dispatcher
90.75849001	Transfer call processed	Sgt
608.4436395	Calls for backup	Officer 2
608.4436395	Calls for backup	Dispatcher
608.4436395	Broadcasts for assistance	Dispatcher
608.4436395	Initiating response	Local PD
608.4436395	Initiating response	Highway Patrol
608.4436395	Initiating response	Sheriff Dept

gression techniques for scientifically guessing the values of the variables (Gosakan 2008). This is an important challenge in simulation modeling of such events.

The simulation output from IMPRINT can be potentially combined with a response surface methodology (Kleijnen 2008) to study the relationships between input parameters, such as the number of responders and the level of training, to output parameters, such as performance time, number of casualties prevented, the number of individuals rescued from site, and the availability of the system. While this is an attractive area for further analysis, to the best of our knowledge, this line of analysis has not been pursued in the literature and forms an attractive avenue for future research.

The second case study that we now present is drawn from a very different domain: civil engineering. Here the role of simulation is in generating key inputs for the subsequent decision-making models. While Monte Carlo simulation is already the preferred tool in the industry for generating these inputs, we also explore opportunities to use discrete-event simulation effectively in the decision-making models themselves.

5.3.2 Preventive Maintenance of Bridges

In 2002, 50% of the national daily traffic in the USA used bridges that were more than 40 years old; about 28% were found to be deficient in some respect and about 14% were found to be structurally deficient (Robelin and Madanat 2007). The US Department of Transportation is interested in developing systems that can help in making maintenance and rehabilitation decisions related to bridges to optimize the available funds. The success of the Arizona Department of Transportation in its pavement management systems (Golabi and Shepard 1997) has further intensified the interest in such systems. In this case study, we will present the modeling details of bridge failure and deterioration. Our analysis will be directed towards devising strategies for timely preventive maintenance that can minimize the probability of devastating failure and lengthen bridge life (Frangopol *et al.* 2001; Kong and Frangopol 2003). We will describe the underlying stochastic process (Markov chain) and the Markov decision process (MDP) model that can be used to devise the strategies of interest to us (Robelin and Madanat 2007). If the MDP is of a reasonable size, i.e., up to a maximum of 1000 states per action, it can be solved via classical dynamic programming techniques, such as value and policy iteration (Bertsekas 1995). On the other hand, if the state space of the MDP becomes too large for dynamic programming to handle, simulation-based techniques called reinforcement learning (Sutton and Barto 1998; Gosavi 2003) or neuro-dynamic programming (Bertsekas and Tsitsiklis 1996) can be employed for solution purposes. The performance measures of interest to us are the probability of bridge failure (related to its reliability), the proportion of time the bridge can be used (availability), and the overall cost of operating the system.

5.3.2.1 Failure Probability and Simulation

To effectively capture the failure dynamics of a bridge, the following deterioration model (Frangopol *et al.* 2004) is popularly used. In its simplest form, a state function, g, is defined as the difference between the structure or the component's resistance, R, and the applied load (also loosely called stress), S:

$$g = R - S \ .$$

The structure is safe if $g \geqslant 0$, which means that the resistance of the component is able to overcome the load applied. Clearly, if $g < 0$, the structure fails. The state g is usually a function of numerous random variables. In particular, if the bridge is supported by girders, then the resistance is a function of the strength of each girder. Also, the resistance is a function of time; usually resistance decays with time. Both R and S tend to be random variables, whose distributions have to be determined by the structural engineer. If both of these random variables are normally distributed, the density function of g can be determined in closed form. If that is not the case, one must use simulation (Melchers 1999). The simulation-based scheme to determine this probability density can be explained as follows.

Random values are generated from their respective distributions for the pair (R, S). If K values are generated for this pair, and out of the K values, if on M occasions, $R < S$, then the probability of failure, p_f, is calculated as

$$p_f = M/K \ .$$

Clearly, this estimate improves as K becomes large, and tends to the correct value as $K \to \infty$. This is called *direct* or *crude* Monte Carlo sampling. While this is intuitively appealing, it turns out that oftentimes K may have to be very large in order to obtain a reliable estimate of the probability of failure. One way to circumvent this difficulty is to use the techniques of importance and directional sampling. The theory of importance and directional sampling is beyond the scope of this chapter, but we refer the interested reader to Melchers (1999). We note that via importance sampling, the estimation of the failure probability can be performed more efficiently, i.e., via fewer samples.

It is usually the case that the failure probability is a function of time. The resistance R of the bridge decays with every subsequent year. This feature of time changing resistance is modeled by dividing the time horizon over which the bridge is to be analyzed, e.g., a period of no more 75 years, into discrete time zones and then using a separate random variable for the resistance in each time zone. Then the sampling procedure must be performed separately for each time zone to determine the individual failure probabilities in all the time zones. For instance if 25 time zones are created, one has a series of resistance values, $R_1, R_2, \ldots, R_{25}$. These values in turn lead to a series of values of the failure probability: $p_{f_1}, p_{f_2}, \ldots, p_{f_{25}}$. Alternatively, the system can be modeled within one discrete-event simulator provided one is able to generate a generic function for the resistance in terms of time. While this is theoretically appealing, we are unaware of any literature that adopts this approach.

The next step is the process of determining when, during the life cycle of the bridge, preventive maintenance should be performed. To this end, at least three models have been presented in the literature. They are: the reliability index model, the MDP model, and the renewal-theoretic model. We will discuss the first two models in some detail. The renewal theoretic model has been discussed in Frangopol *et al.* (2004). It is necessary to point out here that regardless of the nature of the model used by the analyst, estimating the failure probability, which is performed via simulation, is a key input to all of these models. The reliability index model is the easiest to use and has some simple features that appeal to decision-makers. However, it is the MDP model that can be used for large-scale problems in combination with simulation.

Time-dependent reliability of structures has also been discussed in Juan *et al.* (2008b). They use a simulation-based approach to model a complex structure composed of numerous components. They assume the overall system can be in any one of three states: perfect, partially damaged, and collapsed. Transition from the perfect state to other states is triggered by failure of one or more components. They develop a simulation-based algorithm, named SURESIM, which computes the reliability of the structure at any given point of time. They state that the advantages of using simulation over analytical methods are that one does not have to make restrictive simplifying assumptions on structural behavior which makes analytical methods suitable only for simple structure.

5.3.2.2 Data Collection and Distributions

An important issue that we did not discuss above is that of identifying the distributions for the random variables that govern the system's behavior. Oftentimes, the structures in the bridge are composed of numerous components that have been used in other systems before. For such components, one has access to the distribution of times between failures, e.g., a Weibull or gamma distribution. The distributions have to be selected with care after performing statistical tests; with incorrect distributions, it is likely that one will obtain erroneous outputs from the simulations. If the components have never been used, typically, one must perform accelerated life-testing to determine the distribution of the time between failures. Armed with the distributions of the time between failures of all components, one can run simulation programs to generate the reliability of the entire structure at any point in time (Juan *et al.* 2008a).

5.3.2.3 Reliability Index Model

The reliability index model is a simple decision-tree model that has been widely used in industrial applications (Chung *et al.* 2003; Pandey 1998; Sommer *et al.* 1993; Frangopol *et al.* 1997). It builds a simple decision tree using the probability of failure in each time zone, and then exhaustively enumerates all the scenarios possible with the decisions that can be made. The two decisions that can be made at every inspection of a bridge are: maintain (action 1) or do nothing (action 0). If a failure occurs during the time zone at the beginning of which a decision of do nothing is made, then the cost during that time zone is the cost of replacement or repair of the bridge. If the bridge does not fail as a result of the do-nothing action, there is of course no cost. Hence the expected cost of the do-nothing action is the cost of a failure times the probability of failure; the failure probability is already available from the Monte Carlo simulation described in the previous section. On the other hand, if the bridge is maintained (action 1), then the cost incurred during that time zone is that of maintenance. Thus associated with each decision in each time zone, one can compute the expected cost.

See Figure 5.3 for an illustration of the mechanism described above. Three time zones have been considered in the application shown in Figure 5.3; the first zone starts at T_0 and ends at T_1, and so on. Figure 5.3 shows four possible scenarios at the end of time zone starting at T_3. The cost of each scenario is the sum of the costs along each time zone. The scenario that leads to the minimum cost is the optimal one and should be chosen; from this the optimal action at the beginning of each time zone, i.e., the optimal strategy, is determined. Analysis of this kind is also called life cycle analysis via the reliability index. The reliability index, β, is usually a function of the probability that the structure will not fail, i.e., $1 - p_{\mathrm{f}}$. The reliability index can also be expressed as a function of time (t) if the failure probability is a function of time. Further discussion on the reliability index is beyond the scope of this chapter, but the reader is referred to Melchers (1999).

5.3.2.4 Markov Decision Process Model

The MDP model is a more sophisticated model in which instead of using exhaustive evaluation of all the possible decisions, one uses dynamic programming and Markov

Clock	Response Status	Responding Name
45.38585175	Receives Call 1	Dispatcher
45.38585175	Transfers Call 1	Dispatcher
45.38585175	Transfer call received	Sgt
60	Notifies Officer1	Dispatcher
90.75849001	Transfer call processed	Sgt
608.4436395	Calls for Back Up	Officer 2
608.4436395	Call for backup	Dispatcher
608.4436395	Broadcast for Assistance	Dispatcher
608.4436395	Initiating Response	Local PD
608.4436395	Initiating Response	Highway Patrol
608.4436395	Initiating Response	Sheriff Dept

Figure 5.3 A decision tree constructed for the reliability index model

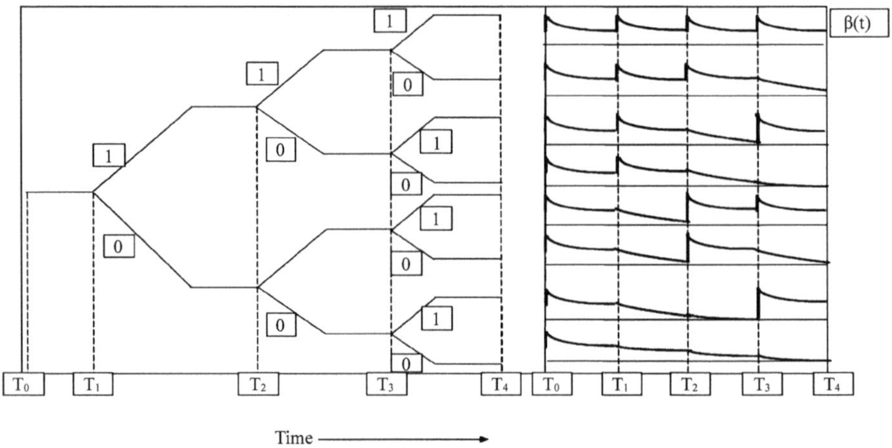

Figure 5.4 The costs underlying a typical MDP model

chains. We will describe this model briefly here. Assume that the state of the bridge is defined by the number of time zones it has survived since the last maintenance or failure. Hence, when we have a new bridge, its state is 0. If it has been just repaired after a failure, its state is also 0. Further, if the bridge has just been maintained, its state is again 0. As the bridge ages, the probability that it will fail, generally, increases. The probability of going from state i to state j under action a is defined by $P_{ij}(a)$. It is required for the MDP model to be useful to show that underlying the transitions of every action, there exists a Markov chain. If this is true, one can use the framework of the Bellman equation (see, e.g., Bertsekas 1995) to solve the problem. The transition probabilities can be calculated from the failure probability, which, we reiterate, is obtained from Monte Carlo simulation discussed previously.

As explained above, we have two actions, do nothing and maintain, and the state of the system takes a value from the set: $\{0, 1, 2, 3, \ldots\}$. During the time the bridge is a time zone, the bridge remains in the same state unless the bridge fails. If the latter occurs, the bridge immediately goes to the state 0. It remains in state 0 until it is ready to be operational. The famous Bellman equation for this system can be stated as follows:

$$V^{t+1}(i) = \min_a \left[c(i,a) + \Sigma_j P_{ij}(a)V^t(j)\right] .$$

In the above, $V(i)$ denotes the value function for state i, and holds the key to the optimal solution; $c(i,a)$ denotes the expected cost when action a is taken in state i. The values for this cost function must be determined from the probability of failure and the costs of the actions chosen. The *value iteration* algorithm can be used to solve for the optimal values of the value function. From the optimal values, the optimal action in state i at time t can be determined as follows:

$$a = \text{argmin}_a \left[c(i,a) + \Sigma_j P_{ij}(a)V^t(j)\right] .$$

See Figure 5.4 for a typical cost curve that results from an MDP model for this problem. The state of the system is typically a function of the time at which inspection (for maintenance) is performed. As one delays the time for inspection, the expected cost of failure increases while the expected cost of inspection falls. The total cost hence has a convex shape. The MDP model zones in on the optimal time of maintenance via the Bellman equation, but Figure 5.4 shows the geometry of the cost as it varies with time.

If the system has a very large number of states, one can use a simulation-based approach called reinforcement learning (Sutton and Barto 1998; Gosavi 2003). In reinforcement learning, the underlying Markov chain is simulated, and function approximation methods such as neural networks are used to approximate the value function. The power of this method becomes more obvious on systems that have a large number of states. Problems of preventive maintenance of *production machines* have been solved in the literature via the reinforcement learning approach (Gosavi 2004). However, this approach has not yet been exploited for the problem of bridge maintenance, and hence this remains an open topic for future research.

5.4 Conclusions

Simulation has become an important tool in applied operations research. This chapter was aimed at showing how it can be used for measuring reliability of complex stochastic systems. We presented some of the main ideas underlying usage of simulation for decision-support systems that are designed for public structures and a homeland security application. In particular, the methods that we discussed above were directed towards (1) measuring system reliability and availability via simulation and (2) role of simulation in helping make the right decisions.

We have attempted to present the key concepts in a tutorial style so that a reader unfamiliar with the use of simulation for these tasks gets a clear overview of this topic. One of our other goals was to familiarize the reader with the literature on these topics. To this end, we have presented numerous references. We hope that the reader will find them beneficial for further reading.

Finally, we would like to emphasize that we have selected two case studies that are somewhat unconventional in the literature on applied simulation modeling, which tends to focus on manufacturing systems such as factories or service systems such as banks and airports. We expect that highlighting the use of simulation in these areas that have not attracted as much attention as conventional applications will lead to further interest in simulation-based research and applications in these domains.

Acknowledgements Funding for this research was received by the second author from the Leonard Wood Institute and Alion Science & Technology. The case study would not have been a possibility without the assistance of the police departments and fire department involved with the actual events. The first author would also like to acknowledge support from a research grant from the National Science Foundation (ECS: 0841055) for partially funding this work.

References

Bertsekas D (1995) Dynamic programming and optimal control, vol 2. Athena, Nashua, NH

Bertsekas D, Tsitsiklis J (1996) Neuro-dynamic programming, Athena, Nashua, NH

Chung HY, Manuel L, Frank KH (2003) Optimal inspection scheduling with alternative fatigue reliability formulations for steel bridges. Applications of statistics and probability in civil engineering. Proceedings of ICASP2003, San Francisco, July 6–9. Millpress, Rotterdam

Frangopol DM, Lin KY, Estes AC (1997) Life-cycle cost design of deteriorating structures. J Struct Eng 123(10):1390–1401

Frangopol DM, Kong JS, Gharaibeh ES (2001) Reliability-based life-cycle management of highway bridges. J Comput Civ Eng 15(1):27–34

Frangopol DM, Kallne M-J, van Noorrtwijk JM (2004) Probabilistic models for life-cycle performance of deteriorating structures: review and future directions. Prog Struct Eng Mater 6:197–212

Fujiwara O, Makjamroen T, Gupta KK (1987) Ambulance deployment analysis: A case study of Bangkok. Eur J Oper Res 31:9–18

Golabi K, Shepard R (1997) Pontis: A system for maintenance optimization and improvement for US bridge networks. Interfaces 27(1):71–88

Gosakan M (2008) Modeling emergency response by building the campus incident case study. Final report. Leonard Wood Institute, MO

Gosavi A (2003) Simulation-based optimization: parametric optimization and reinforcement learning. Kluwer Academic Publishers, Norwell, MA, USA

Gosavi A (2004) Reinforcement learning for long-run average cost. Eur J Oper Res 155:654–674

Harewood SI (2002) Emergency ambulance deployment in Barbados: A multi-objective approach. J Oper Res Soc 53:185–192

Henderson SG, Mason AJ (2004) Ambulance service planning: simulation and data visualisation. In: Brandeau ML, Sainfort F, Pierskalla WP (eds) Operations research and health care: a handbook of methods and applications. Kluwer, Norwell, MA, USA

Hlupic V (2000) Simulation software: An operational research society survey of academic and industrial users. In: Joines JA, Barton RR, Kang K, Fishwick PA (eds) Proceedings of the 2000 winter simulation conference. Society for Computer Simulation International, San Diego, CA, USA, pp 1676–1683

Ingolfsson A, Erkut E, Budge S (2003) Simulation of single start station for Edmonton EMS. J Oper Res Soc 54:736–746

Juan A, Faulin J, Serrat C, Bargueño V (2008a) Improving availability of time-dependent complex systems by using the SAEDES simulation algorithms. Reliab Eng System Saf 93(11):1761–1771

Juan A, Faulin J, Serrat C, Sorroche M, Ferrer A (2008b) A simulation-based algorithm to predict time-dependent structural reliability. In: Rabe M (ed) Advances in simulation for production and logistics applications. Fraunhofer IRB Verlag, Stuttgart, pp 555–564

Kleijnen J (2008) Design and analysis of simulation experiments. Springer, New York, NY, USA

Kolesar P, Swersey A (1985) The deployment of urban emergency units: a survey. TIMS Stud Manag Sci 22:87–119

Kong JS, Frangopol DM (2003) Life-cycle reliability-based maintenance cost optimization of deteriorating structures with emphasis on bridges. J Struct Eng 129(6):818–828

Larson RC (1975) Approximating the performance of urban emergency service systems. Oper Res 23(5):845–868

Law AM, Kelton WD (2000) Simulation modeling and analysis, 3rd edn. McGraw Hill, New York, NY, USA

Martin M (2007) Distraught graduate student brings chaos to campus. Missouri Miner, March 1.

Melchers RE (1999) Structural reliability analysis and prediction, 2nd edn. John Wiley, Cichester, UK

Menk P, Mills M (1999) Domestic operations law handbook. Center for Law and Military Operations, US Army Office of the Judge Advocate General, Charlottesville, VA, USA

Murray S, Ghosh K (2008) Modeling emergency response: a case study. Proceedings of American Society for Engineering management conference, West Point, NY, November. Curran Associates, Red Hook, NY, USA

Pandey MD (1998) Probabilistic models for condition assessment of oil and gas pipelines. NDT&E Int 31(5):349–358

Pidd M, de Silva FN, Eglese RW (1996) A simulation model for emergency evacuation. Eur J Oper Res 90:413–419

Robelin C-A, Madanat S (2007) History-dependent bridge deck maintenance and replacement optimization with Markov decision processes. J Infrastruct Syst 13(3):195–201

Sommer A, Nowak A, Thoft-Cristensen P (1993) Probability-based bridge inspection strategy. J Struct Eng 119(12):3520–3526

Sutton R, Barto A (1998) Reinforcement learning: an introduction. MIT Press, Cambridge, MA, USA

Walker W, Chaiken J, Ignall E (eds) (1979) Fire department deployment analysis. North Holland Press, New York

Chapter 6
Reliability Models for Data Integration Systems

A. Marotta, H. Cancela, V. Peralta, and R. Ruggia

Abstract Data integration systems (DIS) are devoted to providing information by integrating and transforming data extracted from external sources. Examples of DIS are the mediators, data warehouses, federations of databases, and web portals. Data quality is an essential issue in DIS as it concerns the confidence of users in the supplied information. One of the main challenges in this field is to offer rigorous and practical means to evaluate the quality of DIS. In this sense, DIS reliability intends to represent its capability for providing data with a certain level of quality, taking into account not only current quality values but also the changes that may occur in data quality at the external sources. Simulation techniques constitute a non-traditional approach to data quality evaluation, and more specifically for DIS reliability. This chapter presents techniques for DIS reliability evaluation by applying simulation techniques in addition to exact computation models. Simulation enables some important drawbacks of exact techniques to be addressed: the scalability of the reliability computation when the set of data sources grows, and modeling data sources with inter-related (non independent) quality properties.

6.1 Introduction

This chapter presents static reliability models and simulation techniques developed to support data quality-oriented design and management of information systems, specifically data integration systems (DIS).

DIS are devoted to provide large volumes of information, extracted from several (possibly external) sources, which is integrated, transformed, and presented to the users in a unified way. A DIS basically consists of a set of data sources (databases,

A. Marotta · H. Cancela · V. Peralta · R. Ruggia
Universidad de la República, Montevideo, Uruguay

V. Peralta
Laboratoire d'Informatique, Université François Rabelais Tours, France

web services, web sites, etc.), a set of data targets (pre-defined queries, integrated schema, etc.) and a transformation process that, applied to data extracted from the sources, allows the calculation of data targets. Examples of DIS are the mediation systems, data warehousing systems, federations of databases, and web portals.

Information quality is a hot topic for all kinds of information systems. Information quality problems have been reported as critical in several scientific and social areas, e.g., environment [12,25], genetics [17,23], economy [16], and informatics on the web [9]. In the case of DIS, quality problems are aggravated by the heterogeneity of source data and the potential inconsistencies introduced by the integration and transformation processes. Although DIS present an opportunity for delivering more complete and robust responses to user queries (as they collect and contrast data from multiple sources), they may suffer from inconsistencies and quality problems generated by the integration and transformation of poor quality source data. Among the most critical quality problems stressed in the literature are: duplicate data (e.g., the same customers extracted from several sources), inconsistencies and contradictions (e.g., different addresses for the same person), inaccurate values (e.g., wrong contact information for a patient), typing and format errors (e.g., data may be represented differently at each source), obsolete data (e.g., sources may be updated at different periods or not updated at all), and incomplete data (e.g., missing values in a patient file or even missing patients).

Data quality is a wide research area, which involves many different aspects and problems as well as important research challenges. On the other hand, it has an enormous relevance for industry due to its great impact on the usefulness of information systems in all application domains. A great amount of work on data quality can be found in the literature, mostly generated in the last decade. Interesting analyses of the evolution and current state of the field are presented in [18] and [24]. In her work [18], Neely concludes that the problem of quality dimensions has been well researched, while more work still needs to be done with regard to measurement or metrics. There is very little work related to economic resources and operation and assurance costs. She claims that more research is needed in determining how data quality dimensions define the quality of data in terms of the user, and in improving the analysis and design of information systems to include quality constructs. In their work [24], Scannapieco *et al.* focus on the multidimensional characteristic of data quality, as researchers have traditionally done. They precisely define the dimensions: *accuracy*, *completeness*, *currency*, and *consistency*. They say that this core set of dimensions is shared by most proposals in the literature, although the research community is still debating the exact meaning of each dimension, and studying which is the best way to define data quality. In addition, they comment that for specific application domains it may be appropriate to have more specific sets of dimensions.

Data quality assurance is a main issue in DIS, which strongly impacts the confidence of the user in the information system. In order to properly use the DIS information, users should be able to evaluate the quality of the retrieved data, and more generally, the reliability of the DIS.

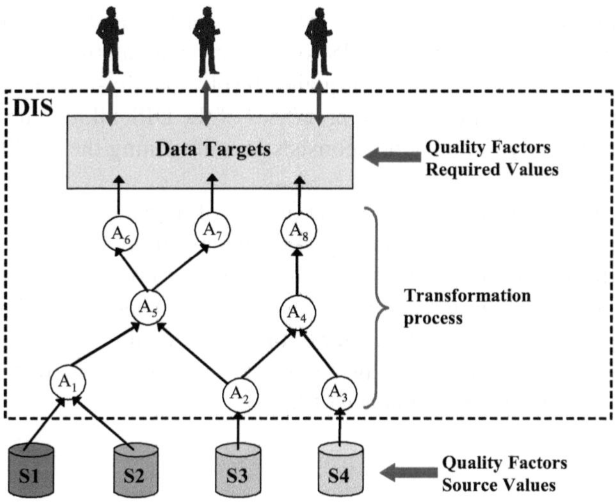

Figure 6.1 DIS and quality factors

Several quality management frameworks have been proposed [8, 10, 13, 20, 21], aiming at the definition and classification of quality factors or dimensions[1] in DIS, the proposition of metrics for assessing these factors and the definition of quality models that include these factors and metrics. Data quality frameworks distinguish between quality-factor values measured at the data sources and the DIS, and quality-factor values required by the users. Figure 6.1 shows a generic architecture of a DIS, which integrates data from four sources (S1, S2, S3, and S4). Quality factors values are associated to data sources and also to data targets.

In this context, the *reliability* of the DIS is considered as its capability for providing data with a certain level of quality. More concretely, the reliability of the DIS is the probability that a certain level of quality, given through the specification of quality requirements, can be attained.

Based on the architecture outlined in Figure 6.1, the complexity of DIS systems mainly lies in three aspects: (1) the potentially large number of data sources participating in the DIS, (2) the heterogeneity and autonomy of these sources, and (3) the complexity of the transformation process that is associated to the business semantics.

As a consequence, data quality management in DIS has to deal with two main characteristics: (1) the heterogeneity and variability of the quality factors at the different sources (each quality factor has particularities that strongly affect its behavior), and (2) the architecture of the DIS, which integrates a number of autonomous and heterogeneous sources through a business-oriented transformation process.

From a management process point of view, quality management in DIS consists of three main tasks: quality evaluation, quality-oriented design, and quality main-

[1] In the solution given in this work we do not differentiate between "quality factor" and "quality dimension."

tenance. Quality evaluation allows the estimation of the quality values associated to the information provided by the DIS to the user, and calculation of the reliability values associated to certain target quality levels. Quality-oriented design is the task of designing the architecture and processes of the DIS, taking into account data quality values. Quality maintenance consists in maintaining the quality of the DIS at a required level.

As DIS quality mainly depends on the quality of its data sources, modeling this dependency is central on a data quality management framework. In the following sections we discuss reliability formulations, based on probabilistic models of the data sources behavior. These models, evaluated by exact and simulation techniques, allows the representation of the behavior of the source quality values and their propagation to the DIS as perceived by the end users, showing the interest of these techniques for the design and maintenance of DIS.

6.2 Data Quality Concepts

In this section we describe some concepts about data quality that are used throughout this chapter. In particular, we present the definitions of some quality factors, an abstract representation of DIS, and some quality evaluation algorithms.

6.2.1 Freshness and Accuracy Definitions

It is widely accepted that data quality is multidimensional. Among the multiple dimensions proposed in the literature, we focus on data freshness and data accuracy, which are two of the most used ones [24]. There are many different aspects (called quality factors) of freshness and accuracy [19]. In order to avoid any ambiguity, but keeping the model as general as possible, we choose one freshness factor (age) and three accuracy factors (semantic correctness, syntactic correctness, and precision) for illustrating our approach.

Age captures how old the data is. It is generally measured as the amount of time elapsed between the moment data is updated at a data source and the moment it is queried at a data target. Age is zero at the moment data is updated, and then it increases as time passes until the following update. Its measurement implies having some knowledge about source updates (e.g., logs) or update frequencies. *Semantic correctness* describes how well data represent real-world states. Its measurement implies a comparison with the real world or with referential tables assumed to be correct; for that reason, measurement methods are very hard to implement. *Syntactic correctness* expresses the degree to which data is free of syntactic errors (typos, format). Its measurement generally consists in verifying format rules or checking belonging to domain dictionaries. *Precision* concerns the level of detail of data representation. Its measurement implies some knowledge on data domains in order to determine the precision of each domain value.

In the remainder of the chapter we refer to freshness and accuracy by considering any of the previous factors, and we refer directly to "factor" and not to "dimension." We make some assumptions about the measurement units and granularity, which fix the context for developing quality models. We assume the following context characteristics:

- *Granularity*. We represent the data in terms of the relational model, and we manage quality values at relation level granularity, i.e., we associate a quality value for each source relation, for each intermediate relation (result of an activity) and for each data target. Some measurement methods (especially those for syntactic correctness) measure quality cell by cell; we assume that these measures are aggregated (e.g., averaged) obtaining a measure for the relation.
- *Measurement units.* Freshness measures are natural numbers that represent units of time (e.g., days, hours, seconds). Accuracy measures (for the three factors) are decimal numbers between 0 and 1, where 1 represents a perfectly accurate value. Freshness and accuracy values are lately discretized for building probabilistic models. For discretizing we must determine the desired precision for quality values and the methods for rounding them. These criteria vary according to the quality factor and the use of the quality values in concrete applications.
- *Information about sources*. We assume that we have the methods for periodically measuring source quality or that we have some source metadata that allows their deduction (for example, knowledge about the completed updates allows estimation of data freshness).
- *User required values*. We assume that users express their required quality values with the same units, precision, and interpretation we presented above.

6.2.2 Data Integration System

A DIS is modeled as a workflow process, which activities perform the different tasks of extraction, transformation, and conveying data to end-users. Each activity takes data from sources or from other activities and produces result data that can be used as input for other activities. Then, data traverses a path from sources to users where it is transformed and processed according to the system logics. This workflow model allows the representation of complex data manipulation activities and facilitates their analysis for quality management.

In order to reason about data quality in DIS, we define the concept of *quality graph*, which is a directed acyclic graph that has the same workflow structure as the DIS and is labeled with additional source and DIS properties that are useful for quality evaluation (e.g., source data quality, DIS processing costs, and DIS policies). The nodes are of three types: (1) *activity nodes* representing the major tasks of a DIS, (2) *source nodes* representing data sources accessed by the DIS, and (3) *target nodes* representing data targets fed by the DIS. Activities consume input data elements and produce output data elements which may persist in repositories. As we work with an acyclic graph, we have arcs, also called edges, connecting pairs of nodes. These edges represent the data flow from sources to activities, from activities to targets and between activities (i.e., the output data of an activity is taken as input by a successor activity). Both nodes and edges can have labels. Labels are *property* = *value* pairs that represent DIS features (costs, delays, policies, strategies, constraints, etc.) or quality measures (a quality value corresponding to a quality factor). Values may belong to simple domains (e.g., numbers, dates, strings), structured domains, lists, or sets.

Formally, a quality graph is a directed acyclic graph $G = (V, E, \rho V, \rho E, \rho G)$ where:

- V is the set of nodes. Vs, Vt, and Va are the sets of source, target, and activity nodes respectively; with $V = Vs \cup Vt \cup Va$.
- $E \subset (Vs \cup Va) \times (Va \cup Vt)$ is the set of edges.
- $\rho_V: V \to L_V$ is a function assigning labels to the nodes. L_V denotes the set of node labels.
- $\rho_E: E \to L_E$ is a function assigning labels to the edges. Analogously, L_E denotes the set of edge labels.
- $\rho_G \in L_G$ is a set of labels associated to the whole graph. Analogously, L_G denotes the set of graph labels.

Figure 6.2 shows the graphical representation of quality graphs.

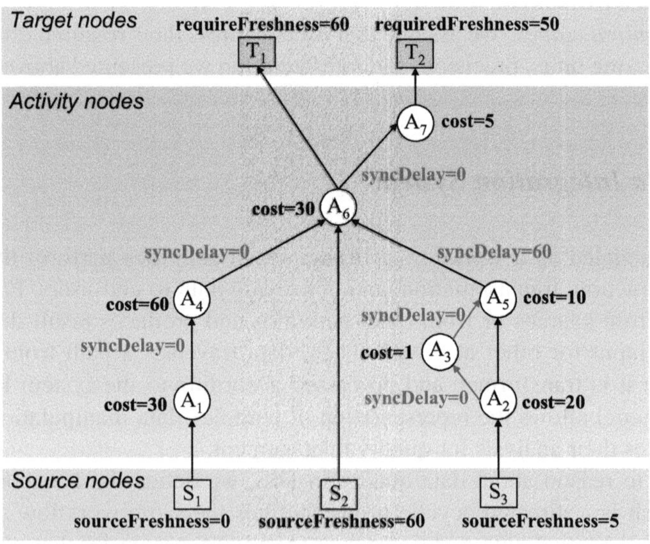

Figure 6.2 Quality graph

6.2.3 Data Integration Systems Quality Evaluation

DIS quality evaluation is based on the quality graph presented above. A set of quality evaluation algorithms, each one specialized in the measurement of a quality factor, compute quality values by combining and aggregating property values (e.g., source quality values or activity costs), which are labels of the quality graph. Evaluation algorithms traverse the graph, node by node, operating with property values and calculating the quality of data outgoing each node. This mechanism for calculating data quality applying operations along the graphs is what we call *quality propagation*.

There are two kinds of quality propagations: (1) propagation of measured values, and (2) propagation of required values. In the former, quality values of source data are propagated along the graph, in the sense of the data flow (from source to target nodes). It serves to calculate the quality of data delivered to users. In the latter, quality values required by users are propagated along the graph but in the opposite sense (from target to source nodes). It serves to constrain source providers on the quality of data. A direct application of this is the comparison of alternative data sources for selecting the one that provides the data with the highest quality.

In the propagation of measured values, the quality of data outgoing each node is calculated as a function of the quality of predecessor nodes and (possibly) other node properties (e.g., costs, delays, policies). Such function, called *composition function*, is defined considering some knowledge about the quality factor being propagated, the nature of data and the application domain. Composition functions may range from simple arithmetic functions (e.g., maximum, sum, or product) to sophisticated decisional formulas.

In the propagation of required values, the quality requirement for a node is decomposed into quality requirements for predecessor nodes. A *decomposition function* must have the inverse effect of the composition function, i.e., it should return all the combinations of requirements for predecessors that, when composed, allows the requirements of the node to be obtained. Consequently, the decomposition function may return an equation system. Applying this propagation towards the sources, we finally obtain a set of *restriction vectors*, where each vector dimension corresponds to a restriction for the quality values of one source. For example, for a DIS with three sources S_1, S_2, and S_3, we obtain a set of vectors for accuracy, each one of the form: $v = \langle acc(S_1) \geqslant a_1, acc(S_2) \geqslant a_2, acc(S_1) \geqslant a_3 \rangle$, where a_1, a_2, and a_3 are accuracy values. All the values combinations that satisfy these restrictions comply to the user quality requirements at the targets.

In the following, we consider freshness and accuracy evaluation algorithms that are simplified versions of those proposed in [19], which allows us to concentrate on solving the probabilistic calculations of the quality values.

The freshness of the data delivered to users depends on the freshness of source data but also on the amount of time needed for executing all the activities (processing costs) as well as on the delays that may exist among activities executions (synchronization delays). The composition function adds such properties; if the node has several predecessors, the maximum value is taken. Formula 6.1 calculates the freshness of node N

$$\text{Freshness}\,(N) = \text{ProcessingCost}\,(N) + \max_{P \in \text{Predecessor}(N)} (\text{Freshness}\,(P) + \text{SyncDelay}\,(P, N)) \tag{6.1}$$

The propagation of freshness required values is similar; processing costs and synchronization delays are subtracted from successor freshness. The decomposition function returns the same value to all predecessors, which has the opposite effect to the maximum operator used in the composition function:

$$\begin{aligned}&\text{RequiredFreshness}\,(P_1 \ldots P_n / P_i \in \text{Predecessor}\,(N)) = \langle v_1 \ldots v_n \rangle / \\ &v_i = \text{RequiredFreshness}\,(N) - \text{ProcessingCost}\,(N) - \text{SyncDelay}\,(P_i, N)\end{aligned} \tag{6.2}$$

The accuracy of the data delivered to users depends on the accuracy of source data but also on the operation semantics. For example, if an activity joins data from two predecessors, the accuracy of resulting data is calculated as the product of the accuracy of operands; however, if the activity performs a union of predecessor data, accuracy is calculated by adding predecessors accuracy weighted by their size. In addition, when the activity may improve or degrade accuracy (e.g., performing data cleaning or losing precision), a correction ratio is added. The composition function may be different for each activity node. An example of composition is

$$\text{Accuracy}\,(N) = \prod_{P \in \text{Predecessor}\,(N)} (\text{Accuracy}\,(P)) \tag{6.3}$$

The propagation of accuracy required values is quite different from freshness ones because it returns an equation system. Precisely, it returns all possible combinations of source values that, multiplied (according to the composition function) allow the node required accuracy to be obtained:

$$\begin{aligned}&\text{RequiredAccuracy}\,(P_1 \ldots P_n / P_i \in \text{Predecessor}\,(N)) = \\ &\{\langle v_1 \ldots v_n \rangle / \text{composition}\,(v_1 \ldots v_n) = \text{RequiredAccuracy}\,(N)\}\end{aligned} \tag{6.4}$$

The quality graph is labeled with all the property values needed to calculate previous formulas (processing costs, synchronization delays, error correction ratio, result size, etc.). These values are estimations of property values obtained from statistics of previous DIS executions. Source measured quality values may also be estimations. Consequently, quality propagation obtains estimations of target data quality and source restrictions vectors. We emphasize that in this way, quality propagation can be performed off-line, without degrading DIS performance. Details on the design of evaluation algorithms and the choice of graph properties can be found in [19].

6.3 Reliability Models for Quality Management in Data Integration Systems

An important objective for a DIS is to satisfy user quality requirements. For achieving this aim, it is necessary to measure the quality it offers, and in addition the DIS must be prepared to endure or detect and manage the changes that affect its quality.

As introduced before, an approach for solving this problem is to have models of the behavior of DIS quality [14, 15]. With this tool, we are able to construct and maintain a DIS that satisfies user quality requirements and that is tolerant to quality changes (minimizing the impact of inevitable changes).

This behavior of DIS quality can be modeled through application of probabilistic techniques. The application of such an approach to the context of DIS has as main advantage the flexibility it provides to the DIS design and maintenance. As a clear example, consider the task of selecting which data sources will participate or continue participating in the system. If we only consider the worst quality values they have, or the current ones, perhaps there are many sources that do not reach the quality requirements of the users. However, by considering probability based values sources can be selected with a fairer criterion and with better overall results.

The following is a case study, which will be used throughout the rest of the chapter for showing the different problems and techniques, as they are presented.

Case Study. The case study is a DIS oriented to people who want to choose a game for playing or who dedicate to creating games for the internet, and need to have trustable information about users' preferences on the existing internet games. The DIS provides this facility to users, extracting and integrating data from different sources, and allowing queries about games ratings.

Source1 is an add-on application developed for a web social network. This application provides an environment for people to play games together. The users agree meetings for playing games on the web. After they play they have the possibility of rating the game. The source registers the total rating (sum of individual ratings) and the rating quantity. Hundreds of users use this application daily, generating the continuous growth and change of the application data.

Source2 is a data source providing qualifications for games given by web users.

In *DataTarget1* the DIS provides the average rating of a game and the oldest date that corresponds to this rating, obtaining this information from a table called *Games* from *Source1* and a table called *Ratings* from *Source2*. Figure 6.3 shows the data processing graph.

Table 6.1 shows a description of each data transformation activity.

In addition to providing the required information in *DataTarget1*, the DIS provides to the users information about the accuracy of the given data. Accuracy is measured in each source, at cell level, and is aggregated obtaining one accuracy value for *Source1* and one accuracy value for *Source2*. In [14] details about these quality measurements can be found.

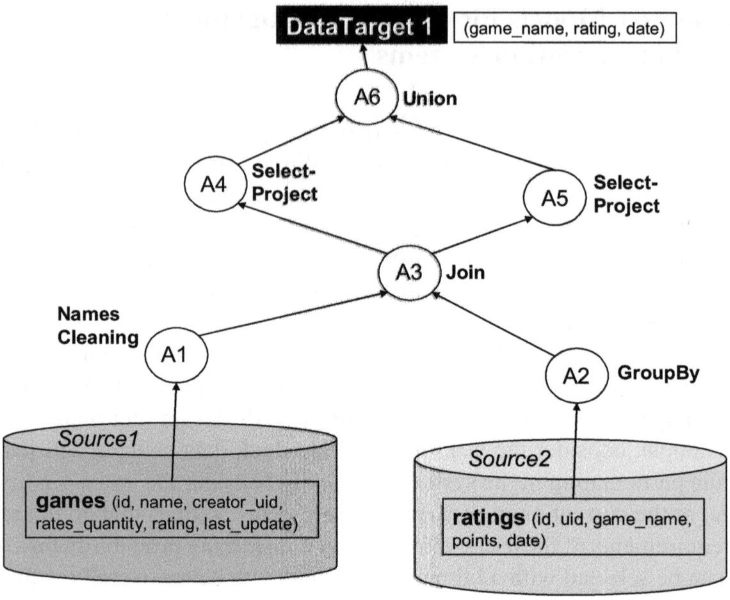

Figure 6.3 Data processing for *DataTarget1*

6.3.1 Single State Quality Evaluation in Data Integration Systems

If the quality values of the sources are known, it is easy to compute quality values of the information provided by the DIS, taking into account the process that is applied to the sources' data until it arrives at the targets. Applying intrinsic properties of each quality factor (such as freshness or accuracy), composition functions are defined in order to estimate the quality values at the data targets, as shown in Section 6.2.3.

Continuing with the case study, we show the calculations that are applied for obtaining the accuracy value provided by the DIS.

Case Study (cont.). For evaluating accuracy in *DataTarget1*, a propagation function is defined for each activity of the transformation graph, and the functions are successively applied starting from the sources accuracy values and ending at the data target. The propagation functions are shown in Table 6.2.

We can calculate *DataTarget1* accuracy value at a given moment, where sources' accuracy values were measured. For example:

$$Acc(Source1) = 0.8$$
$$Acc(Source2) = 0.9$$
$$Acc(DataTarget1) = 0.81$$

Table 6.1 Data transformation activities

Activity	Description
A1: Names Cleaning	This activity performs a cleaning of the attribute "name," which corresponds to names of games. Each value of the attribute is cleaned through a comparison to a referential table of game names
A2: GroupBy	SQL query over Source2.ratings: SELECT game_name, AVG(points) AS points, MAX(date) AS date FROM ratings GROUP BY game_name
A3: Join	SQL join between A1 result and A2 result: SELECT * FROM A1, A2 WHERE A1.name = A2.game_name
A4: Select-Project	SQL query over A3 result: SELECT game_name, ((rating/rates_quantity)+points)/2 AS rating, last_update AS date FROM A3 WHERE last_update < date
A5: Select-Project	SQL query over A3 result: SELECT game_name, ((rating/rates_quantity)+points)/2 AS rating, date FROM A3 WHERE date < last_update
A6: Union	It performs a union between A4 and A5 results.

Table 6.2 Propagation functions for accuracy

Activity	Description
A1: Names Cleaning	$acc(output\_data) = \min(acc(input\_data) + 0.1, 1)$ The cleaning process is estimated to improve the accuracy of the input relation in a 10%
A2: GroupBy	$acc(output\_data) = acc(input\_data)$
A3: Join	$acc(output\_data) = acc(input\_data\_1) * acc(input\_data\_2)$
A4, A5: Select-Project	$acc(output\_data) = acc(input\_data)$
A6: Union	$acc(output\_data) = \frac{acc(input\_data\_1) * \lvert input\_data\_1 \rvert + acc(input\_data\_2) * \lvert input\_data\_2 \rvert}{\lvert input\_data\_1 \rvert + \lvert input\_data\_2 \rvert}$

6.3.2 Reliability-based Quality Behavior Models

Probabilistic models are a useful tool for modeling both uncertainties and dynamic aspects of system quality. *Quality behavior models* are probabilistic models for sources quality, which can be combined to obtain the behavior of DIS quality. These models provide stochastic information about quality-factors' values. A simple model for each source consists in the probability distribution of the quality values. In the case of the DIS, there are two kinds of models: (1) the reliability, which corresponds to the probability that it satisfies a required quality level, and (2) the probability distribution of the quality values that it may provide. The models for DIS quality are calculated from the sources' quality models.

6.3.2.1 Source Models

Instead of considering that each data source has a (deterministic, well known) value for each of the quality factors under study, we model quality behavior through probabilistic models. In particular, for each quality factor we define a sample space, corresponding to the set of all source quality possible values; we will suppose that this sample space is a subset of the real numbers (it may be discrete or continuous), and that there is an order relation between quality values over this sample space (usually the "larger than" or the "smaller than" relation over the real numbers), such that given two different quality values, we can always know which is "better". Then, for each source and for each quality factor, we define a random variable whose value corresponds to the source quality value for this factor.

The source model provides us the probability distribution of the quality values at the source, and also with useful indicators such as the expectation, the mode, the maximum, and the minimum values. In the case of freshness, for example, suppose we have the random variables $X_1, X_2, \ldots, X_n$, so that each one corresponds to one of the n sources of the integration system. X_i represents the freshness value of source i at a given instant. The probability that $\text{freshness}_i = k_i$ is verified (where freshness_i is the current freshness value of source i and k_i is a positive integer number) is $p(X_i = k_i)$, and the probability that $\text{freshness}_i \leqslant k_i$ is verified is $p_i = p(X_i \leqslant k_i) = \sum_{j=0 \ldots ki} p(X_i = j)$.

In practice, we can use empirical distributions computed from collected data; for example, using relative frequencies tables, which are a good estimation of the respective probabilities [2, 3]. Alternatively, we can fit theoretical distributions to this empirically collected data, or if there is enough information about the source behavior (for example, if the mechanism of data actualization is well-known), by exactly computing the probability distribution.

In our case study, for obtaining the accuracy probability distribution of *Source1* and *Source2*, we collected data of successive measurements of the sources' accuracy values.

Case Study (cont.). We measured accuracy of *Source1* during 50 days, and accuracy of *Source2* during 55 days, each 5 days. Then we calculated the probability distribution of each source's accuracy. In Figures 6.4 and 6.5 we show the results.

With this information we can predict the behavior of sources' accuracy, assuming that the error rate will maintain the same if there is not a significant change in the generation of the information.

Measurements:

Date	Source Accuracy
2008-08-05	0.922086
2008-08-10	0.856135
2008-08-15	0.811043
2008-08-20	0.78865
2008-08-25	0.746933
2008-08-30	0.724847
2008-09-04	0.704601
2008-09-09	0.684663
2008-09-14	0.679141
2008-09-19	0.67546
2006-09-24	0.671472

Distribution:

Source Accuracy	Probability
0	0
0.1	0
0.2	0
0.3	0
0.4	0
0.5	0
0.6	0
0.7	0.636364
0.8	0.181818
0.9	0.181818
1	0

Maximum: 0.9
Minimum: 0,7
Expected Value: 0.8
Mode: 0.7

Figure 6.4 Probability distribution of *Source1* accuracy values

Measurements:

Date	Source2 Accuracy
2008-08-05	0.953226
2008-08-10	0.943406
2008-08-15	0.945094
2008-08-20	0.943674
2008-08-25	0.942186
2008-08-31	0.942667
2008-09-06	0.942939
2008-09-12	0.942601
2008-09-15	0.943346
2008-09-18	0.943469
2008-09-21	0.943188
2008-09-24	0.942935
2008-09-27	0.943357

Distribution:

Source2 Accuracy	Probability
0	0
0.1	0
0.2	0
0.3	0
0.4	0
0.5	0
0.6	0
0.7	0
0.8	0
0.9	0.923077
1	0.0769231

Maximum: 1
Minimum: 0.9
Expected Value: 0.9
Mode: 0.9

Figure 6.5 Probability distribution of *Source2* accuracy values

6.3.2.2 Data Integration System Models

After we have the source models, we can build a DIS model to compute the probability that the DIS satisfies the quality requirements. We make the analogy with reliability theory [7], in particular, *structural reliability*, which relates the behavior of the components of a system to the reliability of the whole system. It is based on the *structure function*, which relates the state of the system to the state of its components. The possible states of the system are two, operational (up) and failure (down). The state of component $i = 1, 2, \ldots, n$, is a random variable X_i. The state of the system is determined by the state of its components through the structure function: $\varphi(X)$, where $X = (X_1, \ldots, X_n)$, and $\varphi(X) = 1$ if the system is up and $\varphi(X) = 0$ if the system is down.

In our case, the system is the DIS, the components are the sources, the state of a component is the value of the quality factor of interest, and the system is operational when its quality requirements are being satisfied (actually it is possible to model many quality factors at the same data source, in this case each component will be a quality factor at a given source).

In all cases, we define *DIS reliability* as the probability that the DIS complies with its quality requirements:

$$DR = P(\varphi(X) = 1) = E(\varphi(X)) = \sum_{(x_1,\ldots,x_n)} P(X = (x_1, \ldots, x_n))\, \varphi(X) \quad (6.5)$$

A particular case is when the sources' states can be considered as binary random values X_i such that $X_i = 1$ if the component is up (i.e., the source complies with a given quality requirement) and $X_i = 0$ if the component is down (i.e., the source does not comply with the requirement). We will discuss below a particular case which fits this paradigm, and afterwards the general model.

Case 1: Binary reliability models. Suppose that the DIS structure is such that the quality restriction for the DIS is satisfied if and only if a simple quality restriction is satisfied at each of the sources. For example, suppose a DIS with a single data target, computed as the join of n data sources, taking delay D; and with the requirement that the data target freshness be less or equal to F; this requirement directly translates into a requirement that the freshness of every data source be less or equal to $F - D$.

Considering n sources, we can define a restriction vector $v = \langle r_1, \ldots, r_n \rangle$, and we can define a set of n random variables Y_i, $i = 1, \ldots, n$, each of which is equal to 1 if source S_i is operational (if the quality value X_i at source i satisfies the restriction r_i), and is equal to 0 if S_i is not operational.

When the sources are independent, i.e., their quality values vary independently, probability that DIS satisfies the quality requirements can be exactly calculated.

In this case, the Y_i are statistically independent random variables (each source quality value varies independently from each other).

According to the usual definition in structural reliability [7], a series system is a system that is operational if and only if all of its elements are up. Our system can be considered as a series system, and our structural function is $\varphi(Y) = \prod_{i=1 \ldots n} Y_i$. Therefore we calculate the *DIS reliability*,

$$DR = P(\varphi(Y) = 1) = \prod_{i=1 \ldots n} P(Y_i = 1) \quad (6.6)$$

Case 2: General models. In the general case, there are multiple combinations of quality values at the sources, such that if one of these combinations is satisfied, the quality requirements at the DIS are satisfied. As we saw before, the general formula for computing the DIS reliability is

$$DR = P\left(\varphi(X) = 1\right) = E\left(\varphi(X)\right) = \sum_{(x_1,\ldots,x_n)} P\left(X = (x_1, \ldots, x_n)\right) \varphi\left(X\right) \quad (6.7)$$

The DIS reliability is well defined in terms of the joint probability distribution of the quality values at the sources and of the structure function: $\varphi(X)$. It is then possible to directly apply this formula to compute the DIS reliability measure *DR*; this implies generating all possible values for the vector X; this set of values has size equal to the product of all the numbers of possible values for the quality factor at each source, which means it grows exponentially with the number of quality values and with the number of data sources (and can be very large even for a few data sources, if the quality factors values are computed with much granularity).

When the sources are independent, i.e., their quality values vary independently, the probability distribution $P(X = (x_1, \ldots, x_n))$ is simply the product of the marginal distributions $P(X_i = x_i)$.

In our case study we can easily apply this formula, since there are very few possible accuracy values (accuracy values for which the probability is equal to 0) at the sources and we have only two sources. We show how DIS reliability is calculated in this case.

Case Study (cont.). In this case the only possible vectors, $X^i = (acc(S1), acc(S_2))$, are the following:

$X^1 = (0.7, 0.9)$
$X^2 = (0.7, 1)$
$X^3 = (0.8, 0.9)$
$X^4 = (0.8, 1)$
$X^5 = (0.9, 0.9)$
$X^6 = (0.9, 1)$

Suppose the quality requirement stated for *DataTarget1* is *accuracy = 0.8.*
$\varphi(X^i) = 1$ if $acc(DataTarget1) \geqslant 0.8$ for X^i
$\varphi(X^i) = 0$ otherwise
For X^1:
$P(X^1) = P(acc(Source1) = 0.7) P(acc(Source2) = 0.9) = 0.63 * 0.92 = 0.58$

As we have $acc(DataTarget1) = 0.72$, then for X^1 we have $\varphi(X_1) = 0$.

We calculate analogously $P(X^i)$ and $\varphi(X^i)$ for $i = 2, \ldots, 6$, and then we apply the formula

$$DR = \sum_{(x_1,\ldots,x_n)} P\left(X = (x_1, \ldots, x_n)\right) \varphi\left(X\right) \quad (6.8)$$

The final result is $DR = 0.41$.

Note that if the calculation is done automatically, 121 vectors X^i are considered, since the probabilities in each source are not known in advance.

As said before, if there are many data sources and/or accuracy values are computed with much granularity, the calculation may become very hard. In this case it is possible to apply other exact algorithms (similar to the ones employed for computing network reliability values, see [22]) enabling us to compute the DIS reliability. For example, it is possible to employ inclusion–exclusion formulas, or to employ a variant of the super-cubes method.

The method that applies inclusion–exclusion formula considers *restriction vectors* instead of *values vectors*. As we described in Section 6.2, a set of restriction vectors can be obtained propagating the quality requirements from the data targets to the sources.

Let $v_1, \ldots, v_m$ be the set of restriction vectors, we calculate the DIS reliability

$$\begin{aligned} DR = & \sum_{1 \leqslant i \leqslant m} P(v_i) - \sum_{1 \leqslant i < j \leqslant m} P(v_i \bigcap v_j) + \\ & \sum_{1 \leqslant i < j < k \leqslant m} P(v_i \bigcap v_j \bigcap v_k) - \ldots + (-1)^{m-1} P(v_1 \bigcap v_2 \bigcap \ldots \bigcap v_m) \end{aligned} \tag{6.9}$$

In our case study, the automatic calculation of this formula for a required accuracy value of 0.6 does not finish in a reasonable time. It is not possible to consider more sources or more precision in the accuracy values, which are normal scenarios in these systems.

In the worst case, all the mentioned methods have a runtime complexity of exponential order in the size of the system being evaluated. On the other hand, the independence assumption among quality values in the different sources does not always hold.

6.4 Monte Carlo Simulation for Evaluating Data Integration Systems Reliability

The exact algorithms discussed before can become unpractical when the size and complexity of the system grows; in particular, in the case where multiple vector state combinations lead to target quality values satisfaction, the complexity of evaluation grows quickly with the number of data sources and the number of different quality states at each data source. Moreover, some algorithms depend on the assumption that data source states are independent random variables, and that there does not exist any correlation in their behavior. This is a very strong assumption, which does not usually hold in practice, and greatly limits the applicability of the model.

A powerful alternative consists in employing Monte Carlo simulation techniques to approximately evaluate DIS reliability values.

It is very easy to apply a classical Monte Carlo simulation method for the DIS reliability computation problem. The method consists in generating N random samples X^j, independent and identically distributed, following the same distribution of the DIS data sources' state vector $\boldsymbol{X}$. Then the sample mean $\hat{R}$ gives an unbiased point estimator of the DIS reliability *DR*:

$$\hat{R} = \frac{\sum_{j=1}^{N} \varphi\left(X^j\right)}{N} \tag{6.10}$$

It is also possible to compute an estimation of the variance of this estimator, which is given by

$$\hat{V} = \frac{\hat{R}\left(1 - \hat{R}\right)}{N - 1}$$

We give in Figure 6.6 a pseudo-code of the direct Monte Carlo approach. From this code, it can be seen that the most computationally demanding steps are the sampling of the data source state vector X and the propagation of these quality

```
Input:   Quality graph G=(V, E, ρV, ρE, ρG)
         Probability distribution for source nodes quality values.
         Quality requirements at data targets, Q.
         Sample size N
Output:  R̂ , the estimator of DIS reliability
         V̂ , the estimator of the variance of R̂ .
Begin
         S= 0;
         For   i=1 to N
               Sample X = (X_1, …, X_n) data source state vector, according to their joint prob-
               ability distribution.
               Using the quality graph G, propagate the source quality values given by state
               vector X to compute the corresponding data target quality values V.
               If computed data target quality values V are better (component-wise) to re-
               quired quality values Q, then φ(X)=1; otherwise, φ(X)=0.
               S=S+φ(X)
         End for
         R̂ = S/N
         V̂ = R̂(1− R̂) / (N−1)
End.
```

Figure 6.6 Pseudo-code for direct Monte Carlo

values in the quality graph G to obtain the data target quality values. As discussed in Section 6.3.1, the propagation is done following simple rules (which depend on the quality factor itself), starting from the nodes in the graph corresponding to the data sources, and "going upstream" along the edges of the graph. Then, this process will take time linear in the number of nodes and edges of the graph. Sampling X is also straightforward, if we know the joint distribution function of the quality values of the data sources (in the case of independent data sources, this is even easier, as it suffices to sequentially sample the state of each of the sources).

The straightforward estimation will work quite well when the estimated DIS reliability measure is not too small (near 0) or too large (near 1), and when the precision needed is not too large. A typical indicator of the precision of the estimator is its standard deviation, which has exact value

$$\sigma = \sqrt{\frac{DR\,(1 - DR)}{N}}$$

When the DR is neither too small nor too large, the value of σ will essentially be the square root of the inverse of the number of samples; roughly, having a precision of d digits will require sampling about 10^{2d} independent values X^j.

When a high precision is needed, the required simulation time will grow very quickly. In these cases, or when the DR is too small or too large, and a "rare events" situation arises, variance reduction techniques may be applied, instead of the standard, "crude" Monte Carlo estimator. As the mathematical structure subjacent to the DIS reliability model is quite similar to the one corresponding to static network reliability models, we can take profit from the huge literature on network reliability Monte Carlo simulation (some recent papers proposing highly efficient variance reduction techniques include [4, 6, 11]; a review of such methods can be found in [5]).

It is evident that the structure function of a DIS system differs from the classical network connectivity one. Another difference from the classical model is the need to cope with multiple states for the elementary network components. As we saw both these aspects are easy to include in standard Monte Carlo simulation, and have also been studied in the literature of other sophisticated variance reduction methods. An example is the work by Bulteau and El Khadiri [1], which tackles a flow reliability problem with these characteristics, proposing an efficient variance reduction method employing recursive decomposition.

Case Study (cont.). We did a simple implementation of the Monte Carlo method, in order to be able to conduct simulations for our case study, where the *DataTarget1* required accuracy is 0.8. Using a sample size of $N = 1000$ independent samples, we obtained an estimated DR value of 0.412, with a standard deviation of 0.016. This compares well with the exact value, as computed in Section 6.3.2.2.

Clearly, for such a simple case, exact computation can be done with small computational expenditure, so that there is no need for using Monte Carlo.

Nevertheless, it is very easy to create more realistic examples, where exact computation times are too large, while Monte Carlo simulation has essentially the same performance. If, for example we consider a DIS that obtains data from internet sites, it would be normal to have more than 10 sources. In addition, it could be necessary to make a finer measurement of accuracy, having a granularity of 0.01 for the sources accuracy values.

We also can profit from simulation to compute *DR* values when we have correlations between the sources quality. We show a simple example of this, in the following lines.

Case Study (cont.). We extend our previous example, by adding a new data source, *Source3*, which is a replication of *Source1*, but is updated in one-week cycles, just one day after *Source1*. The accuracy loss during this day is assumed to be of 0.1. Table 6.3 shows the joint distribution of the accuracy values of *Source1* and *Source3* (values not shown in the table have 0 probability of occurrence).

Source3 is integrated into the DIS as an alternative data source for *Source1*. The difference between this DIS and the previous one is that now there is an activity that

Table 6.3 Probability distribution of *Source1* and *Source3* accuracy values

Source1	*Source3*	Joint probability
0.7	0.6	0.090909
0.7	0.7	0.545455
0.8	0.7	0.025974
0.8	0.8	0.155844
0.9	0.8	0.025974
0.9	0.9	0.155844

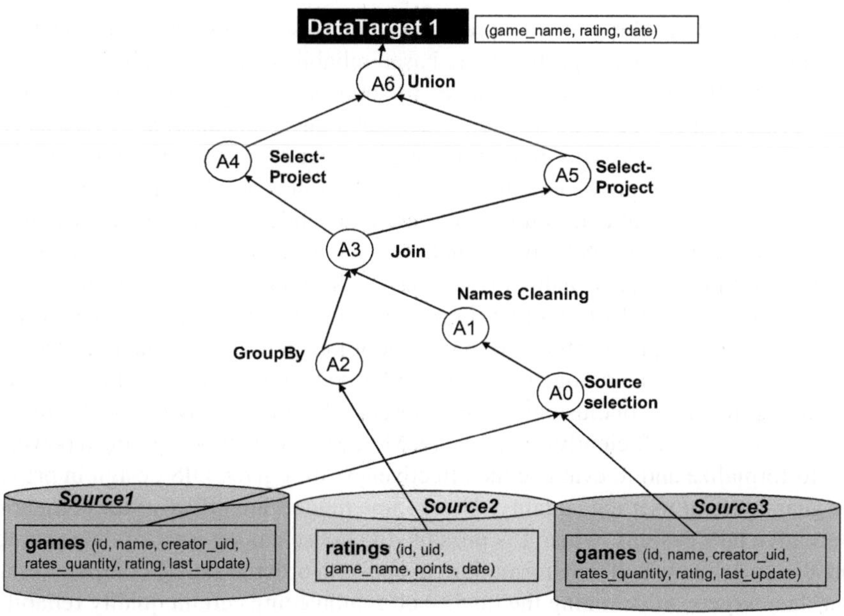

Figure 6.7 Data processing in extended DIS

extracts information from *Source1* or *Source3*, depending on which source responds first (we assume there is a probability of 1/2 for each source responding first). See the new transformation graph in Figure 6.7.

DIS reliability cannot be calculated through the inclusion–exclusion algorithm, since independence between sources does not hold any more.

Again, it was very simple to implement a Monte Carlo method to simulate this modified model. *DataTarget1* required accuracy is 0.8. Using a sample size of $N=1000$ independent samples, we obtained an estimated *DR* value of 0.395, with a standard deviation of 0.015.

Furthermore, the Monte Carlo scheme enables one to easily model the combination of different user quality targets, which can also depend on more than one quality factor, obtaining more comprehensive evaluations. For example, in the previous model, we could also include a model for freshness values at the sources, even correlated with the accuracy values, and add a freshness requirement at *DataTarget1*; the simulation would be just as easy as with the previous examples.

6.5 Conclusions

Data integration has become a necessity in a wide variety of domains (e.g., e-government, life sciences, environmental systems, e-commerce, etc.), in which the increasing availability of data makes possible the construction of DIS. In such scenarios, data quality management constitutes a key issue to ensure the usefulness of the DIS.

DIS reliability enables one to represent how the system satisfies a certain level of data quality. Additionally, probabilistic-based reliability models enable one to take into account the phenomena of changes in sources' data quality. Although reliability evaluation can be done through exact techniques, this approach presents two main drawbacks: (1) As DIS consist of large number of sources, the complexity of reliability evaluation grows and its calculation becomes problematic, and (2) these techniques assume that data source states are independent random variables unable to represent meaningful real-world situations such as inter-related data sources.

Simulation techniques enable one to face the limitations of exact techniques for reliability calculation, by providing simple but potent methods. As shown in the case study, these techniques enable one to treat both independent and inter-related data sources scenarios in a homogeneous way and based on the same simulation model.

Among the main conclusions, we can observe that the proposed models are very flexible and can be efficiently computed by Monte Carlo methods, giving a powerful tool to formalize and to evaluate the effectiveness of a given DIS design in providing guarantees of user data quality. These same models are effective for comparing alternative DIS designs, when it is possible to choose among a set of data sources, and also to define which transformations to apply, in order to satisfy compute certain user data targets, maximizing the quality or guaranteeing certain quality reliability levels.

Simulation methods show promise for carrying out practical application of data quality evaluations in DIS, especially in large-scale ones. This chapter is a step forward in this trend, and forms the basis for further extensions.

References

1. Bulteau S, El Khadiri M (2002) A new importance sampling Monte Carlo method for a flow network reliability problem. Naval Res Logist 49(2):204–228
2. Canavos G (1988) Probabilidad y estadística. Aplicaciones y métodos. McGraw Hill, Madrid, Spain [ISBN: 968-451-856-0]
3. Cho J, Garcia-Molina H (2003) Estimating frequency of change. ACM Trans Internet Technol 3(3):256–290
4. Cancela H, El Khadiri M, Rubino G (2006) An efficient simulation method for K-network reliability problem. In 6th international workshop on rare event simulation (RESIM'2006), Bamberg, Germany
5. Cancela H, El Khadiri M, Rubino G (2009) Rare events analysis by Monte Carlo techniques in static models. In: Rubino G and Tuffin B (eds) Rare event simulation methods using Monte Carlo methods, Chap 7. Wiley, Chichester, UK
6. Cancela H, Murray L, Rubino G (2008) Splitting in source-terminal network reliability estimation. In: 7th international workshop on rare event simulation (RESIM'2008), Rennes, France
7. Gertsbakh I (1989) Statistical reliability theory. Probability: pure and applied. (A series of text books and reference books.) Marcel Dekker, New York, NY, USA [ISBN: 0-8247-8019-1]
8. Gertz M, Tamer Ozsu M, Saake G, Sattler K (1998) Managing data quality and integrity in federated databases. In: 2nd working conference on integrity and internal control in information systems (IICIS'1998), Warrenton, USA, Kluwer, Deventer, The Netherlands
9. Gertz M, Tamer Ozsu M, Saake G, Sattler K (2004) Report on the Dagstuhl seminar: data quality on the web. SIGMOD Rec 33(1), March. vol 33, issue 1 (March 2004) ACM, New York, NY, USA, pp 127–132
10. Helfert M, Herrmann C (2002) Proactive data quality management for data warehouse systems. In: International workshop on design and management of data warehouses (DMDW'2002), Toronto, Canada. University of Toronto Bookstores, Toronto, Canada, pp 97–106
11. Hui K, Bean N, Kraetzl M, Kroese D (2005) The cross-entropy method for network reliability estimation. Oper Res 134:101–118
12. Jankowska M A (2000) The need for environmental information quality. Issues in Science and Technology Librarianship. `http://www.library.ucsb.edu/istl/00-spring/article5.html` (Last modified in 2000.)
13. Jarke M, Vassiliou Y (1997) Data warehouse quality: a review of the DWQ project. In: 2nd conference on information quality (IQ'1997), Cambridge, MA, MIT Pub, Cambridge, MA, USA
14. Marotta A (2008) Data quality maintenance in data integration systems. PhD thesis, University of the Republic, Uruguay
15. Marotta A, Ruggia R (2008) Applying probabilistic models to data quality change management. In: 3rd international conference on software and data technologies (ICSOFT'2008), Porto, Portugal, INSTICC, Setubal, Portugal
16. Mazzi G L, Museux J M, Savio G (2005) Quality measures for economic indicators. Statistical Office of the European Communities, Eurostat, `http://epp.eurostat.ec.europa.eu/cache/ITY_OFFPUB/KS-DT-05-003/EN/KS-DT-05-003-EN.PDF` [ISBN 92-894-8623-6]

17. Müller H, Naumann F (2003) Data quality in genome databases. In: Proceedings of the 8th international conference on information quality (IQ 2003), MIT, Cambridge, MA, USA
18. Neely M (2005) The product approach to data quality and fitness for use: a framework for analysis. In: 10th international conference on information quality (IQ'2005), Cambridge, MA, MIT Pub, Cambridge, MA, USA
19. Peralta V (2006) Data quality evaluation in data integration systems. PhD thesis, University of Versailles, France and University of the Republic, Uruguay.
20. Peralta V, Ruggia R, Bouzeghoub M (2004) Analyzing and evaluating data freshness in data integration systems. Ing Syst Inf 9(5–6):145–162
21. Peralta V, Ruggia R, Kedad Z, Bouzeghoub M (2004) A framework for data quality evaluation in a data integration system. In: 19th Brazilian symposium on databases (SBBD'2004), Brasilia, Brazil, Universidade de Brasilia, Brasilia, Brasil, pp 134–147
22. Rubino G (1999) Network reliability evaluation. In: Walrand J, Bagchi K, Zobrist G (eds) Network performance modeling and simulation. Gordon and Breach Science Publishers, Amsterdam
23. Salanti G, Sanderson S, Higgins J (2005) Obstacles and opportunities in meta-analysis of genetic association studies. Genet Med 7(1):13–20
24. Scannapieco M, Missier P, Batini C (2005) Data quality at a glance. Datenbank-Spektrum 14:6–14
25. US Environment Protection Agency (2004) Increase the availability of quality health and environmental information. Available at `http://www.epa.gov/oei/increase.htm` (last accessed August 2004)

Chapter 7
Power Distribution System Reliability Evaluation Using Both Analytical Reliability Network Equivalent Technique and Time-sequential Simulation Approach

P. Wang and L. Goel

Abstract A power system is usually divided into the subsystems of generation, transmission, and distribution facilities according to their functions. The distribution system is the most important part of a power system, which consists of many step-down transformers, distribution feeders, and customers. Therefore to evaluate the reliability of power distribution systems is a very complicated and tedious process. This chapter illustrates a reliability network equivalent technique for complex radial distribution system reliability evaluation. This method avoids the required procedure of finding the failure modes and their effect on the individual load points and results in a significant reduction in computer solution time. A time-sequential simulation technique is also introduced in this chapter. In the simulation technique, the direct search technique is used and overlapping time is considered. The simulation technique evaluates the reliability indices by a series of trials and therefore the procedure is more complicated and requires longer computer time. The simulation approach can provide both the average values and probability distributions of the load point and system indices. It may be practical therefore to use the analytical technique for basic system evaluation and to use the simulation technique when additional information is required.

7.1 Introduction

Reliability evaluation techniques have been widely used in many industries such as power, nuclear, airspace, etc. Many techniques [1–4,6–20,25–30,32,33,35–43] have been developed for different applications. The basic function of an electric power system is to supply customers with reasonably economical and reliable electricity. To build an absolutely reliable power system is neither practically realizable nor economically justifiable. The reliability of a power system can only be improved

Power Engineering Division, Electrical and Electronic Engineering School, Nanyang Technological University, Singapore

through the increased investment in system equipment during either the planning phase or operating phase. However, over-investment can lead to excessive operating costs, which must be reflected in the tariff structure. Consequently, the economic constraint will be violated although the probability of the system being inadequate may become very small. On the other hand, under-investment leads to the opposite situation. Therefore the reliability and economic constraints should be balanced at both the planning and operating phases.

A power system is usually divided into the subsystems of generation, transmission, and distribution facilities according to their functions. The distribution system is the most important part of a power system, which consists of step-down transformers, distribution feeders, and customers. This chapter introduces the techniques for the reliability evaluation of distribution systems. The basic techniques used in power system reliability evaluation can be divided into the two categories of analytical and simulation methods. The analytical techniques are well developed and have been used in practical applications for many years [1–4, 6–20, 25–28]. Analytical techniques represent the system by mathematical models and evaluate the reliability indices from these models using mathematical solutions. The mathematical equations can become quite complicated and approximations may be required when the system is complicated. Some approximate techniques therefore have been developed to simplify the calculations [1–4, 6–20, 25–28]. Analytical techniques are generally used to evaluate the mean values of the load point and system reliability indices. The mean values are extremely useful and are the primary indices of system adequacy in distribution system reliability evaluation. The mean values have been used for many years to assist power system planners to make planning and operation decisions. A mean value, however, does not provide any information on the variability of the reliability index. The probability distributions of the index, however, provide both a pictorial representation of the way the parameter varies and important information on significant outcomes, which, although they occur very infrequently, can have very serious system effects. These effects, which can easily occur in real life, may be neglected if only average values are available. Probability distributions of the relevant reliability indices can be important for industrial customers with critical processes or commercial customers with nonlinear cost functions. An analytical technique to evaluate the probability distributions associated with distribution system reliability indices is described in [16]. However, this technique can be used to evaluate approximate probability distributions. The technique may be difficult to apply when the distribution system configuration is large or complex.

Time-sequential simulation techniques [1–4, 11, 12, 19, 20, 26–28, 30, 32, 35, 36, 39–43] can be used to estimate the reliability indices by directly simulating the actual process and the random behavior of the system and its components for both power systems and other systems. These techniques can be used to simulate any system and component characteristics that can be recognized. The sequential method simulates component and system behavior in chronological time and the system and component states in a given hour are dependent on the behavior in the previous hour. Time-sequential simulation techniques can be used to evaluate both the mean values of the reliability indices and their probability distributions without excessive complications due to the probability distributions of the element parameters and the network configuration complexity. Simulation can be used to provide useful information on both the mean and the distribution of an index and in general to provide information that would not otherwise be possible to obtain analytically. The disadvantage of the simulation technique is that the solution time can be extensive.

Basic distribution system reliability indices are introduced first in this chapter. An analytical reliability network equivalent approach [21, 22] is presented to simplify the evaluation procedure. A test distribution system is analyzed to illustrate the technique and the results are presented. This chapter also briefly illustrates a time-sequential Monte Carlo simulation technique. The procedure of this technique in distribution system reliability evaluation is described. The technique is used to analyze the reliability of a test distribution system.

7.2 Basic Distribution System Reliability Indices

The basic function of a distribution system is to supply electrical energy from a substation to the customer load points. Service continuity is, therefore, an important criterion in a distribution system, Service continuity can be described by three basic load point indices and a series of system indices [6].

7.2.1 Basic Load Point Indices

In distribution system reliability evaluation, three basic load point indices are usually used to measure load point reliability [6]. These are average failure rate, average outage time, and average annual outage time. For a series system, the average failure rate λ_i, average annual outage time U_i, and average outage time r_i for the load point i can be calculated using the following equations:

$$\lambda_i = +\sum_{j=1}^{n} \lambda_j \tag{7.1}$$

$$U_i = \sum_{j=1}^{n} \lambda_j r_j \tag{7.2}$$

$$r_i = \frac{U_i}{\lambda_i} \tag{7.3}$$

where n is the total number of components which affect load point i, λ_j is the average failure rate of element j, and r_j is the average restoration time to restore load point i due to the failure of component j.

7.2.2 Basic System Indices

The three primary load point indices are fundamentally important parameters. They can be aggregated to provide an appreciation of the system performance using a series of system indices. The additional indices that are most commonly used [6] are defined in the following sections.

7.2.2.1 Customer-oriented Indices

System average interruption frequency index (*SAIFI*) [6]:

$$SAIFI = \frac{\text{total number of customer interruptions}}{\text{total number of customers served}} = \frac{\sum \lambda_i N_i}{\sum N_i} \tag{7.4}$$

where λ_i is the failure rate and N_i is the number of customers at load points i.

System average interruption duration index (*SAIDI*) [6]:

$$SAIDI = \frac{\text{sum of customer interruption durations}}{\text{total number of customers}} = \frac{\sum U_i N_i}{\sum N_i} \tag{7.5}$$

where U_i is the annual outage time and N_i is the number of customers at load point i.

Customer average interruption duration index (*CAIDI*) [6]:

$$CAIDA = \frac{\text{sum of customer interruption durations}}{\text{total number of customer interruptions}} = \frac{\sum U_i N_i}{\sum N_i \lambda_i} \tag{7.6}$$

where λ_i is the failure rate, U_i is the annual outage time, and N_i is the number of customers at load point i.

Average service availability index (*ASAI*) [6]:

$$ASAI = \frac{\text{customer hours of available service}}{\text{customer hours demanded}} = \frac{\sum N_i \times 8760 - \sum N_i U_i}{\sum N_i \times 8760} \tag{7.7}$$

Average service unavailability index (*ASUI*) [6]:

$$ASUI = 1 - ASAI \tag{7.8}$$

where λ_i is the failure rate, U_i is the annual outage time, N_i is the number of customers at load point i, and 8760 is the number of hours in a calendar year.

7.2.2.2 Load- and Energy-oriented Indices

Energy not supplied index (*ENS*):

$$ENS = \text{total energy not supplied by the system} = \sum L_{a(i)} U_i \tag{7.9}$$

where $L_{a(i)}$ is the average load connected to load point i.

Average energy not supplied index (*AENS*):

$$AENS = \frac{\text{total energy not supplied}}{\text{total number of customers served}} = \frac{\sum L_{a(i)} U_i}{\sum N_i} \tag{7.10}$$

where $L_{a(i)}$ is the average load connected to load point i.

7.3 Analytical Reliability Network Equivalent Technique

The analytical techniques required for distribution system reliability evaluation are highly developed. Many of the published concepts and techniques are presented and summarized in [6–8]. Conventional techniques for distribution system reliability evaluation are generally based on failure mode and effect analysis (FMEA) [6, 23, 31]. This is an inductive approach that systematically details, on a component-by-component basis, all possible failure modes and identifies their resulting effects on the system. Possible failure events or malfunctions of each component in the distribution system are identified and analyzed to determine the effect on surrounding load points. A final list of failure events is formed to evaluate the basic load point indices. The FMEA technique has been used to evaluate a wide range of radial distribution systems. In systems with complicated configurations and a wide variety of components and element operating modes, the list of basic failure events can become quite lengthy and can include thousands of basic failure events. This requires considerable analysis when the FMEA technique is used. It is therefore difficult to directly use FMEA to evaluate a complex radial distribution system. A reliability network equivalent approach is introduced in this section to simplify the analytical process. The main principle in this approach is using an equivalent element to replace a portion of the distribution network and therefore decompose a large distribution system into a series of simpler distribution systems. This approach provides a repetitive and sequential process to evaluate the individual load point reliability indices.

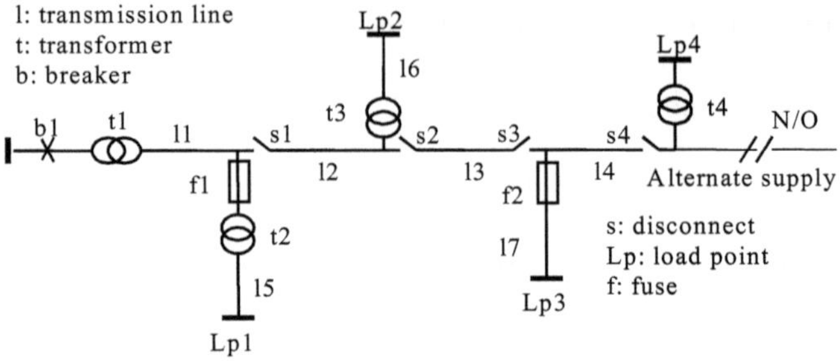

Figure 7.1 A simple distribution system

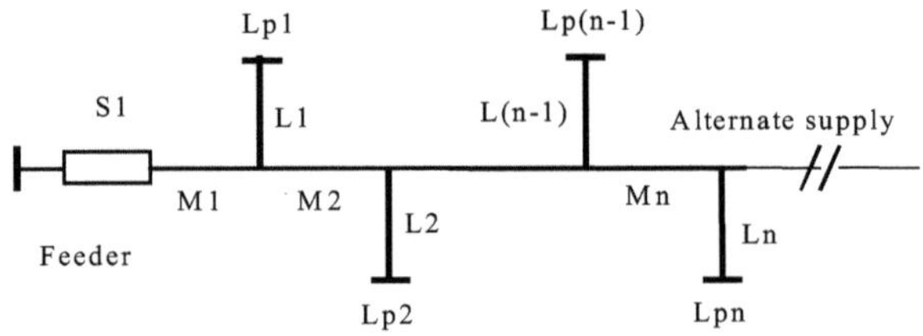

Figure 7.2 General feeder

7.3.1 Definition of a General Feeder

Figure 7.1 shows a simple radial distribution system consisting of transformers, transmission lines (or feeders), breakers, fuses, and disconnects. Disconnects and transmission lines such as s1 and l2 are designated as a main section. The main sections deliver energy to the different power supply points. An individual load point is normally connected to a power supply point through a transformer, fuse and lateral transmission line. A combination such as f1, t2, and l5 is called a lateral section.

A simple distribution system is usually represented by a general feeder which consists of n main sections, n lateral sections, and a series component as shown in Figure 7.2. In this feeder, Si, Li, Mi, and Lpi represent series component i, lateral section i, main section i, and load point i, respectively. Li could be a transmission line, a line with a fuse or a line with a fuse and a transformer. Mi can be a line, a line with one disconnect switch, or a line with disconnect switches on both ends.

7.3.2 Basic Formulas for a General Feeder

Based on the element data (λ_i, λ_k, λ_s, r_i, r_k, r_s, p_k) and the configuration of the general feeder, a set of general formulas for calculating the three basic load point indices of load point failure rate λ_j, average outage duration r_j and average annual outage time U_j for load point j of a general feeder is as follows:

$$\lambda_j = \lambda_{sj} + \sum_{i=1}^{n} \lambda_{ij} + \sum_{k=1}^{n} p_{kj}\lambda_{kj} \text{ (1)} \tag{7.11}$$

$$U_j = \lambda_{sj} r_{sj} + \sum_{i=1}^{n} \lambda_{ij} r_{ij} + \sum_{k=1}^{n} p_{kj}\lambda_{kj} r_{kj} \tag{7.12}$$

$$r_j = \frac{U_j}{\lambda_j} \tag{7.13}$$

where is the control parameter of lateral section k that depends on the fuse operating model. It can be 1 or 0 corresponding to no fuse or a 100% reliable fuse respectively and a value between 0 and 1 for a fuse which has a probability of unsuccessful operation of p_{kj}. The parametersλ_{ij}, λ_{kj}, and λ_{sj} are the failure rates of the main section i, lateral section k and series element s respectively, and r_{ij}, r_{kj}, and r_{sj} are the outage durations (switching time or repair time) for the three elements respectively.

The r_{ij}, r_{kj}, and r_{sj} data have different values for different load points when different alternate supply operating modes are used and disconnect switches are installed in different locations on the feeder. This is illustrated in the following three cases.

7.3.2.1 Case 1: No Alternate Supply

In this case, r_s is the repair time of the series element s and r_i is the switching time for those load points that can be isolated by disconnection from the failure main section i or the repair time for those load points that cannot be isolated from a failure of the main section i. In this case, r_k is the switching time for those load points that can be isolated by disconnection from a failure on a lateral section k or the repair time for those load points that cannot be isolated from a failure on a lateral section k.

7.3.2.2 Case 2: 100% Reliable Alternate Supply

In this case, r_i and r_k take the same values as in Case 1. The parameter r_s is the switching time for those load points that are isolated from the failure of a series element by disconnection or the repair time for those load points not isolated from the failure of a series element s.

7.3.2.3 Case 3: Alternate Supply with Availability

In this case, r_i is the repair time (r_1) for those load points not isolated by disconnection from the failure of main section i, the switching time (r_2) for those load points supplied by the main supply and isolated from the failure of the main section i, or $r_2 p_a + (1 - p_a) r_1$ for those load points supplied by an alternate supply and isolated from the failure of the main section i. The parameter r_k is the repair time r_1 for those load points not isolated by disconnection from the failure of lateral section k, the switching time r_2 for those load points supplied by the main supply and isolated from the failure of lateral section k or $r_2 p_a + (1 - p_a) r_1$ for those load points supplied by an alternate supply and isolated from the failure of a lateral section k. r_s is the same as in Case 2.

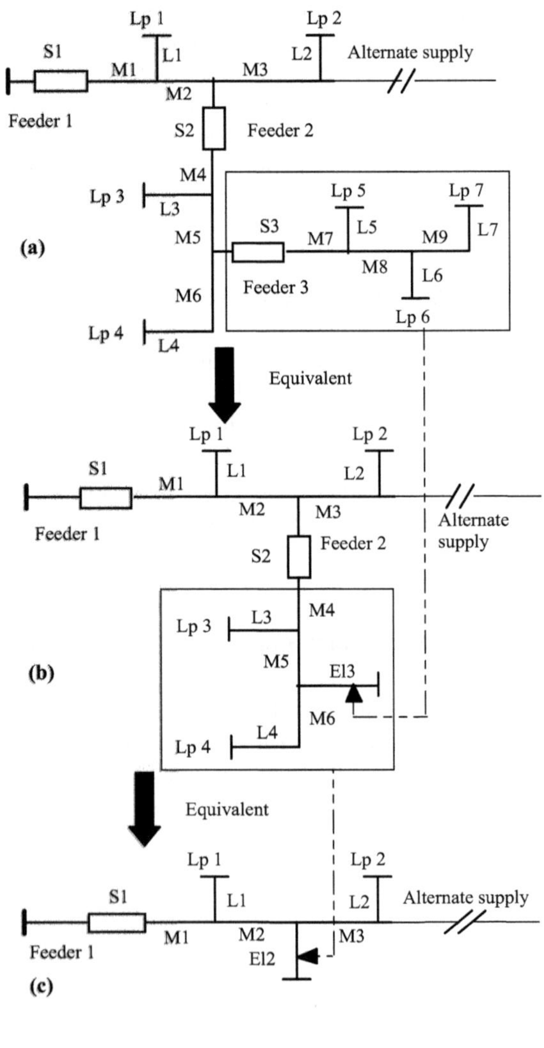

Figure 7.3 Reliability network equivalent: (a) original configuration; (b) and (c) successive equivalents

7.3.3 Network Reliability Equivalent

A practical distribution system is usually a relatively complex configuration that consists of a main feeder and subfeeders as shown in Figure 7.3. The main feeder is connected to a bus station. A subfeeder is a feeder connected such as Feeder 2 and Feeder 3 in Figure 7.3. The three basic equations presented earlier cannot be used directly to evaluate the reliability indices of this system. The reliability network equivalent approach, however, provides a practical technique to solve this problem. The basic concepts in this approach can be illustrated using the distribution system shown in Figure 7.3. The original configuration is given in Figure 7.3a and successive equivalents are shown in Figure 7.3b and c. The procedure involves the development of equivalent lateral sections and associated series sections.

7.3.3.1 Equivalent Lateral Sections

The failure of an element in Feeder 3 will affect load points not only in Feeder 3 but also in Feeder 1 and Feeder 2. The effect of Feeder 3 on Feeder 1 and Feeder 2 is similar to the effect of a lateral section on Feeder 2. Feeder 3 can be replaced using the equivalent lateral section (El 3) shown in Figure 7.3b. The equivalent must include the effect of the failures of all elements in Feeder 3. The equivalent lateral section (El 2) of Feeder 2 can then be developed as shown in Figure 7.3c. The contributions of the failures of different elements to parameters of an equivalent lateral section will depend on the location of the disconnect switches. The reliability parameters of an equivalent lateral section can be divided into two groups and obtained using the following equations:

$$\lambda_{e1} = \sum_{i=1}^{m} \lambda_i \tag{7.14}$$

$$U_{e1} = \sum_{i=1}^{m} \lambda_i r_i \tag{7.15}$$

$$r_{e1} = \frac{U_{e1}}{\lambda_{e1}} \tag{7.16}$$

$$\lambda_{e2} = \sum_{i=1}^{n} \lambda_i \tag{7.17}$$

$$U_{e2} = \sum_{i=1}^{n} \lambda_i r_i \tag{7.18}$$

$$r_{e2} = \frac{U_{e2}}{\lambda_{e2}} \tag{7.19}$$

where λ_{e1} and λ_{e1} are the total failure rate and restoration time of the failed components that are not isolated by disconnects in the subfeeder and m is the total number of these elements. The effect of this equivalent lateral section on the load points in the prior supply feeder (designated as upfeeder) depends on the configuration and operating mode of the upfeeder elements. The parameters λ_{e1} and λ_{e1} are the total equivalent failure rate and the switching time of those failed elements that can be isolated by disconnects in the branch and n is the total number of these elements. They do not depend on the configuration and operating modes of the upfeeders. The equivalent parameters do not depend on alternate supplies in the subfeeders.

7.3.3.2 Equivalent Series Components

Using successive network equivalents, the system is reduced to a general distribution system in the form shown in Figure 7.3c. Only Feeder 1 remains in the system. The basic formulas can now be used to evaluate the load point indices of Feeder 1. On the other hand, the failure of elements in Feeder 1 also affects the load points in Feeder 2 and Feeder 3. These effects are equivalent to those of a series element S2 in Feeder 2. The parameters of the equivalent series component S2 are obtained as the load point indices of Feeder 1 are calculated. Feeder 2 becomes a general distribution system after the equivalent series element is calculated. The load point indices of Feeder 2 and the parameters of the equivalent series element S3 are then calculated in the same way as with Feeder 1. Finally, the load point indices of Feeder 3 are evaluated. The reliability parameters of an equivalent series component can be calculated using the same method used for the load point indices. The only difference is that the equivalent parameters should be divided into two groups. The effect of one group on the load points of a subfeeder is independent of the alternate supplies in subfeeders; the effect of the other group depends on the alternate supplies in the subfeeders.

7.3.4 Evaluation Procedure

The procedure described in the previous section for calculating the reliability indices in a complex distribution system using the reliability network equivalent approach can be summarized by two processes. A bottom-up process is used to search all the subfeeders and to determine the corresponding equivalent lateral sections. As shown in Figure 7.3, the equivalent lateral section El 3 is first found, followed by El 2. The system then is reduced to a general distribution system. Following the bottom-up process, a top-down procedure is then used to evaluate the load point indices of each feeder and equivalent series components for the corresponding subfeeders until all the load point indices of feeders and subfeeders are evaluated. The load point indices and the equivalent parameters of the series components are calculated using Equations 7.1–7.3. Referring to Figure 7.3, the load point indices in Feeder 1 and the equivalent series element S2 for Feeder 2 are first calculated, followed by the load

Table 7.1 Load point indices for Case 1

Load point	Failure rate (occ/yr)	Outage Duration (h)	Unavailability (h/yr)
1	0.3303	2.4716	0.8163
10	0.3595	2.2434	0.8065
20	3.4769	4.1915	14.5735
25	3.4769	5.0216	17.4595
30	3.3586	5.0223	16.8680
35	3.6498	4.2298	15.4380
40	3.8734	5.0194	19.4420

point indices in Feeder 2 and S3. The load point indices in Feeder 3 are finally calculated. After all the individual load point indices are calculated, the final step is to obtain the feeder and system indices. The example presented in Figure 7.3a considers a single alternate supply. The procedure can be extended, however, to consider more than one supply to a general feeder.

7.3.5 Example

A small but practical test system known as the RBTS [5,24] developed at the University of Saskatchewan. Figure 7.4 shows the distribution system connected to bus 6. The distribution system contains 4 feeders, 3 subfeeders, 42 main sections, 42 lateral sections and 40 load points. Each system segment consists of a mixture of components. The disconnect switches, fuses, and alternate supplies can operate in the different modes described earlier. The data used in these studies is given in [5, 24]. The existing disconnect switches are shown in Figure 7.4, but additional switches can be added at any location. System analysis has been carried out for three different operating conditions. The detailed procedure followed in the reliability network equivalent approach is illustrated in Case 1.

7.3.5.1 Case 1

In order to illustrate the reliability network equivalent approach in a general sense, breakers b6, b7, and b8 are assumed to be 80% reliable with no alternate supply to main Feeder 4. There are three subfeeders in main Feeder 4. The first step is to find the equivalent lateral sections of subfeeders F5, F6, and F7.

After finding the equivalent lateral sections of subfeeders F5, F6, and F7, Feeder 4 becomes a general feeder. The next step is to calculate the load point indices in Feeder 4. After determining the parameters of the three equivalent series elements, the indices of load points connected to the three subfeeders can be calculated. Table 7.1 shows the representative load point reliability indices for Feeder 4.

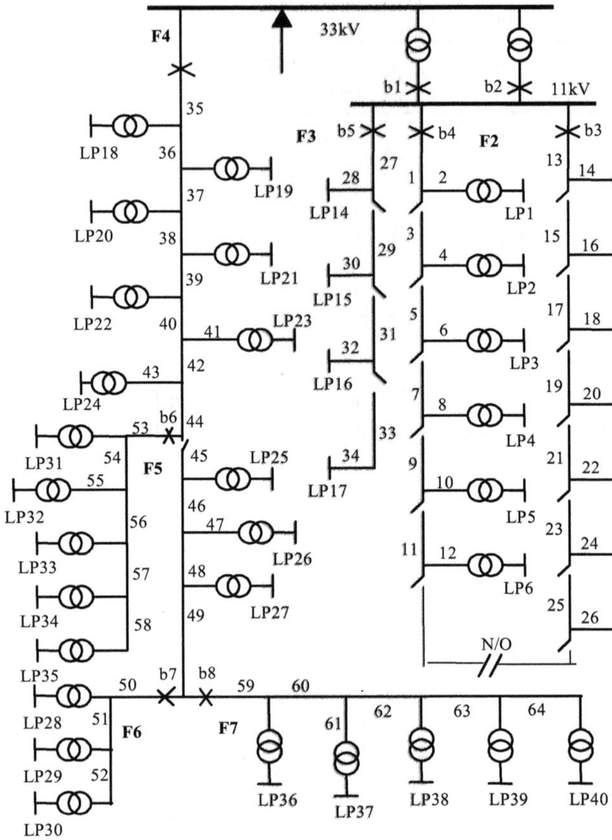

Figure 7.4 Distribution system of RBTS

Table 7.2 System indices for Case 1

SAIFI (interruptions/customer yr)	1.6365
SAIDI (hours/customer yr)	6.9695
CAIDI (hours/customer interruption)	4.2588
ASAI	0.9992
ASUI	0.0008
ENS (MWh/yr)	83.9738
AENS (kWh/customer yr)	0.02858

The system indices for Feeder 4 can be evaluated using the load point indices and are shown in Table 7.2.

Table 7.3 System indices for Case 2

SAIFI (interruptions/customer yr)	1.0065
SAIDI (hours/customer yr)	3.8197
CAIDI (hours/customer interruption)	3.7949
ASAI	0.999956
ASUI	0.00044
ENS (MWh/yr)	48.3691
AENS (kWh/customer yr)	0.01646

Table 7.4 System indices for Case 3

SAIFI (interruptions/customer yr)	1.6365
SAIDI (hours/customer yr)	4.8478
CAIDI (hours/customer interruption)	2.9623
ASAI	0.99945
ASUI	0.00055
ENS (MWh/yr)	57.8922
AENS (kWh/customer yr)	0.0197

7.3.5.2 Case 2

In this case, breakers 6, 7, and 8 are assumed to be 100% reliable and no alternative supply is available to Feeder 4. The system indices are shown in Table 7.3.

7.3.5.3 Case 3

In this case, breakers 6, 7, and 8 are assumed to be 80% reliable and alternative supply is available to Feeder 4 at the point between the two breakers in F6 and F7. The system indices are shown in Table 7.4.

It can be seen by comparing the results of Case 2 with those of Case 1 that the probability of successful operation of breakers 6, 7, and 8 is important for the reliability of the whole distribution system. Comparing the results of Case 1 and Case 3, it can be seen that the reliability of the overall system is greatly increased by providing the alternate supply in Feeder 4.

These conclusions can obviously be determined by other techniques such as the standard FMEA approach. The reliability network equivalent method is a novel approach to this problem which uses a repetitive and sequential process to evaluate the individual load point and subsequently the overall system indices.

7.4 Time-sequential Simulation Technique

Monte Carlo simulation has been used in reliability evaluation of generating systems, transmission systems, substations, switching stations, and distribution systems. The behavior pattern of n identical systems in real time will all be different in varying degrees, including the number of failures, times to failure, restoration times, etc. This is due to the random nature of the processes involved. The behavior of a particular system could follow any of these behavior patterns. The time-sequential simulation process can be used to examine and predict behavior patterns in simulated time, to obtain the probability distributions of the various reliability parameters and to estimate the expected or average value of these parameters.

In a time-sequential simulation, an artificial history that shows the up and down times of the system elements is generated in chronological order using random number generators and the probability distributions of the element failure and restoration parameters. A sequence of operating-repair cycles of the system is obtained from the generated component histories using the relationships between the element states and system states. The system reliability indices and their probability distributions can be obtained from the artificial history of the system.

7.4.1 Element Models and Parameters

The essential requirement in time-sequential simulation is to generate realistic artificial operating/restoration histories of the relevant elements. These artificial histories depend on the system operating/restoration modes and the reliability parameters of the elements. Distribution system elements include basic transmission equipment such as transmission lines and transformers, and protection elements such as disconnect switches, fuses, breakers, and alternate supplies.

Transmission equipment can generally be represented by the two-state model shown in Figure 7.5 where the up state indicates that the element is in the operating state and the down state implies that the element is inoperable due to failure.

The time during which the element remains in the up state is called the time to failure (TTF) or failure time (FT). The time during which the element is in the down state is called the restoration time that can be either the time to repair (TTR) or the time to replace (TTR). The process of transiting from the up state to the down state is the failure process. Transition from an up state to a down state can be caused by

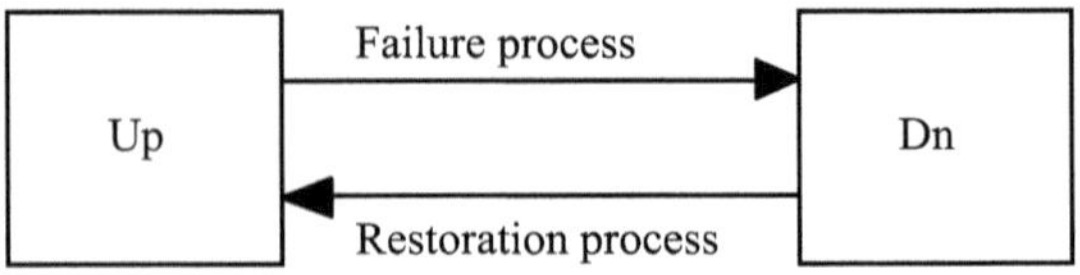

Figure 7.5 State-space diagram of element

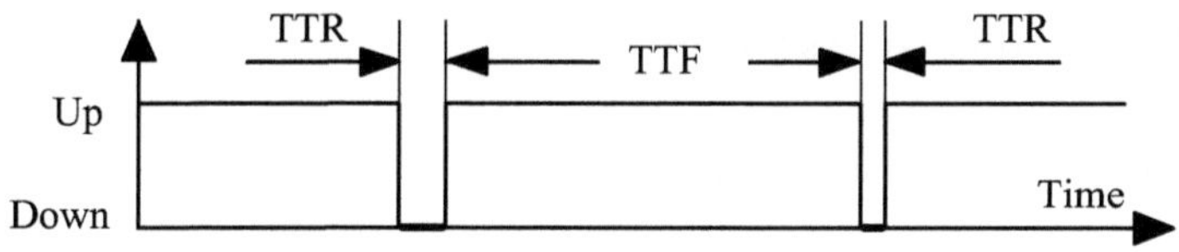

Figure 7.6 Element operating/repair history

the failure of an element or by the removal of elements for maintenance. Figure 7.6 shows the simulated element operating/restoration history of an element.

The parameters TTF, TTR are random variables and may have different probability distributions. The probability distributions used to simulate these times are exponential, gamma, normal, lognormal, and Poisson distributions.

Protection elements are used to automatically isolate failed elements or failed areas from healthy areas when one or more failures occur in system. They can exist in either functioning or failed states which can be described in terms of their probabilities. Alternative supply situations can be described by probabilities that alternative supplies are available. A uniform distribution is used to simulate these probabilities.

7.4.2 Probability Distributions of the Element Parameters

The parameters that describe the operating/restoration sequences of the elements such as TTF, TTR, repair time (RT), and switching time (ST) are random variables, and may have different probability distributions. The most useful probability distributions in distribution system reliability evaluation are given in the following sections.

7.4.2.1 Uniform Distribution

The probability density function (p.d.f.) of a uniform distribution is

$$f_U(u) = \begin{cases} 1 & 0 \leqslant u \leqslant 1 \\ 0 & \text{otherwise} \end{cases} \tag{7.20}$$

The availability of an alternate supply and the probability that a fuse or breaker operates successfully can be obtained directly from this distribution.

7.4.3 Exponential Distribution

The p.d.f. of an exponential distribution is

$$f_T(t) = \begin{cases} \lambda e^{-\lambda t} & 0 < t < \infty \\ 0 & \text{otherwise} \end{cases} \tag{7.21}$$

Many studies indicate that time to failure is reasonably described by an exponential distribution.

7.4.3.1 Gamma Distribution

A random variable T has a gamma distribution if the p.d.f. is defined as

$$f_T(t) = \begin{cases} \frac{t^{\alpha-1} e^{-t/\beta}}{\beta^{\alpha}\Gamma(\alpha)} & 0 \leqslant t \leqslant \infty \\ 0 & \text{otherwise} \end{cases} \tag{7.22}$$

7.4.3.2 Normal Distribution

A random variable T has a normal distribution if the p.d.f. is

$$f_T(t) = \frac{1}{\sigma\sqrt{2\pi}} \exp\left[-\frac{(t-\mu)^2}{2\sigma^2}\right] \tag{7.23}$$

and is denoted by $N(\mu, \sigma^2)$, where μ is the mean and σ^2 is the variance.

7.4.3.3 Lognormal Distribution

Let T be from $N(\mu, \sigma^2)$, then $Y = e^T$ has the lognormal distribution with p.d.f.

$$f_T(t) = \begin{cases} \frac{1}{\sqrt{2\pi}\sigma t} \exp\left[-\frac{(\ln t-\mu)^2}{2\sigma^2}\right] & 0 \leqslant t \leqslant \infty \\ 0 & \text{otherwise} \end{cases} \tag{7.24}$$

7.4.3.4 Poisson Distribution

A random variable x has a Poisson distribution if the probability mass function is

$$p_x = \frac{\lambda^x e^{-\lambda}}{x!} \quad x = 0, 1, \ldots, \quad \lambda > 0 \tag{7.25}$$

Studies show that the number of element failures in a year is Poisson distributed. The TTR, TTF, RT, and ST in the operating/restoration history of the elements and load point can be described by any one of these distributions.

7.4.4 Generation of Random Numbers

As described earlier, the uniform distribution can be generated directly by a uniform random number generator. The random variables from other distributions are converted from the generated uniform number. The three basic methods are the inverse transform, composition, and acceptance–rejection techniques. These methods are discussed in detail in [23, 34]. The following example shows how to convert the uniform distribution into an exponential distribution using the inverse transform method.

The cumulative probability distribution function for the exponential distribution 7.18 is

$$U = F_T(t) = 1 - \mathrm{e}^{-\lambda t} \tag{7.26}$$

where U is a uniformly distributed random variable over the interval [0,1].

Solving for T:

$$T = -\frac{1}{\lambda}\ln(1-U) \tag{7.27}$$

Since $(1-U)$ is distributed in the same way as U, then

$$T = -\frac{1}{\lambda}\ln U \tag{7.28}$$

U is uniformly distributed and T is exponentially distributed.

7.4.5 Determination of Failed Load Point

The function of a distribution system is to supply electric power to individual customers. Element failures may affect one or more load points. The most difficult problem in the simulation is to find the load points and their failure durations affected by the failure of an element, which are dependent on the network configuration, the system protection and the maintenance philosophy. In order to create a structured approach, the distribution system can be broken down into general segments. A complex radial distribution system can be divided into the combination of a main feeder (a feeder is connected to a switch station) and subfeeders (a subfeeder is a branch connected to a main feeder or to other subfeeders [21]. The direct search procedure for determining the failed load points and their operating-restoration histories is as follows:

Step 1. Determine the type (main section, lateral section, or series element). If the failed element is a lateral section, go to step 2. If the failed element is a main section or a series element, go to step 3.

Step 2. Determine the state of the corresponding lateral fuse If the failed element is a lateral section line. If the lateral fuse is in a functioning state, the load point connected to this lateral section is the only failed load point and the search procedure is stopped. If the lateral fuse is in a malfunction state, go to the next step.

Step 3. Determine the location of the failed element, that is, the failed element number and the feeder that the failed element is connected to. If the failed feeder is the main feeder, all the load points connected to this main feeder are the failed load points and the search procedure is stopped. If the failed feeder is a subfeeder, go to step 4.

Step 4. Determine the subfeeders which are the downstream feeders connected to the failed subfeeder and all the load points connected to these subfeeders are the failed load points.

Step 5. Determine the breaker state of the failed subfeeder. If the breaker is in a functioning state, the search procedure is stopped. If not, go to step 6.

Step 6. Determine the upfeeder which is the upstream feeder to which the failed subfeeder is connected. All the load points in the upfeeder are the failed load points. The upfeeder becomes the new failed subfeeder.

Step 7. Repeat steps 5 to 6 until the main feeder is reached and all the failed load points are found.

Some failed load points can be restored to service by switching action. The failure duration therefore is the switching time that is the time to isolate the failed element from the system. Others can only be restored by repairing the failed elements. In this case, the failure duration is the repair time of the failed element. The failure durations of the load points, are determined based on the system configuration and operating scheme for the disconnect switches in the system.

The operating/restoration history of a load point is shown in Figure 7.7 and is conceptually similar to that of a component as shown in Figure 7.6. In this case, however, it is based on the operating/restoration histories of the pertinent elements, the system configuration and protection scheme. The TTR is the time to restoration, which can be the repair time or the switching time.

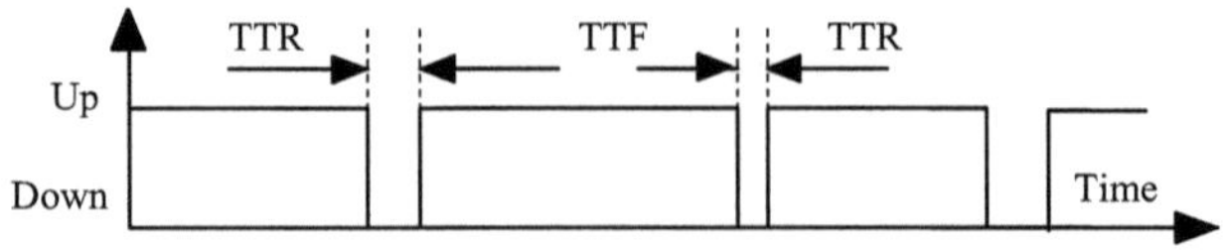

Figure 7.7 Load point operating/restoration history

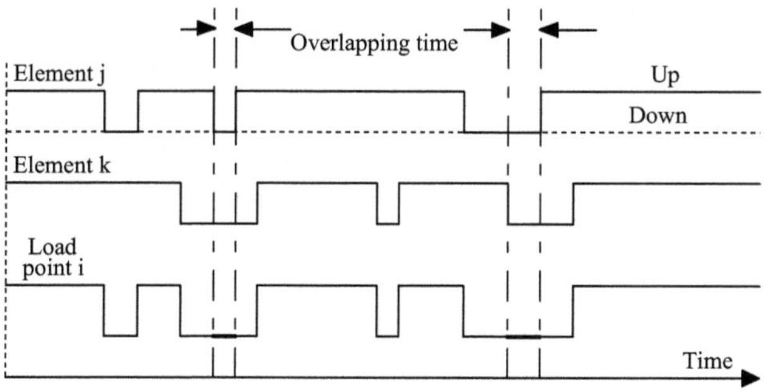

Figure 7.8 Overlapping time of element failures

7.4.6 Consideration of Overlapping Times

The failure of one element can overlap that of another element. The duration of this event is called overlapping time and can occur with more than one element. Overlapping time can affect load point failure duration as illustrated in Figure 7.8. The artificial histories of the elements j, k, and the load point i are shown in Figure 7.8 where the failures of both elements j and k affect load point i. It is usually assumed in radial distribution system reliability evaluation, that the restoration time is very short compared with the operating time which means that the probability of two elements or more elements being failed at the same time is very small. This is not true if all the elements have similar failure rates and the deviations in TTF are large. The effects of overlapping times on the load point indices are considered in the simulation program.

7.4.7 Reliability Indices and Their Distributions

Distribution system reliability can be expressed in terms of load point and system indices. Both the average values and the probability distributions of these indices can be calculated from the load point operating/restoration histories. The average values of the three basic load point indices for load point j can be calculated from the load point up–down operating history using the following formulae:

$$\lambda_j = \frac{N_j}{\sum T_{uj}} \tag{7.29}$$

$$r_j = \frac{\sum T_{dj}}{N_j} \tag{7.30}$$

$$U_j = \frac{\sum T_{dj}}{\sum T_{uj} + \sum T_{dj}} \tag{7.31}$$

where $\sum T_{uj}$ and $\sum T_{dj}$ are the respective summations of all the up times T_u and all the down times T_d and N_j is the number of failures during the total sampled years.

In order to determine the probability distributions of the load point failure frequency, the period values k of this index are calculated for each sample year. The number of years $m(k)$ in which the load point outage frequency equals k is counted. The probability distribution $p(k)$ of the load point failure frequency can be calculated using

$$p(k) = \frac{m(k)}{M} \quad k = 0, 1, 2 \ldots \tag{7.32}$$

where M is the total sample years. The probability distribution of the load point unavailability can be calculated in a similar manner. To calculate the probability distribution of outage duration, the failure number $n(i)$ with outage duration between $i - 1$ and i is counted. The probability distribution $p(i)$ is

$$p(i) = \frac{n(i)}{N} \quad i = 1, 2, 3, \ldots \tag{7.33}$$

where N is the total failures in the sampled years.

The system indices can be calculated from the basic load point indices as system indices are basically weighted averages of the individual load point values. Distributions of the system indices therefore can also be obtained from the period load point indices.

7.4.8 Simulation Procedure

The process used to evaluate the distribution system reliability indices using time-sequential simulation consists of the following steps:

Step 1. Generate a random number for each element in the system.

Step 2. Convert these random numbers into TTFs corresponding to the probability distribution of the element parameters.

Step 3. Generate a random number and convert this number into the RT of the element with minimum TTF according to the probability distribution of the repair time.

Step 4. Generate another random number and convert this number into an ST according to the probability distribution of the switching time if this action is possible.

Step 5. Utilize the procedure described earlier under determination of load point failures (Section 7.4.4) and record the outage duration for each failed load point.

Step 6. Generate a new random number for the failed element and convert it into new TTF, and return to step 3 if the simulation time is less than one year. If the simulation time is greater than one year, go to step 9.

Step 7. Calculate the number and duration of failures for each load point for each year.

Step 8. Calculate the average value of the load point failure rate and failure duration for the sample years.

Step 9. Calculate the system indices of *SAIFI*, *SAIDI*, *CAIDI*, *ASAI*, *ASUI*, *ENS*, and *AENS* and record these indices for each year.

Step 10. Calculate the average values of these system indices.

Step 11. Return to step 3 if the simulation time is less than the specified total simulation years, otherwise output the results.

7.4.9 Stopping Rules

For the time-sequential techniques, a large number of simulation (sampling) years are required to obtain relatively accurate results. There are two stopping rules used in sequential simulation. One is to stop the simulation using the specific simulation time which is usually obtained from simulation experiences of the program users based on the accuracy required. One is to stop the simulation using the given accuracy between two simulation years. The later one will increase the simulation time.

7.4.10 Example

The developed program has been used to evaluate a range of distribution systems. The following illustrates an application to the system shown in Figure 7.4. The failure rate of each element is assumed to be constant. The repair and switching times are assumed to be lognormally distributed. It is assumed that the standard deviations of the transmission line repair time, transformer replace time, and switching time of all elements are 1 hour, 10 hours, and 0.4 hours, respectively. The simulation was performed for a period of 15,000 years in order to obtain specific accuracy. The following shows the simulation results.

7.4.11 Load Point and System Indices

The average values of the load point and system indices can be calculated using both the analytical and simulation techniques. Table 7.5 shows representative results of the load point indices obtained using the analytical (A) and simulation (S) techniques. The average values of system indices are shown in Table 7.6 for two approaches.

The results from both simulation and analytical approaches are very close. The maximum difference in the load point indices is 7.95% at load point 20. The max-

Table 7.5 Comparison of the load point indices

Load Point (i)	Failure rate (occ/yr) (i)	(S)	Difference (%)	Unavailability (h/yr) (A)	(S)	Difference (%)
1	0.3303	0.3340	−1.11	0.8163	0.8310	−1.77
5	0.3400	0.3460	−1.73	0.8260	0.8520	−3.05
10	0.3595	0.3570	0.07	0.8065	0.8170	−1.29
20	1.6274	1.7680	−7.95	5.5515	5.5919	−0.07
25	1.6725	1.7681	−5.41	8.4375	8.8573	−4.73
35	2.5370	2.6008	−2.45	9.8740	9.7233	1.55
40	2.5110	2.5593	−1.88	12.6300	12.7872	−1.23

Table 7.6 Comparison of the system indices

Indices	(S)	(A)	Difference (%)
SAIFI (interruption/customer yr)	1.03872	1.00655	3.18
SAIDI (h/customer yr)	3.86350	3.81970	1.15
CAIDI (h/customer interruption)	3.71951	3.79485	−1.98
ASAI	0.99956	0.99956	0
ASUI	0.00044	0.00044	0
ENS (MWh/yr)	48.85556	48.36910	1.00
AENS (kWh/customer yr)	0.01663	0.01646	1.03

imum error in the system indices is 3.18% for *SAIFI*. The analytical approach provides a direct and practical technique for radial distribution system evaluation and is quite adequate if only the average values of the load point and system indices are required.

7.4.12 Probability Distributions of the Load Point Indices

The probability distributions of the annual failure frequency and failure duration for each load point in the distribution system have been evaluated. Figures 7.9 and 7.10 present the histograms of the failure frequency for load point 1 and load point 30.

The probability distribution of the failure frequency clearly shows the probability of having a different number of load point failures in each year for each load point. It can be seen in Figure 7.9 that the probability of having zero failures per year at load point 1 is more than 0.9. The probability of having one failure per year is less than 0.1 and the probability of two failures per year is less than 0.01. It can be seen from Figure 7.10 that the probability of zero failures per year is about 0.02 at load point 30 and the probability of having six or more outages per year is very small. The additional information provided by the probability distributions can be very important for those customers which have special reliability requirements.

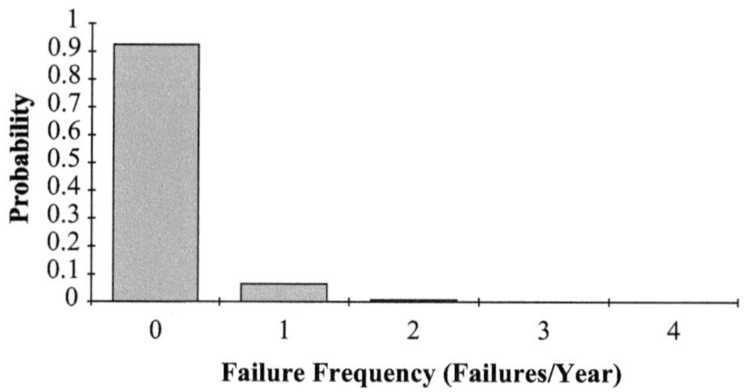

Figure 7.9 Failure frequency histogram, load point 1

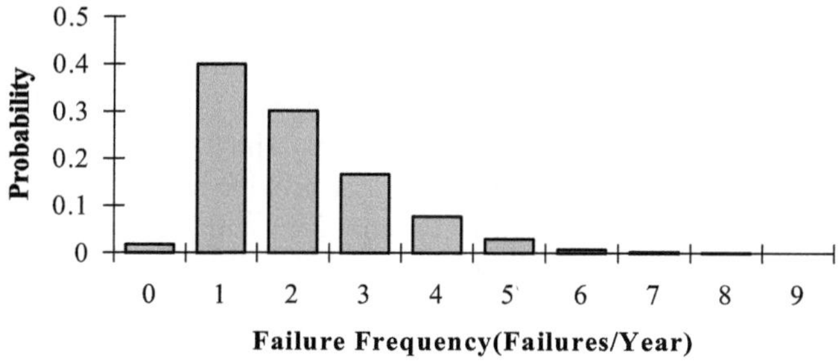

Figure 7.10 Failure frequency histogram, load point 30

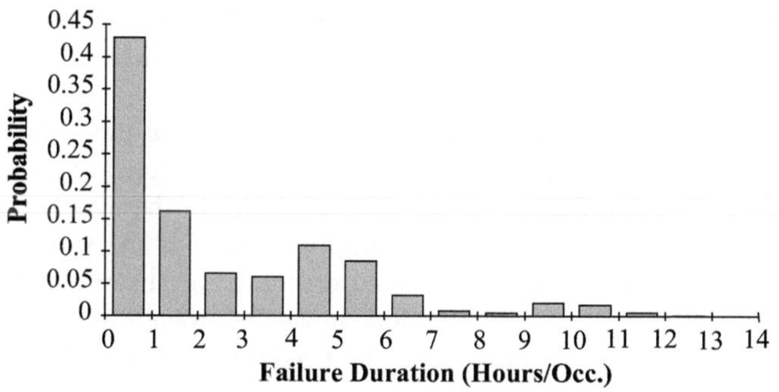

Figure 7.11 Failure duration histogram, load point 1

The probability distributions of failure durations for load point 1 and 30 are shown in Figures 7.11 and 7.12. A class interval width of 1 hour has been used in this example. It can be seen from Table 7.4 that failure durations between 0 and 1 hour at load point 1 have the largest probability. The durations between 1 and 2 hours has the second largest probability and the duration with the third largest probability is between 4 and 5 hours. Durations in excess of 12 hours have a very small possibility. For load point 30, outage durations between 4 and 5 hours have the largest probability 0.38. The durations are mainly distributed between 0 and 12

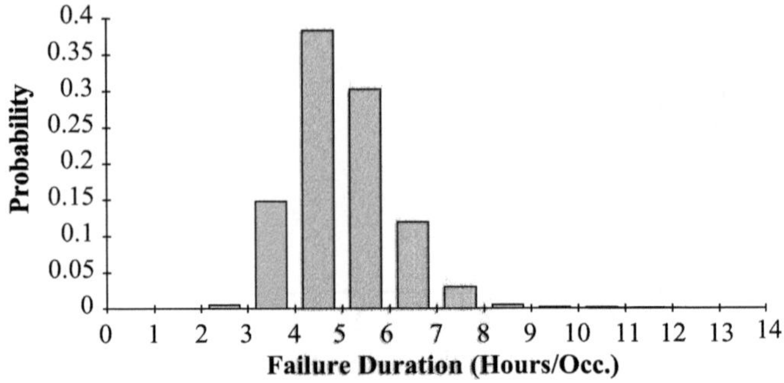

Figure 7.12 Failure duration histogram, load point 30

hours. The longest duration is about 12 hours. The information provided by these probability distributions is very useful for reliability worth/cost analysis for customers with nonlinear customer damage functions. The 2.488-hour average failure duration from the analytical technique does not provide any distribution information. A 1-hour class interval is used in Figures 7.11 and 7.12. Any class interval, however, can be used in the simulation.

7.4.12.1 Probability Distributions of System Indices

The probability distributions of all seven system indices for each feeder were also evaluated. Figures 7.13–7.19 show the probability distributions of *SAIFI*, *SAIDI*, *CAIDI*, *ASAI*, *ASUI*, *ENS*, and *AENS* for Feeder 4.

The probability distribution of *SAIFI* is a combination of the failure frequency distribution weighted by the percentage of customers connected to the corresponding load points. The distribution shows the variability in the average annual customer interruption frequency. The distribution of *SAIDI* is the summation of the unavailability distribution weighted by the percentage of customers connected to corresponding load points. The distribution shows the probabilities of different average annual customer failure durations. The *CAIDI* distribution shows the probability of different failure durations for each customer interruption in each year. The

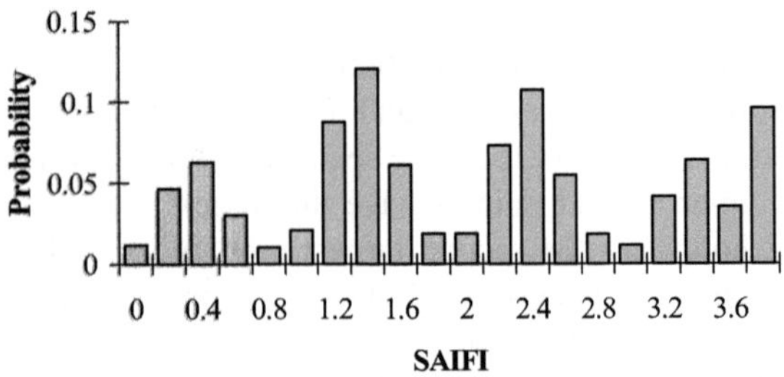

Figure 7.13 Histogram of *SAIFI*, Feeder 4

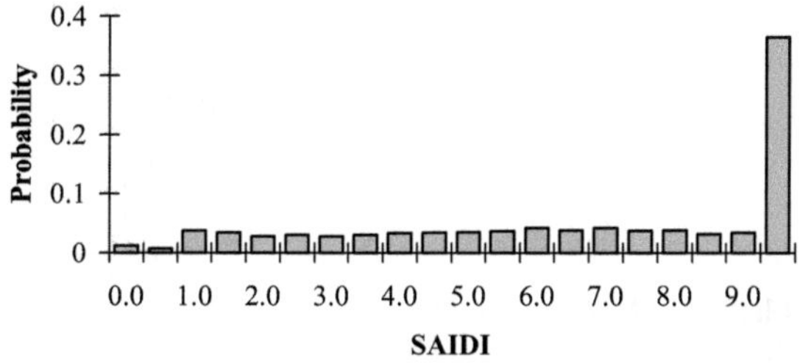

Figure 7.14 Histogram of *SAIDI*, Feeder 4

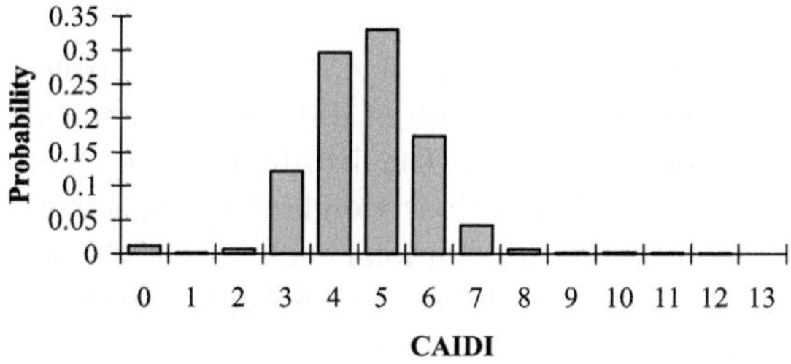

Figure 7.15 Histogram of *CAIDI*, Feeder 4

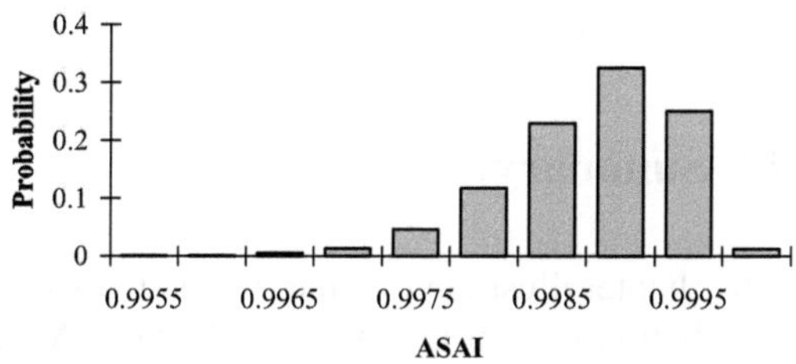

Figure 7.16 Histogram of *ASAI*, Feeder 4

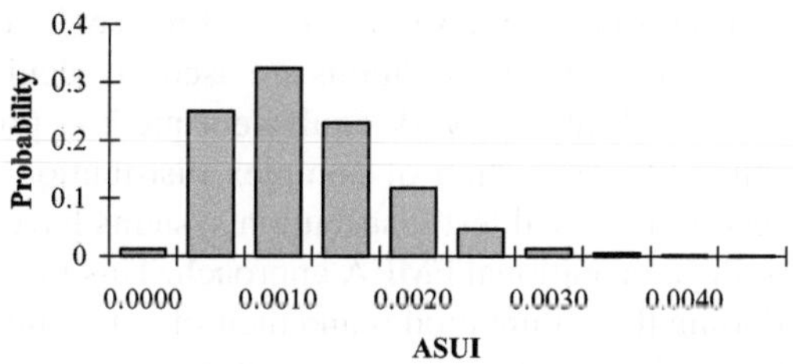

Figure 7.17 Histogram of *ASUI*, Feeder 4

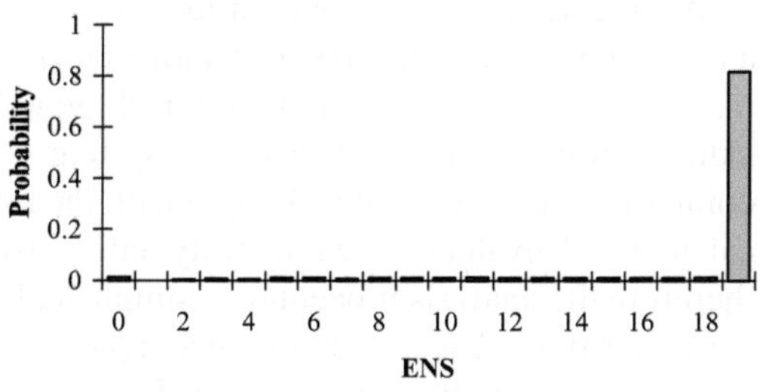

Figure 7.18 Histogram of *ENS*, Feeder 4

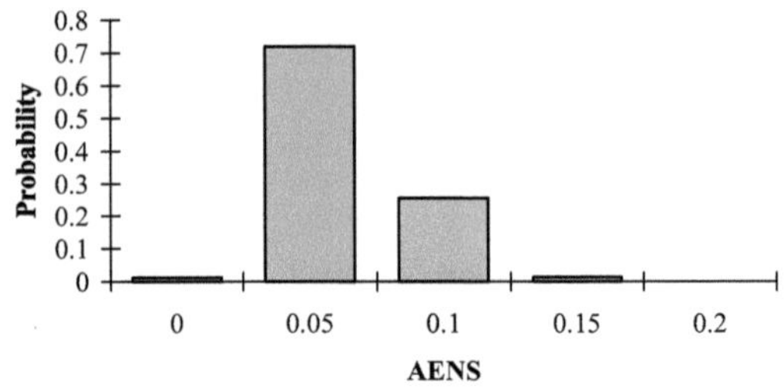

Figure 7.19 Histogram of *AENS*, Feeder 4

probability distribution of *ASUI* mainly depends on the distribution of *SAIDI* and provides the probability of different percentages of unavailable customer hours in each simulation year. The distribution of *ENS* is a summation of the load point unavailability distributions weighted by the corresponding load level and shows the probability of different total energies not supplied in each year. The distribution of *AENS* is the distribution of *ENS* per customer. These indices provide a complete picture based on the number of customers, the energy level, duration hours, and the number of interruptions.

7.5 Summary

This chapter illustrates a reliability network equivalent technique for complex radial distribution system reliability evaluation. A general feeder is defined and a set of basic equations is developed based on a general feeder concept. A complex radial distribution system is reduced to a series of general feeders using reliability network equivalents. Basic equations are used to calculate the individual load point indices. The reliability network equivalent method provides a simplified approach to the reliability evaluation of complex distribution systems. Reliability evaluations for several practical test distribution systems have shown this technique to be superior to the conventional FMEA approach. This method avoids the required procedure of finding the failure modes and their effect on the individual load points and results in a significant reduction in computer solution time.

A time-sequential simulation technique is also introduced in this chapter, and a computer program has been developed using the simulation approach. In the simulation technique, the direct search technique is used and overlapping time is considered. A practical test distribution system was evaluated using this technique. In comparing the analytical technique with the time-sequential technique, the analytical approach evaluates the reliability indices by a set of mathematical equations and therefore the analysis procedure is simple and requires a relatively small amount of computer time. The simulation technique evaluates the reliability indices by a series of trials and therefore the procedure is more complicated and requires a longer computer time. The simulation approach can provide information on the load point and system indices that the analytical techniques cannot provide. It may be practical therefore to use the analytical technique for basic system evaluation and to use the simulation technique when additional information is required.

References

1. Allan RN, Billinton R, Lee SH (1984) Bibliography on the application of probability methods in power system reliability evaluation. IEEE Trans PAS 103(2):275–282
2. Allan RN, Billinton R, Shahidehpour SM, Singh C (1988) Bibliography on the application of probability methods in power system reliability evaluation. IEEE Trans Power Syst 3(4):1555–1564
3. Allan RN, Billinton R, Breipohl AM, Grigg CH (1994) Bibliography on the application of probability methods in power system reliability evaluation. IEEE Trans Power Syst 9(1):41–49
4. Allan RN, Bhuiyan MR (1993) Effects of failure and repair process distribution on composite system adequacy indices in sequential Monte Carlo simulation. Proceedings of the joint international IEEE power conference, Power Tech. IEEE, Los Alamitos, CA, USA, pp 622–628
5. Allan RN, Billinton R, Sjarief I, Goel L et al. (1991) A reliability test system for educational purpose-basic distribution system data and results. IEEE Trans Power Syst 6(2):823–831
6. Billinton R, Allan RN (1996) Reliability evaluation of power systems, 2nd edn. Plenum Press, New York
7. Billinton R, Allan RN (1990) Basic power system reliability concepts. Reliab Eng Syst Saf 27:365–384
8. Billinton R, Allan RN, Salvaderi L (1988) Applied reliability assessment in electric power systems. Institute of Electrical and Electronics Engineers, New York
9. Billinton R, Wang P (1998) Distribution system reliability cost/worth analysis using analytical and sequential simulation techniques. IEEE Trans Power Syst 13(4):1245–1250
10. Billinton R, Billinton JE (1989) Distribution system reliability indices. IEEE Trans Power Deliv 4(1):561–568
11. Billinton R, Wacker G, Wojczynski E (1983) Comprehensive bibliography of electrical service interruption costs. IEEE Trans PAS 102:1831–1837
12. Billinton R (1972) Bibliography on the application of probability methods in power system reliability evaluation. IEEE Trans PAS 91(2):649–660
13. Billinton R, Grover MS (1975) Reliability assessment of transmission system and distribution systems. IEEE Trans PAS 94(3):724–732
14. Billinton R, Grover MS (1975) Quantitative evaluation of permanent outages in distribution systems. IEEE Trans PAS 94(3):733–741
15. Billinton R, Grover MS (1974) A computerized approach to substation and switching station reliability evaluation. IEEE Trans PAS 93(5):1488–1497
16. Billinton R, Goel R (1986) An analytical approach to evaluate probability distribution associated with the reliability indices of electrical distribution systems. IEEE Trans Power Deliv 1(3):145–251
17. Billinton R, Wojczynski E (1985) Distribution variation of distribution system reliability indices. IEEE Trans PAS 104:3152–3160
18. Billinton R, Wang P (1995) A generalized method for distribution system reliability evaluation. Conference proceedings, IEEE WESCANEX. IEEE, Los Alamitos, CA, USA, pp 349–354
19. Billinton R, Wang P (1999) Teaching distribution system reliability evaluation using Monte Carlo simulation. IEEE Trans Power Syst 14(2):397–403
20. Billinton R, Cui L, Pan Z, Wang P (2002) Probability distribution development in distribution system reliability evaluation. Electric Power Compon Syst 30(9):907–916
21. Billinton R, Wang P (1998) Reliability-network-equivalent approach to distribution system reliability evaluation. IEE Proc Gener Transm Distrib 145(2):149–153

22. Billinton R, Wang P (1999) Deregulated power system planning using a reliability network equivalent technique. IEE Proc Gener Transm Distrib 146(1):25–30
23. Billinton R, Allan RN (1984) Reliability evaluation of engineering systems. Plenum Press, New York
24. Billinton R, Jonnavithula S (1997) A test system for teaching overall power system reliability assessment. IEEE Trans Power Syst 11(4):1670–1676
25. Brown RE, Hanson AP (2001) Impact of two-stage service restoration on distribution reliability. IEEE Trans Power Syst 16(4):624–629
26. Ding Y, Wang P, Goel L, Billinton R, Karki R (2007) Reliability assessment of restructured power systems using reliability network equivalent and pseudo-sequential simulation techniques. Electric Power Syst Res 77(12):1665–1671
27. Durga Rao K, Kushwaha HS, Verma AK, Srividya A (2007) Simulation based reliability evaluation of AC power supply system of Indian nuclear power plant. Int J Qual Reliab Manag 24(6):628–642
28. Durga Rao K, Gopika V, Rao VVSS et al. (2009) Dynamic fault tree analysis using Monte Carlo simulation in probabilistic safety assessment. Reliab Eng Syst Saf 94(4):872–883
29. Goel L, Billinton R (1991) Evaluation of interrupted energy assessment rates in distribution systems. IEEE Trans Power Deliv 6(4):1876–1882
30. Goel L, Ren S, Wang P (2001) Modeling station-originated outages in composite system using state duration sampling simulation approach. Comput Electr Eng 27(2):119–132
31. Henley EJ, Hiromitsu K (1981) Reliability engineering and risk assessment. Prentice-Hall, Englewood Cliffs, NJ
32. IEEE Subcommittee on the Application of Probability Methods. Power System Engineering Committee (1978) Bibliography on the application of probability methods in power system reliability evaluation. IEEE Trans PAS 97(6):2235–2242
33. Li W, Wang P, Li Z, Liu Y (2004) reliability evaluation of complex radial distribution systems considering restoration sequence and network constraints. IEEE Trans Power Deliv 19(2):753–758
34. Rubinstein RY (1981) Simulation and Monte Carlo method. Wiley, New York
35. Tollefson G, Billinton R, Wacker G (1991) Comprehensive bibliography on reliability worth and electrical service interruption costs. IEEE Trans Power Syst 6(4):1980–1990
36. Ubeda R, Allan RN (1992) Sequential simulation applied to composite system reliability evaluation. IEE Proc C 139(2):81–86
37. Wang P, Li W (2007) Reliability evaluation of distribution systems considering optimal restoration sequence and variable restoration times. IET Proc Gener Transm Distrib 1(4):688–695
38. Wang P, Billinton R (2001) Impacts of station-related failures on distribution system reliability. Electr Mach Power Syst 29:965–975
39. Wang P, Billinton R (2002) Reliability cost/worth assessment of distribution system incorporating time varying weather conditions and restoration resources. IEEE Trans Power Deliv 17(1):260–265
40. Wang P, Billinton R (1999) Time-sequential distribution system reliability worth analysis considering time varying load and cost models. IEEE Trans Power Deliv 14(3):1046–1051
41. Wang P, Billinton R (2001) Time-sequential simulation technique for rural distribution system reliability cost/worth evaluation including wind generation as an alternative supply. IEE Proc Gener Transm Distrib 148(4):355–360
42. Wang P, Goel L, Billinton R (2000) Evaluation of probability distributions of distribution system reliability indices considering WTG as alternative supply. Electr Mach Power Syst 28:901–913
43. Zio E, Marella M, Podollini L (2007) A Monte Carlo simulation approach to the availability assessment of multi-state systems with operational dependencies. Reliab Eng Syst Saf 92:871–882

Chapter 8
Application of Reliability, Availability, and Maintainability Simulation to Process Industries: a Case Study

Aijaz Shaikh and Adamantios Mettas

Abstract This chapter demonstrates the application of RAM (reliability, availability, and maintainability) analysis to process industries by providing a case study of a natural-gas processing plant. The goal of the chapter is to present RAM analysis as a link between the widely researched theoretical concepts related to reliability simulation and their application to complex industrial systems. It is hoped that the concepts and techniques illustrated in the chapter will help spawn new ideas to tackle real-world problems faced by practitioners of various industries, particularly the process industry.

8.1 Introduction

Reliability, availability, and maintainability (RAM) have become the focus of all industries in the present times. Growing competition, tighter budgets, shorter cycle times, and the ever-increasing demand for better, cheaper, and faster products have created greater awareness about the benefits of using the various tools offered by the discipline of reliability engineering. With increasing complexity of industrial systems and the widespread use of powerful computers, reliability simulation is becoming the preferred option to deal with the challenging real-world problems of the modern age that would otherwise be either too difficult or sometimes even impossible to solve using analytical approaches. One such simulation-based approach that is now being regarded by many industries as a standard tool of reliability engineering is RAM analysis.

This chapter illustrates the applicability and benefits of conducting RAM analyses of industrial systems by considering the example of a natural-gas processing plant. The goal of the chapter is to present RAM analysis as a link between the widely researched theoretical concepts related to reliability simulation and their application to complex industrial systems. It is hoped that the concepts and techniques

ReliaSoft Corporation, Tucson, AZ, USA

P. Faulin, A. Juan, S. Martorell, and J.E. Ramírez-Márquez (eds), *Simulation Methods for Reliability and Availability of Complex Systems*.

illustrated in the chapter will help spawn new ideas to tackle real-world problems faced by practitioners of various industries, particularly the process industry.

8.2 Reliability, Availability, and Maintainability Analysis

The most commonly used approach for conducting a RAM analysis on a complex system involves the use of reliability block diagrams (RBDs) to represent the reliability-wise interdependencies of the components of the system under consideration. In RBDs (ReliaSoft 2007), the system is represented as a series of blocks, each block symbolizing a component of the system. The corresponding failure and repair distributions of the component are tied to this block. With the various inputs of the system specified as probabilistic distributions in this manner, Monte Carlo simulation is then used to model the behavior of the system for a large number of life cycles. This provides statistical estimates of various system parameters such as reliability and availability. Details on Monte Carlo simulation can be found in Armstadter (1971) and Fishman (1996).

8.3 Reliability Engineering in the Process Industry

The term process industry is used to refer to a large classification of industries such as oil and gas processing, petrochemicals, general chemicals, pharmaceuticals, cement, iron and steel, food processing, and so on. Suzuki (1994) lists a number of characteristics that distinguish the process industry from other industries. From a reliability engineering perspective, the following features are unique to the process industry:

1. diverse equipment that includes rotating equipment (e.g., pumps, compressors, motors, turbines), static equipment (e.g., vessels, heat exchangers, columns, furnaces), and piping and instrumentation equipment that is used to connect and monitor the rest of the equipment;
2. round-the-clock operation during normal production periods with the use of standby and bypass units to ensure continuous operation;
3. harsh operation environment that exposes equipment to high temperature, high pressure, vibrations, and toxic chemicals;
4. high accident and pollution risk because of the nature of the manufacturing processes and the materials involved;
5. periodic shutdown of the plants to evaluate the condition of all equipment and take preventive measures to mitigate failures.

It is obvious that process industries are highly complex systems, and achieving high reliability, availability, and maintainability in these industries is a very crucial and challenging task. With so much at stake in terms of meeting production, safety, and environmental goals, these industries have rigorous maintenance programs that require considerable planning and devotion of significant amount of resources. To streamline operation and maintenance and to address the reliability issues mentioned above, many of the industries have adopted the maintenance management philosophy of total productive maintenance (TPM). TPM (Suzuki 1994; Wireman 2004) is an integrated approach to maintenance and production where the responsibility of maintenance is shared by all employees. A number of maintenance concepts (Waeyenbergh *et al.* 2000) may be used under the TPM philosophy, the most popular being reliability-centered maintenance (RCM). RCM (Moubray 1997; Smith 1993) involves the selection of the most appropriate maintenance task (corrective, preventive – time or condition based, failure finding or redesign) for each equipment based on the failure consequences. With recent technological advancements, condition-based maintenance tasks, also referred to as predictive maintenance, are gaining popularity (Mobley 2002). Predictive maintenance uses a surveillance system to continuously monitor equipment deterioration and predict equipment failure using mathematical models. The information of the impending failure is used to decide an optimum time to carry out preventive maintenance.

8.4 Applicability of RAM Analysis to the Process Industry

Maintenance concepts such as RCM have been successfully applied in the process industry to reduce unnecessary preventive maintenance actions and come up with a systematic and efficient maintenance plan. RCM and other maintenance concepts involve the selection of maintenance actions at the equipment level. RAM analysis goes one step further and provides a quantitative assessment of how different maintenance tasks would affect performance at the system level. It is a tool that can be used to model and compare complex maintenance strategies with results available both at the equipment and the plant level.

There is also a trend towards greater integration of different functional aspects in the process industry, particularly with the implementation of management philosophies such as TPM. These integration efforts can benefit immensely from an analysis that takes into consideration all aspects of these industries (such as maintenance policies, resources used such as spare parts and crews, layout of the plant and production levels) and quantifies the effects of the different available options. The quantitative predictions obtained can assist plant management in making informed decisions to achieve the goals of the plant. Such an analysis can also be important to win the confidence of engineers, operators, and maintenance personnel so that changes in procedures and policies are readily accepted. RAM analysis is an ideal tool in this regard. It can play an important role in complementing the efforts of philosophies such as TPM by integrating all the functions such as reliability, availability, maintainability, production, and logistics into a single analysis and providing forecasts in terms of quantitative measures. In the following sections the application of RAM analysis to achieve these benefits is illustrated through a case study of a natural-gas plant.

8.5 Features of the Present Work

In recent times, with the widespread awareness of the benefits of RAM, there is an increase in the efforts to apply this tool to process plants. For example, Herder *et al.* (2008) have presented the application of RAM to a plastics plant to assess two key decisions regarding operation and shutdown policies. Racioppi *et al.* (2007) have performed a RAM analysis to evaluate the availability of a sour-gas injection plant. Lee *et al.* (2004) have investigated a subsea production system to verify if the required availability goal is met and suggest improvements. Sikos and Klemeš (2009) have used RAM analysis to provide quantitative forecasts on availability and other performance measurements of a waste management plant. While these publications used the RBD approach to carry out the RAM analysis, Zio *et al.* (2006) have used fault tree diagrams (see Bahr 1997 for details on fault tree diagrams) to assess availability of an offshore plant and Marquez *et al.* (2005) have discussed a general approach of using continuous time Monte Carlo simulations to assess availability using the example of cogeneration plants. A comparative study of the publications indicates that the RBD approach is the most intuitive approach for industrial practitioners.

The publications mentioned previously represent significant efforts towards incorporation of RAM to the process industry. The present chapter is an attempt to add to the work accomplished thus far by including one or more of the following features that are found lacking in the aforementioned publications:

1. RBD modeling of the entire plant with emphasis on all aspects including modeling of standby and bypass equipment;
2. modeling of real-world maintenance policies such as failure finding inspections to detect hidden failures and predictive maintenance;
3. integration of production into the analysis and illustration of throughput modeling for all equipment taking into consideration the product flow while preserving the reliability configuration;
4. integration of resources such as maintenance crews into the analysis;
5. modeling of shutdown and other phases of production;
6. incorporation of variation in throughput before normal steady production is reached;
7. presentation of results in terms of availability, production efficiency, and cost figures to enable informed decision making.

The case study presented in this chapter includes sufficient details that are essential to grasp a thorough understanding of the approach employed. The details include several interesting situations that may appear in process industries which need proper modeling, such as linking maintenance repairs for different equipment and the modeling of throughput. These examples can be of interest to a wide range of practitioners.

8.5.1 Software Used

The case study presented in this chapter uses ReliaSoft Corporation's BlockSim software. BlockSim offers the advantage of advanced modeling capabilities together with reliable results and ease of use. The software has been widely used in many industries since 1998. For process industries, analyses using BlockSim has been conducted by Herder *et al.* (2008), Racioppi *et al.* (2007), Sikos and Klemeš (2009), and Calixto and Rocha (2007), to name a few. A comparison of commercially available RAM packages is found in Brall *et al.* (2007). The present study uses version 7 of BlockSim, which includes the ability to model reliability phase diagrams (RPDs). An RPD is a representation of the changes in the configuration or properties of the system RBD during different periods of time. For the process industry, RPDs can be used to model different phases in the operation of the plant including the periodic shutdowns. A complete description of RPDs and other models and analyses available in BlockSim is found in ReliaSoft (2007).

8.6 Case Study

The following sections present a case study from the natural-gas processing industry. The purpose of this study is to demonstrate the application of RAM analysis. The study is not intended to illustrate results based on the analysis of an actual natural-gas processing facility. The information for this study is taken from a number of sources including Wheeler and Whited (1985), Giuliano (1989), and Peebles (1992). The key objectives of this RAM analysis are to:

1. predict availability and production efficiency of the natural-gas processing facility under consideration;
2. identify the bad actors or the key components responsible for losses in availability and production;
3. conduct a cost analysis to estimate the loss of revenue due to unavailability;
4. identify recommended actions to improve performance;
5. estimate expected availability and production if the recommended actions are implemented.

8.6.1 Natural-gas Processing Plant Reliability Block Diagram Modeling

Natural gas used by consumers is mostly methane. However, raw natural gas occurring in nature is not pure and needs to be processed. Raw natural gas may occur along with a semi-liquid hydrocarbon condensate and liquid water. It also can exist as a mixture consisting of other hydrocarbons such as ethane, propane, butane, and pentanes. These hydrocarbons are valuable by-products of natural-gas processing and are collectively referred to as *natural-gas liquids* (NGLs). Natural gas containing significant amounts of NGL is referred to as *rich gas*. Natural gas also contains impurities such as hydrogen sulfide, carbon dioxide, water vapor, nitrogen, helium, and mercury. Before natural gas is deemed fit to be utilized by the consumers (called *pipeline quality gas*), it has to be processed to remove the impurities. This processing is done at a natural-gas processing plant as described next.

Figure 8.1 shows the RBD of a natural-gas plant that gets two streams of gas – a medium-pressure (MP) stream and a high-pressure (HP) stream. It is assumed that the volume of both the gas streams is 50 MMSCF (million standard cubic feet) per day resulting in a total input of 100 MMSCF per day. The RBD shown in Figure 8.1 is created using process flow diagrams (PFDs) and piping and instrumentation diagrams (PIDs) of the plant together with the reliability-wise relationships of the equipment and systems. Note that an RBD may not necessarily match the physical layout of the plant.

The units modeled as a part of this plant are described next. Please note that some of the plant equipment, such as valves and control systems, plant utility systems, and nitrogen, mercury, and helium treatment units, have not been included in the model to keep the RBD from becoming exceedingly complex and beyond the scope of the present chapter.

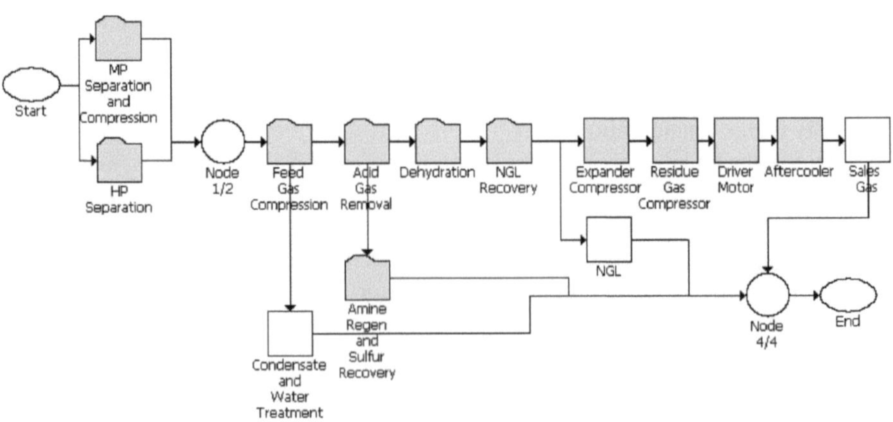

Figure 8.1 RBD of the natural-gas plant

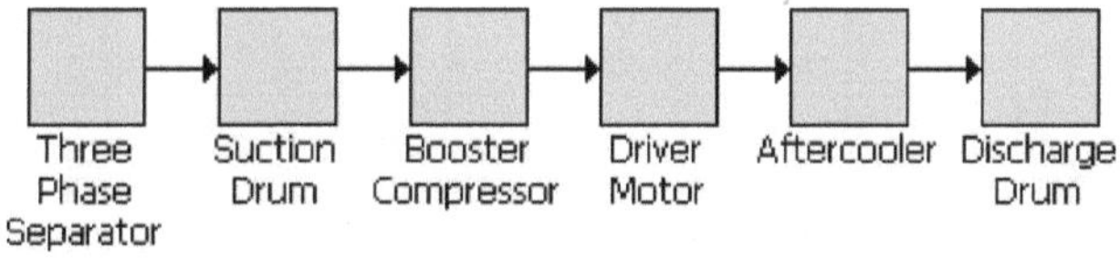

Figure 8.2 RBD of the MP separation and compression unit

8.6.1.1 Medium-pressure Separation and Compression Unit

The first step in natural-gas processing is to separate out condensate and water using vessels called *separators*. As shown in Figure 8.1, the MP gas stream is sent to the MP separation and compression unit while the HP gas stream is sent to the HP separation unit. These units are represented as subdiagrams in Figure 8.1, while Figures 8.2 and 8.3 show the equipment included in the analysis for these units.

The MP gas first goes to the three-phase separator where condensate and free water are removed (see Figure 8.2). The separated condensate and free water are sent to the condensate treatment unit and the water treatment unit respectively. To keep the RBD simple, these units are modeled as one block in Figure 8.1 assuming that the outage of these units will not affect production. The block is shown in white, indicating that the block is not assigned any failure properties.

After the three-phase separator, the MP gas stream is compressed to HP by a single compression train. The MP gas stream enters the suction drum where any entrained liquid is separated. The separated liquid condensate is sent to the condensate treatment unit while the gas is compressed in the booster compressor. The booster compressor is driven by an electric motor. An aftercooler is provided after the compressor to cool the compressed gas before it mixes with the HP gas stream after passing through the discharge drum.

8.6.1.2 High-pressure Separation Unit

Figure 8.3 shows that the HP gas is received at the slug catcher. The slug catcher removes high-velocity liquid slugs that may otherwise damage the piping system by high-energy hydraulic shocks. From here, the gas is sent to the three-phase separator to remove condensate and free water.

The HP gas is then mixed with compressed MP gas and sent to the feed-gas compression unit. Note that a "node" is used in the RBD of Figure 8.1 as the junction of the two gas streams. If either of the gas streams is interrupted due to equipment

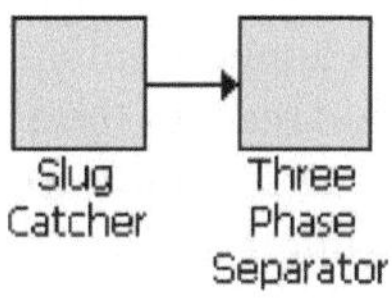

Figure 8.3 RBD of the HP separation unit

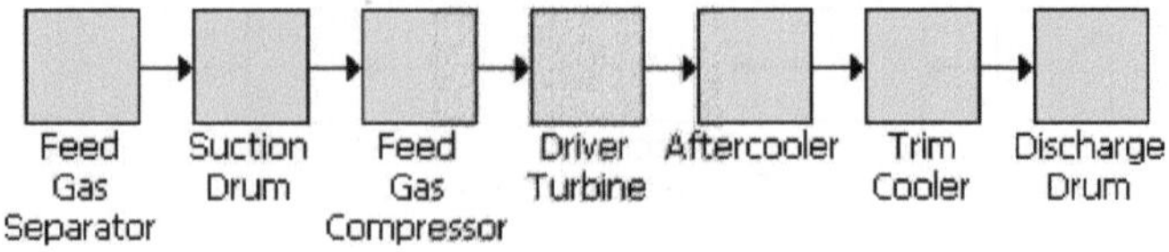

Figure 8.4 RBD of the feed-gas compression unit

failure the plant will continue to function with the remaining stream. As a result, the properties of the node are set to require only one of the two paths coming into the node to be operational.

8.6.1.3 Feed Gas Compression Unit

The combined gas streams from the MP separation and compression unit and the HP separation unit are sent to the feed-gas compressor through the feed-gas separator and the suction drum (see Figure 8.4). The feed-gas compressor is driven by a steam turbine. After compression the gas is cooled in the aftercooler and further cooled using cooling water in the trim cooler. The compressed feed gas is finally sent to the acid-gas removal unit after passing through the discharge drum.

Any condensate or water removed in the feed-gas compression unit is sent to the condensate treatment unit and the water treatment unit respectively. As stated earlier, these units are represented by a single non-failing block in Figure 8.1. The block collectively represents condensate and water removed from the MP separation and compression unit, the HP separation unit, and the feed-gas compression unit. It is assumed that the condensate and water account for 5% of the total gas volume entering the facility. As a result, the throughput of the RBD is split up and five units of throughput are directed to the condensate and water treatment block while 95 units (representing 95 MMSCF per day of gas) go to the acid-gas removal block.

8.6.1.4 Acid-gas Removal Unit

The next step in natural-gas processing is to remove the highly corrosive hydrogen sulfide and carbon dioxide gases called *acid gases*. It is assumed that in this case the carbon dioxide composition of the natural gas is within the pipeline specifications. However, the hydrogen sulfide content is 3% and requires treatment. Natural gas consisting of a significant amount of hydrogen sulfide is termed *sour gas*, while the gas free from these impurities is called *sweet gas*.

The removal of hydrogen sulfide is done by bringing the sour gas in contact with an amine solution in a tower called the absorber. Gas from the feed-gas compression unit reaches the absorber after passing through the sour-gas knock-out drum and the filter separator (see Figure 8.5). The knock-out drum separates any entrained liquid while the filter separator removes ultra fine liquid and solid particles to prevent

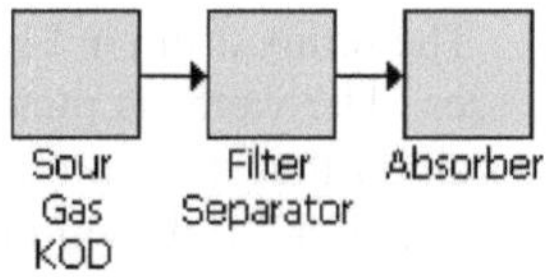

Figure 8.5 RBD of the acid-gas removal unit

contamination of the amine solution. In the absorber, the amine solution absorbs the hydrogen sulfide and sweet natural gas is removed from the top of the vessel. The sweet gas is sent to the dehydration unit while the amine solution is sent to the amine regeneration and sulfur recovery units (SRUs). As a result of the removal of hydrogen sulfide from the gas, 3 units of throughput are sent to the amine regen and sulfur recovery block while the remaining 92 units of throughput (representing 92 MMSCF per day of gas) move on in the RBD to the dehydration block.

8.6.1.5 Amine Regeneration and Sulfur Recovery Units

The amine solution from the absorber containing hydrogen sulfide is called *rich amine* while the regenerated amine is called *lean amine*. The rich amine is sent for regeneration to the amine regeneration unit so that it can then be reused in the absorber. The rich-amine flash drum is used to remove any entrained gas from the rich amine (see Figure 8.6). The rich amine then goes through the rich-amine/lean-amine exchanger where it gets preheated by the regenerated lean amine. Then the rich amine is sent to the regenerator. The overhead gas from the regenerator is mostly hydrogen sulfide. This gas stream is sent to the SRUs through the overhead condenser and the reflux drum. The SRUs convert hydrogen sulfide into elemental sulfur using the Claus process (Kohl and Nielsen 1997). The units are modeled as a single non-failing block for this analysis. The sulfur from the SRUs is sent for storage while the gas is sent to the tail-gas treatment units (TGTUs) and then incinerated. Again, the TGTUs are modeled using a single non-failing block.

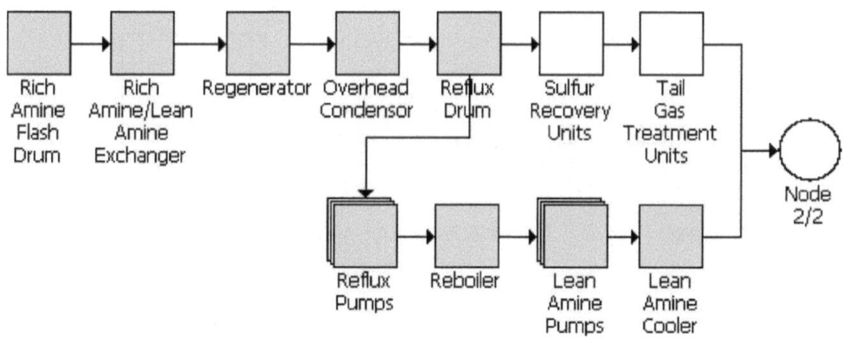

Figure 8.6 RBD of the amine regeneration and sulfur recovery units

The reflux drum in Figure 8.6 separates the reflux water and water-saturated acid gases. The water is pumped back to the regenerator using the reflux pumps. It is assumed that there are two full-capacity reflux pumps (2 × 100%). As a result, a "multiblock" representing two blocks in parallel is used to model the pumps in Figure 8.6.

The reboiler is an exchanger that provides steam to heat and strip the amine from the regenerator to a lean condition. The lean amine is pumped back to the absorber using the lean-amine pumps after going through the lean-amine cooler. It is assumed that two lean-amine pumps are used. Each of these pumps is half-capacity (2×50%) and thus both the pumps need to be in an operational state. In the RBD of Figure 8.6, a multiblock representing two blocks in parallel but requiring both blocks to be functional (2-out-of-2 configuration) is used to model the lean-amine pumps. A 2/2 node is used to specify that both the paths coming into the node (the sulfur recovery path and the amine regeneration path) are needed to be in an operating condition for the plant to function.

8.6.1.6 Dehydration Unit

The dehydration unit removes water vapor from the natural-gas stream using molecular-sieve adsorption beds. The sweet gas from the acid-gas removal unit is first cooled in the sweet-gas cooler and then sent to the sweet-gas knock-out drum. The drum removes any entrained amine thereby preventing downstream problems in processing the treated gas. The gas is then sent to the feed-gas prechiller where it is cooled. The condensed liquids are separated in the feed-gas separator. The treatment of these liquids is excluded from the present analysis. The gas from the feed-gas separator is sent to the inlet-gas filter to prevent liquid contaminants from entering the molecular-sieve beds. As the wet natural gas passes through these beds, water vapor is adsorbed and dry gas is obtained. After the beds get saturated they are regenerated using heated residue gas. A number of beds are used so that some beds are on-line while the others are being regenerated.

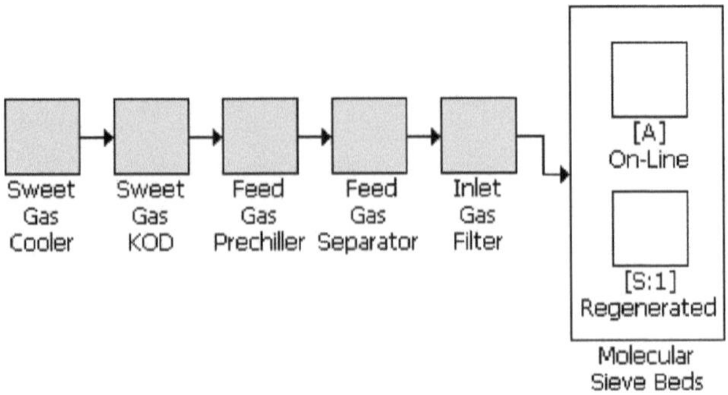

Figure 8.7 RBD of the dehydration unit

The functioning of the molecular-sieve beds can be modeled in BlockSim using the standby container construct as shown in Figure 8.7. The container allows blocks to be specified as active or standby. Thus, the on-line beds can be modeled as active blocks and the regenerated beds can be modeled as standby blocks. Once an on-line bed is saturated, BlockSim will automatically switch to a regenerated bed. The time to regenerate the beds can be modeled as maintenance downtime. A switch-delay time is also available to model any other standby or delay times associated with the beds.

8.6.1.7 Natural-gas Liquids Recovery Unit

Dry natural gas from the dehydration unit is sent to the NGL recovery unit to separate out the NGLs. This is done by the cryogenic expansion process using a turbo expander. As shown in Figure 8.8, the gas is first sent to the feed-gas filter to ensure that molecular-sieve particles are not carried over along with the gas. The gas then goes through the feed-gas/residue-gas exchanger where it is cooled by residue gas from the demethanizer. Condensed liquids are separated in the feed-gas separator and the gas goes to the feed-gas chiller where it is cooled using propane refrigeration. Condensed liquids are separated in the cold separator. The gas then goes to the turbo expander where it expands rapidly, causing the temperature to drop significantly. This rapid temperature drop condenses ethane and other hydrocarbons. These NGLs are separated in the demethanizer as the bottom product. The overhead residue gas obtained from the demethanizer is the processed natural gas. The energy released during the expansion of the gas in the turbo expander is used to drive the expander compressor (see Figure 8.1). The compressor compresses the residue gas, which finally goes to the sales-gas pipeline.

The turbo expander of the natural-gas plant is assumed to have a Joule–Thomson (JT) bypass valve in case the expander goes off-line. This setup can be modeled in BlockSim using the standby container construct (see Figure 8.8). The turbo ex-

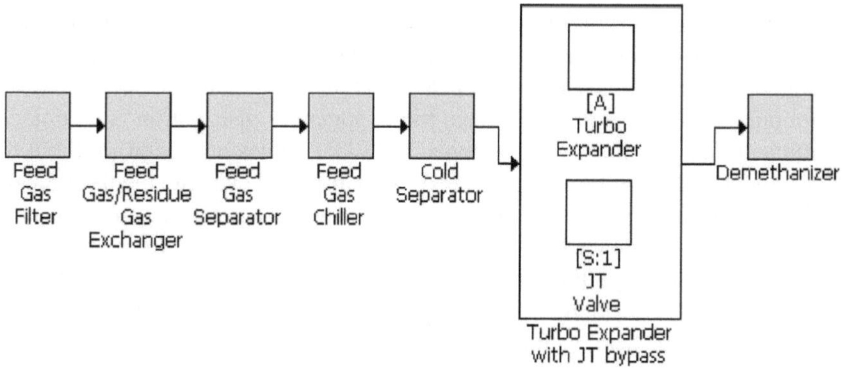

Figure 8.8 RBD of the NGL recovery unit

pander is modeled as the active block of the container, while the JT valve is modeled as the standby block. In the case of failure of the turbo expander, the container will switch to the JT valve.

Note that for the present analysis no failure properties have been assigned to the turbo expander setup. Some of the equipment associated with the demethanizer (such as pumps, reflux drum, and reboiler) and equipment related to propane refrigeration also have not been included in the analysis.

It is assumed that the constitution of NGL in the natural gas is 7%. As a result, seven units of throughput are directed to the NGL block (see Figure 8.1) while the remaining 85 units (representing 85 MMSCF of gas per day) move on in the RBD as the processed natural gas.

8.6.1.8 Residue Gas Compression

Processed natural gas from the NGL recovery unit is compressed by the expander compressor and further compressed by the residue-gas compressor, cooled in the aftercooler and sent to the sales-gas pipeline (see Figure 8.1). The sales-gas block is a non-failing block that is used to track the throughput of this gas.

A node is used as the junction of the four paths coming from the condensate and water treatment block, amine regen and sulfur recovery block, NGL block, and sales-gas block. Since all of these paths are critical for the functioning of the natural-gas processing plant, a setting of four required paths is used on the node.

8.6.2 Failure and Repair Data

Most process plants have computerized maintenance management systems or CMMS (Mather 2003) that can be used to obtain historical performance data for various equipment. Statistical fit to the data can be performed using maximum likelihood estimation or regression techniques (Meekar and Escober 1998). The two-parameter Weibull distribution is selected in this study to model equipment failure as the Weibull distribution is a flexible distribution that can model increasing, decreasing, or constant failure rates. Table 8.1 lists the shape and scale parameters for the equipment failures. For repair data, the exponential distribution is assumed to be sufficient and the mean time to repair (MTTR) values are listed in Table 8.1. Software such as Weibull++ can be used to perform and evaluate the fit of the distributions (Herder *et al.* 2008; ReliaSoft 2005). The parameter values listed in the table are at 50% confidence level and the variation in the parameters is not included in the analysis. Although the uncertainty associated with parameters is ignored in the analysis, this is considered acceptable for the present case. Note that due to proprietary nature of the data, values presented in Table 8.1 do not represent equipment data from an actual natural-gas plant.

Table 8.1 Failure and repair data used for the analysis

Equipment	Failure			Repair	
	Distribution	Parameters		Distribution	Parameter
		Shape	Scale (Days)		MTTR (Days)
Separators, Slug Catcher	Weibull	1.5	2700	Exponential	5
Suction Drums, Discharge Drums, Reflux Drums	Weibull	1.5	2300	Exponential	3
Compressors	Weibull	1.4	650	Exponential	0.75
Motors	Weibull	1.1	700	Exponential	0.5
Turbines	Weibull	1.4	750	Exponential	0.65
Pumps	Weibull	1.2	500	Exponential	0.55
Exchangers, Condensors, Aftercoolers, Trim Coolers, Chillers, Reboilers	Weibull	1.5	12000	Exponential	6
Absorbers, Regenerators	Weibull	1.5	5000	Exponential	7
Filters	Weibull	1.5	2500	Exponential	2
Demethanizer	Weibull	1.5	7000	Exponential	15

8.6.3 Phase Diagram and Variable Throughput

Natural-gas plants have periodic overhauls (referred to as *turnarounds* or *shutdowns*) during which all production is stopped and preventive maintenance actions are carried out on the equipment to minimize failure occurrences during days of normal production. The overhauls also provide opportunities to carry out corrective maintenance on *hidden* or *degraded* failures. Hidden failures are equipment failures that do not cause any loss of production. Degraded failures are failures during which the equipment continues to function with a lower rate of production. These failures may not be corrected until a major overhaul of the facility in order to avoid disruption of production during normal production periods. Periodic overhauls of plants can be modeled in BlockSim using phase diagrams (see Figure 8.9).

After a total shutdown of the plant during a periodic overhaul, normal production is not resumed immediately. Instead, the facility is slowly ramped up to full produc-

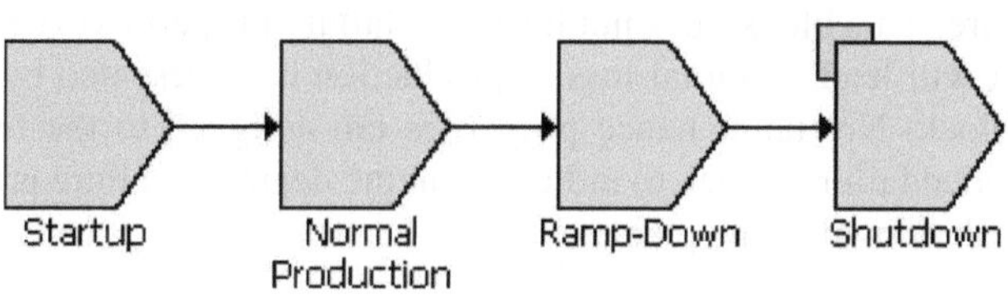

Figure 8.9 Reliability phase diagram for the natural-gas plant

tion over a period of a few days. This variation in production can be modeled in BlockSim using the variable throughput option available with the phase diagrams.

Figure 8.9 illustrates the application of phase diagrams along with the use of variable throughput for the natural-gas facility under consideration. It is assumed that the facility undergoes a periodic overhaul of 15 days every 3 years. After the shutdown, the facility takes 5 days to ramp up to normal production. The first block in the phase diagram, startup, represents this period. It is assumed that the ramping up of production is linear and can be modeled using the equation $y = 20x$. After the startup phase, the facility begins normal production for a period of 1073 days. This is modeled using the normal production block. The facility is then prepared for the upcoming shutdown by ramping down the production over a period of 2 days. This is represented using the ramp-down block, assuming a linear decrease of production following the equation $y = 100-50x$. The final phase is represented by the shutdown block during which the facility is shut down and there is no production.

8.6.4 Hidden and Degraded Failures Modeling

As mentioned previously, equipment in a natural-gas plant may experience hidden or degraded failures in addition to the usual failures that lead to a total loss of production.

Modeling of hidden failures in BlockSim is illustrated using the reflux pumps of the amine regeneration and sulfur recovery units. Recall that the two pumps are in a parallel configuration. Therefore, failure of one of the pumps will not cause any loss of production and is a hidden failure. To model the hidden failure a "corrective maintenance policy" of "upon inspection" can be specified for the two pumps (see Figure 8.10). This means that the failure will only be discovered when an inspection is carried out on the pump. The frequency of these inspections is specified as 30 days for the present analysis.

Figure 8.11 illustrates one of the ways used by practitioners to model degraded failures. It is assumed that the compressor of the feed-gas compression unit may undergo a failure mode, as a result of which it functions at a degraded level of 90% production. This failure is assumed to be a random failure that occurs with a mean time of 1300 days. The feed-gas compressor (degraded failure) block in Figure 8.11 models this failure mode. The throughput of this block is 10 units, representing a loss of 10 MMSCF of production per day. The feed-gas compressor (degraded production) block represents the production that is continued after the occurrence of the degraded failure. This block does not have any failure properties. The original failure mode that will lead to a total loss of production is represented by the feed-gas compressor block. No maintenance properties are assigned to the feed-gas compressor (degraded failure) block to indicate that the degraded failure is not corrected until the next shutdown of the plant.

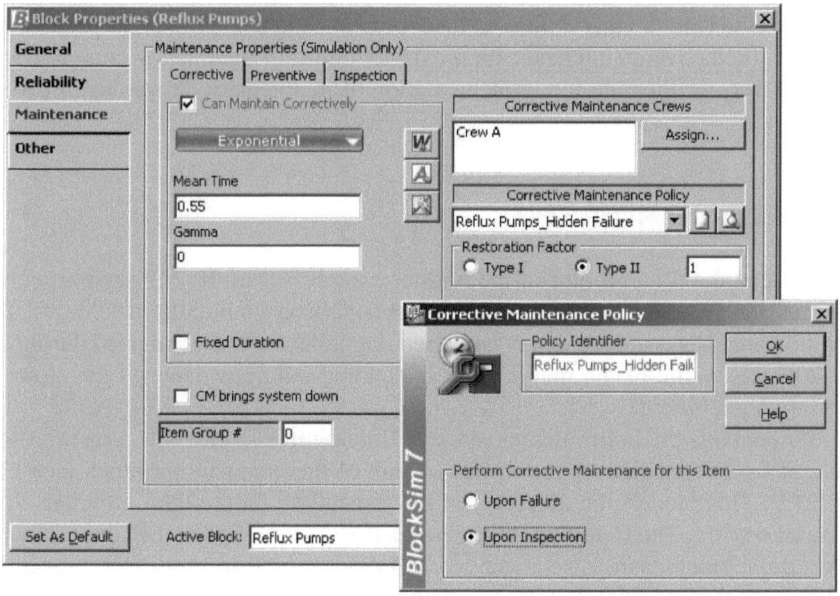

Figure 8.10 Modeling hidden failure of the reflux pumps

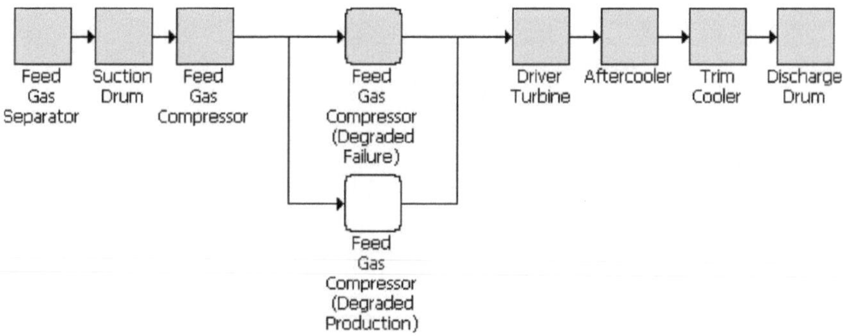

Figure 8.11 Modeling degraded failure of the feed-gas compressor

8.6.5 Maintenance Modeling

The following paragraphs explain the maintenance models used in the present analysis.

8.6.5.1 Normal Production

It is assumed that only corrective maintenance is carried out during the production periods. Preventive maintenance actions are not carried out at these times as these actions would result in disruption of plant operation and result in loss of production. However, predictive maintenance on the driver motor of the booster compressor is modeled next to illustrate the implementation of maintenance strategies in BlockSim.

8.6.5.2 Predictive Maintenance

Assume that the driver motor of the booster compressor of the MP separation and compression unit is subjected to vibration analysis every 6 months (180 days). The vibration analysis is able to detect an impending failure if it is conducted during the last 20% of the life of the motor. If an impending failure is detected the motor is preventively replaced.

To model this predictive maintenance in BlockSim, a failure detection threshold value of 0.8 is specified on the inspection tab of the driver motor block (see Figure 8.12). This models the detection of the impending failure during the last 20% of the motor life. The frequency of the inspections is specified as 180 days. Finally, a preventive maintenance frequency of 180 days is used to model the preventive replacement that is initiated if the results from the vibration test are positive.

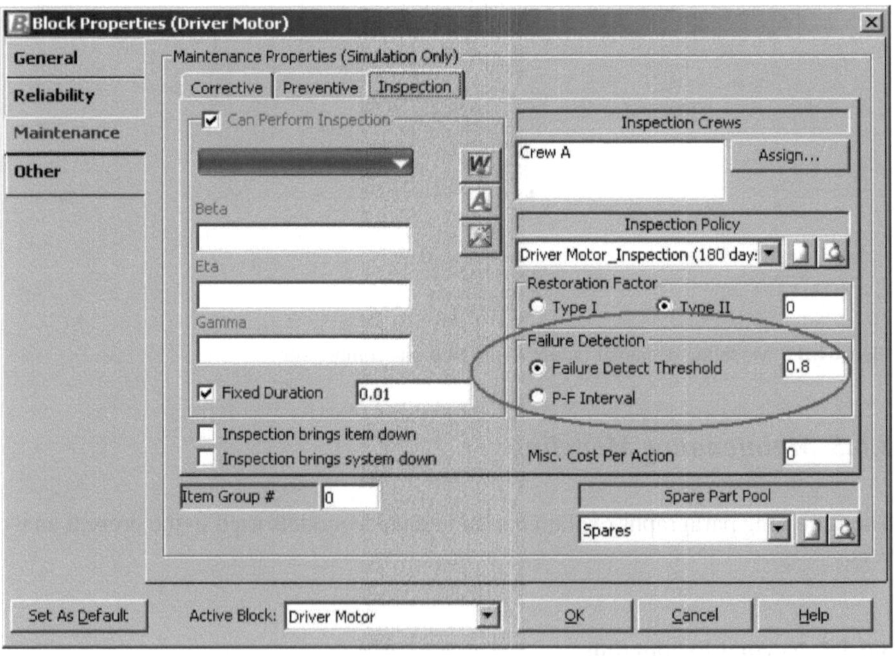

Figure 8.12 Modeling predictive maintenance of the MP compressor's driver motor

8.6.5.3 Complex Maintenance Strategies

It is realized that the predictive maintenance carried out on the driver motor of MP separation and compression unit is not an efficient strategy as the MP gas processing stops every time the motor is replaced. It is decided to carry out the predictive maintenance only when the booster compressor associated with the driver motor fails. By doing this, no additional disruption of the plant is caused if the vibration analysis indicates an impending failure and the motor has to be replaced.

To model this maintenance scenario in BlockSim, the booster compressor and the driver motor are first linked together by specifying a common item group number for both pieces of equipment (i.e., Item Group # = 1 as shown in Figure 8.13). Then the frequency of inspections is changed to the "upon maintenance of another group item" option. This models the fact that vibration tests are done on the motor only when corrective maintenance is performed on the compressor. Finally the preventive maintenance frequency is also changed to the "upon maintenance of another group item" option to model the preventive replacement that is initiated if the results from the vibration test are positive.

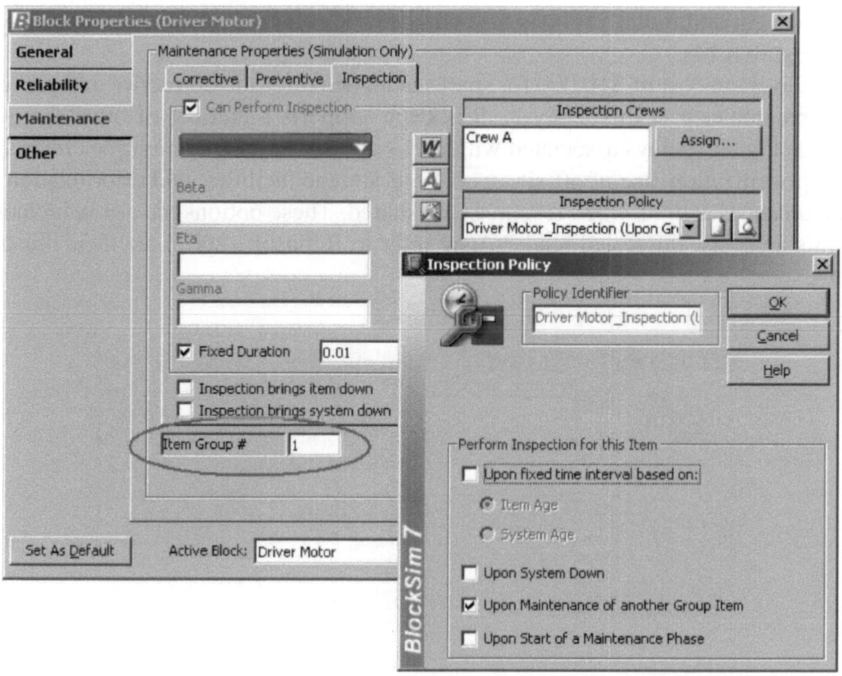

Figure 8.13 Modeling complex maintenance strategies

8.6.5.4 Shutdown

During the shutdown period preventive maintenance is carried out on all equipment. It is assumed that the preventive maintenance actions restore equipment by 90%. As a result, a type II restoration factor of 0.9 is used in BlockSim (for details refer to Reliasoft 2007). In addition to the preventive maintenance actions on all equipment, corrective maintenance is carried out on any equipment that enters the shutdown period in a failed state (such as the equipment failing in degraded mode).

8.6.6 Crews and Spares Resources

To illustrate the modeling of crew resources it is assumed that two different crews are available to perform maintenance work on the natural-gas plant. Crew A is an internal crew that is used to perform maintenance actions during normal production periods. The charges incurred by this crew are $10,000 per day. Maintenance actions during the shutdown period are performed by Crew B, an external contractor called only during the periods of shutdown. This crew charges $13,000 per day for their services. An additional $5000 is also charged for every call answered by this crew (see Figure 8.14).

An average cost of $10,000 for spares is modeled for the corrective and preventive maintenance in this analysis. BlockSim supports a number of other features such as logistic delays associated with crews and spares, use of multiple crews and their prioritization, use of off-site spare parts storage facilities, and prioritization of maintenances actions when resources are shared. These options are not in included in this analysis. Interested readers may refer to Reliasoft (2007) for their illustrations.

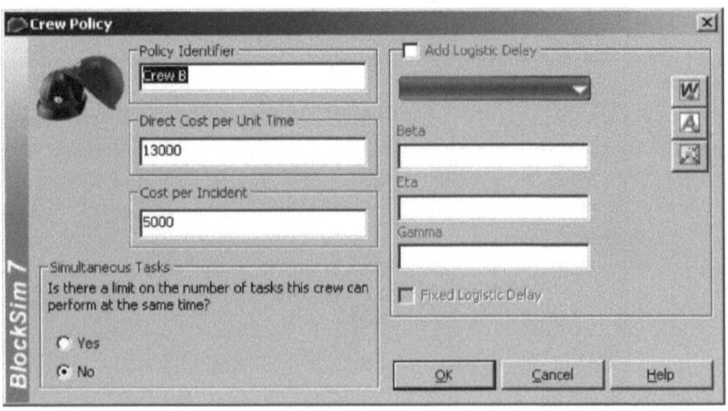

Figure 8.14 Modeling crew resources

8.6.7 Results

Five hundred simulations are run on the natural-gas plant model using an end time of 1095 days to simulate the behavior of the plant for a period of 3 years. The results are explained next. The discussion presented is limited to mean values, the procedure to obtain confidence bounds is available in ReliaSoft (2009).

The availability of the plant at the end of the 3-year period is predicted to be 96.38% (see Figure 8.15). Another metric more relevant to oil and gas, petrochemicals and other process industries is the *production efficiency*. Production efficiency is the ratio of the actual (simulated) production to the ideal production (when the plant is assumed to have no downtime). The actual natural-gas production is known by looking at the throughput of the sales-gas block. This is obtained as 86,273 MMSCF. The ideal production is obtained by running a simulation on the model having no failure and maintenance properties for any of the equipment. The ideal production for the sales-gas block is obtained as 91,503 MMSCF. Therefore the production efficiency is 94.3% and there is an expected production loss of 5230 MMSCF of gas during the 3-year plant operation.

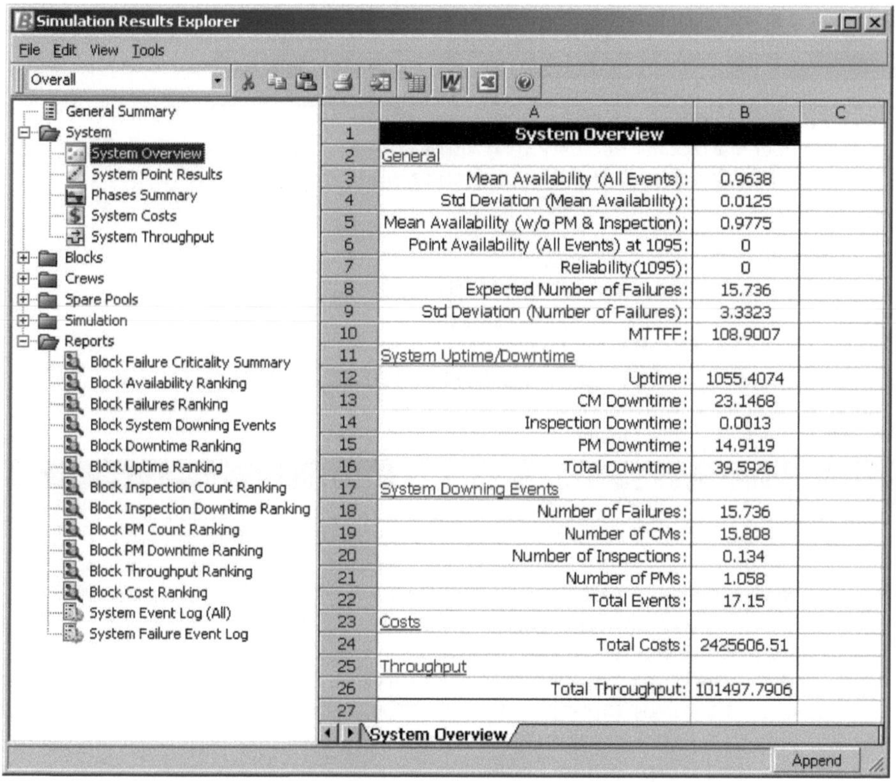

Figure 8.15 Expected availability for 3 years of operation

8.6.8 Bad Actors Identification

Table 8.2 shows a portion of the block failure criticality summary report obtained from BlockSim. The report ranks equipment by RS FCI (ReliaSoft's failure criticality index), which is the percentage of times the system failure was caused by the failure of the particular equipment. Therefore, the top rankers of this table are responsible for the largest losses to the plant availability. The table shows that the top five bad actors are the two lean-amine pumps (that operate in a 2-out-of-2 configuration), the feed-gas compressor, the driver motor of the residue gas compressor and the expander compressor. From the layout of the plant RBD, it can be seen that all these equipment are single point failures. They operate in a series configuration and their failure leads to a disruption of the operation of the plant.

Since the interest in natural-gas plants is in loss of production, an additional metric to look at for the present analysis is equipment downtime. Equipment downtime is directly linked to loss of production. It may or may not be tied to plant availability depending on whether or not the equipment is a single point failure. Table 8.3 shows a portion of BlockSim's block downtime ranking report. The report identifies

Table 8.2 Failure criticality ranking

Block Failure Criticality Summary	
Block Name	**RS FCI**
Amine Regen and Sulfur Recovery:Lean Amine Pumps	13.45%
Amine Regen and Sulfur Recovery:Lean Amine Pumps	12.94%
Feed Gas Compression:Feed Gas Compressor	9.80%
Driver Motor	9.76%
Expander Compressor	9.75%
Residue Gas Compressor	9.74%
Feed Gas Compression:Driver Turbine	8.43%
Dehydration:Sweet Gas KOD	2.07%
NGL Recovery:Feed Gas Filter	2.00%
Amine Regen and Sulfur Recovery:Rich Amine Flash Drum	1.88%

Table 8.3 Block downtime ranking

Block Downtime Ranking	
Block Name (Diagram)	**Block Downtime**
Feed Gas Compressor (Degraded Failure) {Feed Gas Compression}	355.0931
Reflux Pumps {Amine Regen and Sulfur Recovery}	32.9369
Reflux Pumps {Amine Regen and Sulfur Recovery}	32.4961
Feed Gas Separator {Deydration}	1.6778
Slug Catcher {HP Separation}	1.3258
Feed Gas Separator {NGL Recovery}	1.2206
Lean Amine Pumps {Amine Regen and Sulfur Recovery}	1.2128
Cold Separator {NGL Recovery}	1.1861
Lean Amine Pumps {Amine Regen and Sulfur Recovery}	1.159
Residue Gas Compressor	1.1512

the degraded failure of the feed-gas compressor as the cause of the largest equipment downtime in the plant. This result can be explained by the fact that the model assumed that the degraded failure is not corrected until the next shutdown. Due to the large downtime it can be concluded that the degraded failure of the feed-gas compressor is responsible for the largest loss in production. Downtime of the reflux pumps shown in the table is not as significant because these pumps operate in a parallel configuration.

8.6.9 Cost Analysis

8.6.9.1 Maintenance

For the present analysis, the maintenance cost for running the natural-gas plant for a period of 3 years consists of the cost of maintenance actions carried out by the repair crews and the cost of spares. The overall cost for the 3-year operation of the plant is available in the system cost summary table (Table 8.4) as $2,425,607. A major portion of this cost, approximately 76%, is incurred by preventive maintenance actions.

The total cost can also be broken down as crew cost ($1,753,387) and spares cost ($677,920). The crew summary table gives the break-down of the crew cost by the type of crew used (see Table 8.5). Crew A, the internal crew, costs $330,286 while crew B, the external contractor, costs $1,423,101. It can be seen that the charges of crew B are much higher than that of crew A. Although the calls answered by this

Table 8.4 System cost summary

System Cost Summary		
Misc. Corrective Costs:	0	
Costs for Parts (CM):	240240	
Costs for Crews (CM):	324155.8	
Total CM Costs:		564395.8
Misc. Preventive Costs:	0	
Costs for Parts (PM):	431980	
Costs for Crews (PM):	1422072	
Total PM Costs:		1854052
Misc. Inspection Costs:	0	
Costs for Crews (IN):	7158.2	
Total Inspection Costs:		7158.2
Indirect Pool Costs:		0
Total Costs:		2425607

Table 8.5 Crew summary

Maintenance Crew Summary					
Crew Policy	Calls	Accepted	Rejected	Time Used	Cost
Crew A	95.706	95.706	0	33.0286	330285.9
Crew B	43.098	43.098	0	92.8931	1423101

crew are almost half that of crew A, the duration of the calls is almost three times that of crew A.

8.6.9.2 Production Loss Cost

It was seen previously that the expected production efficiency of the natural-gas plant for the 3 years of operation is 94.3%. Production loss of 5.7%, translating to 5230 MMSCF of natural gas, occurs due to downtime of various equipment of the plant. Assuming that the cost of natural gas is $5 per million BTUs and that there are 1030 BTUs in one cubic foot of gas, the cost of the lost production is $26,934,500.

8.6.9.3 Total Losses

The total revenue loss due to various equipment failures for the 3-year plant operation can be obtained by addition of the maintenance cost ($2,425,607) and cost of lost production ($26,934,500). This comes out to be $29,360,107.

8.6.9.4 Life Cycle Costs

A life cycle costs (LCC) analysis can be performed at this point, based on the maintenance and production loss costs. In addition to these costs, a complete LCC analysis will include acquisition costs and other capital expenses (Kumar *et al.* 2000). These can easily be added to the costs computed here, but are beyond the focus of this chapter and thus omitted.

8.6.10 Sensitivity Analysis

After a review of the analysis results a number of recommended actions are usually put forward to improve plant operation and decrease cost. A sensitivity analysis can be conducted to study the expected effect as a result of the implementation of the recommended actions. Assume that a number of modifications to the feed-gas compressor are recommended that will result in a decrease in the occurrence

Table 8.6 Sensitivity analysis

Item	Availability (%)	Production Efficiency (%)	Production Loss MMSCF	Production Loss ($)	Maintenance Cost ($) Crews	Maintenance Cost ($) Spares	Cost Savings ($)
Base Case	96.38	94.29	5230	26,934,500	1,753,387	677,920	0
Feed Gas Compressor	96.42	96.24	3445	17,741,750	1,755,254	676,640	9,192,163
Lean Amine Pumps	96.62	94.52	5017	25,837,550	1,756,534	673,120	1,098,603
Expander Compressor	96.54	94.67	4879	25,126,850	1,734,044	662,400	1,842,513
Crew	96.38	94.29	5230	26,934,500	1,264,494	677,920	488,893
All	96.75	96.67	3049	15,702,350	1,245,499	656,460	11,761,498

of the degraded failure. With these modifications it is expected that the degraded failure will occur with a mean time of 4000 days instead of the previous 1300 days. Table 8.6 shows the analysis results when the model is simulated with this change. It can be seen that there is a marked increase in production efficiency while the availability increase is comparatively smaller.

Similarly, it is decided to investigate the effect of having two full-capacity lean amine pumps instead of the two half-capacity pumps. Modifications of the expander compressor are also proposed that will lead to a failure distribution with the original shape parameter of 1.4 but a new scale parameter of 2000 days instead of the original 650 days. A proposal to use the internal crew for all maintenance activities instead of using the external contractor crew is also considered. Table 8.6 summarizes the results of these scenarios. The last row represents the expected performance when all the four actions are implemented. A similar approach can be used to investigate the effect of expanding existing facilities, using different maintenance strategies and many other "what-if" scenarios.

8.7 Conclusion

This chapter demonstrates the application of RAM analysis to process industries by providing a case study of a natural-gas processing plant. The approach employed in the chapter can be used to compare maintenance strategies, evaluate equipment performance, decide on appropriate spare inventory levels, plan for manpower requirements, prepare for turnarounds and evaluate their effectiveness, predict production, and obtain assistance in budgeting. The approach can also play an important role in the expansion of existing plants and in the design of new plants. As demonstrated by Sunavala (2008), RAM is becoming an important part of the front-end engineering design phase of these plants. RAM can also be used to drive further analysis related to reliability and availability, for example, RAM can be integrated into process synthesis for chemical industries (Yin and Smith 2008) or it can be integrated in a non-linear model to optimize availability (Calixto and Rocha 2007). The applications of RAM are many and it is set to become a standard reliability tool in the process industry. It is hoped that the concepts and techniques illustrated in the chapter would help spawn new ideas to tackle real-world problems faced by practitioners of various industries particularly the process industry.

References

Armstadter BL (1971) Reliability mathematics: fundamentals, practices, procedures. McGraw-Hill Book Company, New York

Bahr NJ (1997) System safety engineering and risk assessment: a practical approach. Taylor & Francis, Washington

Brall A, Hagen W, Tran H (2007) Reliability block diagram modeling – comparisons of three software packages. In Reliability and maintainability symposium. IEEE, Piscataway, NJ, pp 119–124

Calixto E, Rocha R (2007) The non-linear optimization methodology model: the refinery plant availability optimization study case. In: Proceedings of the European safety and reliability conference 2007 – Risk, reliability and societal safety. Taylor and Francis, London, UK, pp 503–510

Fishman GS (1996) Monte Carlo: concepts, algorithms and applications. Springer, New York

Giuliano FA (ed) (1989) Introduction to oil and gas technology, 3rd edn. Prentice Hall, Englewood Cliffs, NJ

Herder PM, Van Luijk JA, Bruijnooge J (2008) Industrial application of RAM modeling. Development and implementation of a RAM simulation model for the Lexan® Plant at GE Industrial, Plastics. Reliab Eng Syst Saf 93(4):501–508

Kohl AL, Nielsen RB (1997) Gas purifications, 5th edn. Gulf Professional Publishing, Oxford

Kumar UD, Crocker J, Knezevic J, El-Haram M (2000) Reliability, maintenance and logistic support: a life cycle approach. Kluwer Academic Publishers, Boston, MA

Lee D, Nam K, Kim J, Min J, Chang K, Lee S (2004) RAM study on the subsea production system of an offshore oil and gas platform. Proceedings of the international offshore and polar engineering conference. ISOPE, Cupertino, CA, pp 514–519

Marquez AC, Heguedas AS, Iung B (2005) Monte Carlo-based assessment of system availability. A case study for cogeneration plants. Reliab Eng Syst Saf 88(3):273–289

Mather D (2003) CMMS: a timesaving implementation process. CRC Press, Boca Raton, FL

Meeker WQ, Escobar LA (1998) Statistical methods for reliability data. John Wiley & Sons, New York

Mobley RK (2002) An introduction to predictive maintenance, 2nd edn. Butterworth-Heinemann, New York

Moubray J (1997) Reliability-centered maintenance, 2nd edn. Industrial Press, New York

Peebles MWH (1992) Natural-gas fundamentals. Shell International Gas, London

Racioppi G, Monaci G, Michelassi C, Saccardi D, Borgia O, De Carlo F (2007) A methodology to assess the availability of a sour gas injection plant for the production of oil. In: Proceedings of the European safety and reliability conference 2007 – risk, reliability and societal safety, Vol 1, June. Taylor and Francis, London, UK, pp 543–549

ReliaSoft Corporation (2005) Life data analysis reference. ReliaSoft Publishing, Tucson, AZ

ReliaSoft Corporation (2007) System analysis reference: reliability, availability & optimization. ReliaSoft Publishing, Tucson, AZ

ReliaSoft Corporation (2009) An application of BlockSim's log of simulations. Reliab HotWire no 97, March. ReliaSoft Publishing, Tucson, AZ

Sikos L, Klemeš J (2009) RAMS contribution to efficient waste minimisation and management. J Clean Prod 17(10):932–939

Smith AM (1993) Reliability-centered maintenance. McGraw-Hill, New York

Sunavala KP (2008) The value of RAM. ABB Rev (Special report – Process automation services & capabilities):74–78

Suzuki T (ed.) (1994) TPM in process industries. Productivity Press, Portland, OR

Waeyenbergh G, Pintelon L, Gelders L (2000) JIT and maintenance. In: Ben-Daya M, Duffuaa SO, Raouf A (eds) Maintenance, modeling and optimization. Kluwer Academic Publishers, Boston, MA, pp 439–470

Wheeler RR, Whited M (1985) Oil – from prospect to pipeline, 5th edn. Gulf Publishing Company, Houston, TX

Wireman T (2004) Total productive maintenance. Industrial Press, New York

Yin QS, Smith R (2008) Incorporating reliability, availability and maintainability (RAM) into process synthesis. AIChE Annual Meeting, Conference Proceedings. AIChE, New York, NY

Zio E, Baraldi P, Patelli E (2006) Assessment of the availability of an offshore installation by Monte Carlo simulation. Int J Press Vessels Pip 83(4):312–320

Chapter 9
Potential Applications of Discrete-event Simulation and Fuzzy Rule-based Systems to Structural Reliability and Availability

A. Juan, A. Ferrer, C. Serrat, J. Faulin, G. Beliakov, and J. Hester

Abstract This chapter discusses and illustrates some potential applications of discrete-event simulation (DES) techniques in structural reliability and availability analysis, emphasizing the convenience of using probabilistic approaches in modern building and civil engineering practices. After reviewing existing literature on the topic, some advantages of probabilistic techniques over analytical ones are highlighted. Then, we introduce a general framework for performing structural reliability and availability analysis through DES. Our methodology proposes the use of statistical distributions and techniques – such as survival analysis – to model component-level reliability. Then, using failure- and repair-time distributions and information about the structural logical topology (which allows determination of the structural state from their components' state), structural reliability, and availability information can be inferred. Two numerical examples illustrate some potential applications of the proposed methodology to achieving more reliable and structural designs. Finally, an alternative approach to model uncertainty at component level is also introduced as ongoing work. This new approach is based on the use of fuzzy rule-based systems and it allows the introduction of experts' opinions and evaluations in our methodology.

A. Juan · J. Hester
Dept. of Computer Sciences, Multimedia and Telecommunication, IN3 – Open University of Catalonia, Spain

A. Ferrer · C. Serrat
Institute of Statistics and Mathematics Applied to the Building Construction, EPSEB – Technical University of Catalonia, Spain

J. Faulin
Dept. of Statistics and Operations Research, Public University of Navarre, Spain

G. Beliakov
School of Engineering and Information Technology, Deakin University, Australia

P. Faulin, A. Juan, S. Martorell, and J.E. Ramírez-Márquez (eds), *Simulation Methods for Reliability and Availability of Complex Systems*. © Springer 2010

9.1 Introduction

Some building and civil engineering structures such as bridges, wind turbines, and off-shore platforms are exposed to abrupt natural forces and constant stresses. As a consequence of this, they suffer from age-related degradation in the form of deterioration, fatigue, deformation, etc., and also from the effect of external factors such as corrosion, overloading, or environmental hazards. Thus, the state of these structures should not be considered constant – as often happens in structural literature – but rather as being variable through time. For instance, reinforced concrete structures are frequently subject to the effect of aggressive environments [29]. According to Li [18] there are three major ways in which structural concrete may deteriorate, namely: (1) surface deterioration of the concrete, (2) internal degradation of the concrete, and (3) corrosion of reinforcing steel in concrete. Of these, reinforcing-steel corrosion is the most common form of deterioration in concrete structures and is the main target for the durability requirements prescribed in most design codes for concrete structures [24]. In other words, these structures suffer from different degrees of resistance deterioration due to aggressive environments and, therefore, reliability problems associated with these structures should always consider the structure's evolution through time.

In this chapter we propose the use of non-deterministic approaches – specifically those based on discrete-event simulation (DES) and fuzzy rule-based systems – as the most natural way to deal with uncertainties in time-dependent structural reliability and availability (R&A) analysis. With this goal in mind, we first discuss why these approaches should be preferred to others in structural R&A issues, especially in those structures that can be considered time-dependent systems, i.e., sets of individual time-dependent components connected by an underlying logical topology, which allows determining the actual structural state from the components' states. We also review some previous works that promote the use of simulation techniques – mainly Monte Carlo simulation – in the structural reliability arena. Then, our DES approach is introduced and discussed. This approach can be employed to offer solutions to structural R&A problems in complex scenarios, i.e., it can help decision-makers develop more reliable and cost-efficient structural designs. Some potential applications of our approach to structural R&A analysis are illustrated through two numerical examples. Finally, an alternative approach for modeling component-level uncertainty is also proposed. This later approach relies upon the use of fuzzy rule-based systems, and in our opinion it represents a promising line of research in the structural reliability arena.

9.2 Basic Concepts on Structural Reliability

For any given structure, it is possible to define a set of limit states [23]. Violation of any of those limit states can be considered a structural failure of a particular magnitude or type and represents an undesirable condition for the structure. In this sense, Structural reliability is an engineering discipline that provides a series of concepts, methods and tools to predict and/or determine the reliability, availability and safety of buildings, bridges, industrial plants, off-shore platforms, and other structures, both during their design stage and during their useful life. Structural reliability should be understood as the structure's ability to satisfy its design goals for some specified time period. From a formal perspective, structural reliability is defined as the probability that a structure will not achieve each specified limit state (i.e., will not suffer a failure of certain type) during a specified period of time [30]. For each identified failure mode, the failure probability of a structure is a function of operating time, t, and may be expressed in terms of the distribution function, $F(t)$, depending on the time-to-failure random variable, T. The reliability or survival function, $R(t)$, which is the probability that the structure will not have achieved the corresponding limit state at time $t > 0$, is then given by $R(t) = 1 - F(t) = P(T > t)$. According to Petryna and Krätzig [26], interest in structural reliability analysis has been increasing in recent years, and today it can be considered a primary issue in civil engineering. From a reliability point of view, one of the main targets of structural reliability is to provide an assembly of components which, when acting together, will perform satisfactorily (i.e., without suffering critical or relevant failures) for some specified time period, either with or without maintenance policies.

9.3 Component-level Versus Structural-level Reliability

In most cases, a structure can be viewed as a system of components (or individual elements) linked together by an underlying logical topology that describes the interactions and dependencies among the components. Each of these components deteriorates according to an analytical degradation or survival function and, therefore, the structural reliability is a function of each component's reliability function and the logical topology. Thus it seems reasonable to assess the probability of failure of the structure based upon its elements' failure probability information [4, 19]. As noticed by Frangopol and Maute [9], depending on the structure's topology, material behavior, statistical correlation, and variability in loads and strengths, the reliability of a structural system can be significantly different from the reliability of its components. Therefore, the reliability of a structural system may be estimated at two levels: component level and system or structural level. At the component level, limit-state formulations and efficient analytical and simulation procedures have been developed for reliability estimation [25]. In particular, if a new structure will likely have some components that have been used in other structural designs, chances are that there will be plenty of available data; on the other hand, if a new structure uses components about which no historical data exists, then survival analysis methods, such as accelerated life testing, can be used to obtain information about component reliability behavior [22]. Also, fuzzy sets theory can be used as a natural and alternative way to model individual component behavior [14, 27]. Component failures may be modeled as ductile (full residual capacity after failure), brittle (no residual capacity after failure), or semi-brittle (partial residual capacity after failure). Structural-level analysis, on the other hand, addresses two types of issues: (1) multiple performance criteria or multiple structural states, and (2) multiple paths or sequences of individual component failures leading to overall structural failure. Notice that sometimes it will be necessary to consider possible interactions among structural components, i.e., to study possible dependencies among component failure-times.

9.4 Contribution of Probabilistic-based Approaches

In most countries, structural design must agree with codes of practice. These structural codes used to have a deterministic format and describe what are considered to be the minimum design and construction standards for each type of structure. In contrast to this, structural reliability analysis worries about the rational treatment of uncertainties in structural design and the corresponding decision making. As noticed by Lertwongkornkit *et al.* [17], it is becoming increasingly common to design buildings and other civil infrastructure systems with an underlying "performance-based" objective which might consider more than just two structural states (collapsed or not collapsed). This makes it necessary to use techniques other than just design codes in order to account for uncertainty on key random variables affecting structural behavior. According to other authors [20, 31], standards for structural design are basically a summary of the current "state of knowledge" but offer only limited information about the real evolution of the structure through time. Therefore, these authors strongly recommend the use of probabilistic techniques, which require fewer assumptions. Camarinopoulos *et al.* [3] do also recommend the use of probabilistic methods as a more rational approach to deal with safety problems in structural engineering. In their words, "these [probabilistic] methods provide basic tools for evaluating structural safety quantitatively."

9.5 Analytical Versus Simulation-based Approaches

As Park *et al.* [25] point out, it is difficult to calculate probabilities for each limit-state of a structural system. Structural reliability analysis can be performed using analytical methods or simulation-based methods [19]. A detailed and up-to-date description of most available methods can be found at [5]. On one hand, analytical methods tend to be complex and generally involve restrictive simplifying assumptions about structural behavior, which makes them difficult to apply in real scenarios. On the other hand, simulation-based methods can also incorporate realistic structural behavior [2, 15, 20]. Traditionally, simulation-based methods have been considered to be computationally expensive, especially when dealing with highly reliable structures [21]. This is because when there is a low failure rate, a large number of simulations are needed in order to get accurate estimates – this is usually known as the "rare-event problem." Under these circumstances, use of variance reduction techniques (such as importance sampling) are usually recommended. Nevertheless, in our opinion these computational concerns can now be considered mostly obsolete due to outstanding improvement in processing power experienced in recent years. This is especially true when the goal – as in our case – is to estimate time-dependent structural R&A functions, where the rare-event problem is not a major issue.

9.6 Use of Simulation in Structural Reliability

There is some confusion in structural reliability literature about the differences between Monte Carlo simulation and DES. They are often used as if they were the same thing when, in fact, they are not [16]. Monte Carlo simulation has frequently been used to estimate failure probability and to verify the results of other reliability analysis methods. In this technique, the random loads and random resistance of a structure are simulated and these simulated data are then used to find out if the structure fails or not, according to predetermined limit states. The probability of failure is the relative ratio between the number of failure occurrences and the total number of simulations. Monte Carlo simulation has been applied in structural reliability analysis for at least three decades now. Fagan and Wilson [6] presented a Monte Carlo simulation procedure to test, compare, and verify the results obtained by analytical methods. Stewart and Rosowsky [29] developed a structural deterioration reliability model to calculate probabilities of structural failure for a typical reinforced concrete continuous slab bridge. Kamal and Ayyub [13] were probably the first to use DES for reliability assessment of structural systems that would account for correlation among failure modes and component failures. Recently, Song and Kang [28] presented a numerical method based on subset simulation to analyze the reliability sensitivity. Following Juan and Vila [12], Faulin *et al.* [7], and Marquez *et al.* [21], the basic idea behind the use of DES in structural reliability problems is to model uncertainty by means of statistical distributions which are then used to generate random discrete events in a computer model so that a structural lifetime is generated by simulation. After running some thousands or millions of these structural lifetimes, which can be attained in just a few seconds with a standard personal computer, confidence interval estimates can be calculated for the desired measures of performance. These estimates can be obtained using inference techniques, since each replication can be seen as a single observation randomly selected from the population of all possible structural lifetimes. Notice that, apart from obtaining estimates for several performance measures, DES also facilitates obtaining detailed knowledge on the lifetime evolution of the analyzed structure.

9.7 Our Approach to the Structural Reliability Problem

Consider a structure with several components which are connected together according to a known logical topology, that is, a set of minimal paths describing combinations of components that must be operating in order to avoid a structural failure of some kind. Assume also that time-dependent reliability/availability functions are known at the component-level, i.e., each component failure- and/or repair-time distribution is known. As discussed before, this information might have been obtained from historical records or, alternatively, from survival analysis techniques (e.g., accelerated life tests) on individual components. Therefore, at any moment in time the structure will be in one of the following states: (1) perfect condition, i.e., all components are in perfect condition and thus the structure is fully operational; (2) slight damage, i.e., some components have experienced failures but this has not affected the structural operability in a significant way; (3) severe damage, i.e., some components have failed and this has significantly limited the structural operability; and (4) collapsed, i.e., some components have failed and this might imply structural collapse. Notice that, under these circumstances, there are three possible types of structural failures depending upon the state that the structure has reached. Of course, the most relevant – and hopefully least frequent – of these structural failures is structural collapse, but sometimes it might also be interesting to be able to estimate the reliability or availability functions associated with other structural failures as well. To attain this goal, DES can be used to artificially generate a random sample of structural lifecycles (Figure 9.1).

In effect, as explained in [8], component-level failure- and repair-time distributions can be used to randomly schedule component-level failures and repairs. Therefore, it is possible to track the current state of each individual component at each

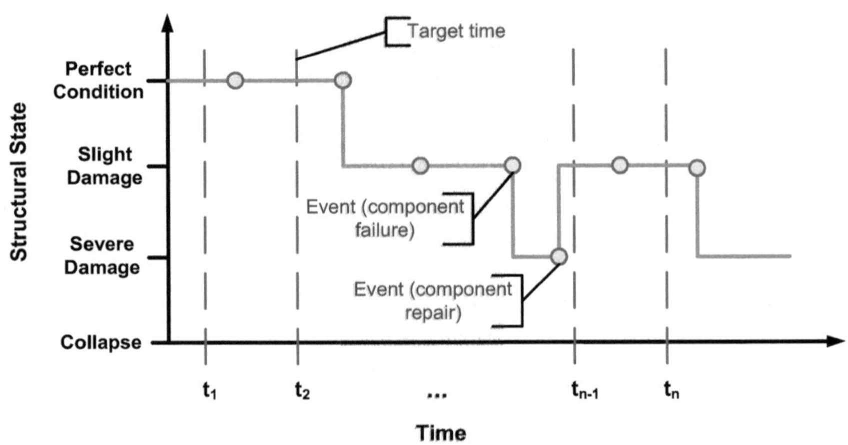

Figure 9.1 Using DES to generate a structural lifecycle

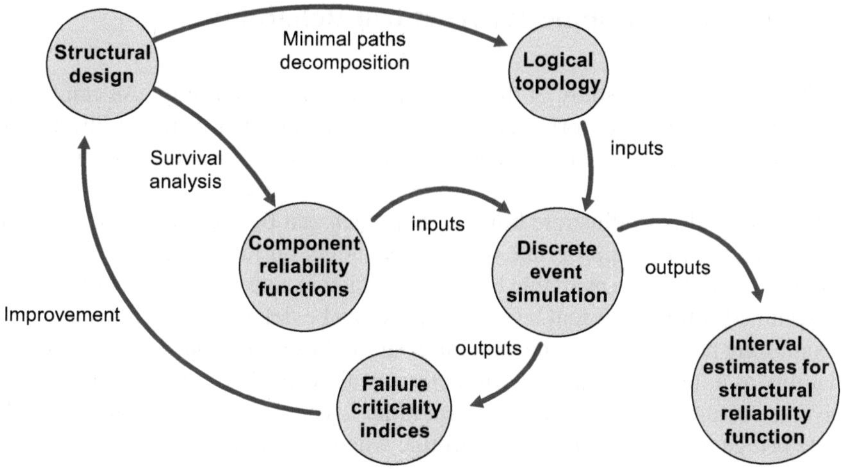

Figure 9.2 Scheme of our approach

target time. This information is then combined with the structural logical topology to infer the structural state at each target time.

By repeating this process, a set of randomly generated lifecycles is provided for the given structure. Each of these lifecycles provides observations of the structural state at each target time. Therefore, once a sufficient number of iterations has been run, accurate point and interval estimates can be calculated for the structural reliability at each target time [12]. Also, additional information can be obtained from these runs: which components are more likely to fail, which component failures are more likely to cause structural failures (failure criticality indices), which structural failures occur more frequently, etc. [11].

Moreover, notice that DES could also be employed to analyze different scenarios (what-if analysis), i.e., to study the effects of a different logical topology on structural reliability, the effects of adding some redundant components on structural reliability, or even the effects of improving reliability of some individual components (Figure 9.2).

Finally, DES also allows for considering the effect of dependencies among component failures and/or repairs. It is usually the case that a component failure or repair affects the failure or repair rate of other components. In other words, component failure- and repair-times are not independent in most real situations. Again, discrete-event simulation can handle this complexity by simply updating the failure- or repair-time distributions of each component each time a new component failure or repair takes place [8]. This way, dependencies can be also introduced in the model. Notice that this represents a major difference between our approach and other approaches, mainly analytical ones, where dependencies among components, repair-times or multi-state structures are difficult to consider.

9.8 Numerical Example 1: Structural Reliability

We present here a case study of three possible designs for a bridge. As can be seen in Figure 9.3, there is an original design (case A) and two different alternatives, one with redundant components (case B) and another with reinforced components (case C).

Our first goal is to illustrate how our approach can be used in the design phase to help pick the most appropriate design, depending on factors such as the desired structural reliability, the available budget (cost factor), and other project restrictions. As explained before, different levels of failure can be defined for each structure, and in examining how and when the structures fail in these ways, one can measure their reliability as a function of time. Different survival functions can then be obtained for a given structure, one for each structural failure type. By comparing the reliability of one bridge to another, one can determine whether a certain increase in structural robustness – either via redundancy or via reinforcement – is worthwhile according to the engineer's utility function. As can be deduced from Figure 9.3, the three possible bridges are the same length and height, but the second one (case B) has three more trusses connecting the top and bottom beam and is thus more structurally redundant. If the trusses have the same dimensions, the second bridge should have higher reliability than the first one (case A) for a longer period of time. Regardless of how failure is defined for the first bridge, a similar failure should take longer to occur in the second bridge. Analogously, the third bridge design (case C) is likely to be more reliable than the first one (case A), since it uses reinforced components with improved individual reliability (in particular, components $1'$, $2'$, $5'$, $6'$, $9'$, $10'$, and $13'$ are more reliable than their corresponding components in case A).

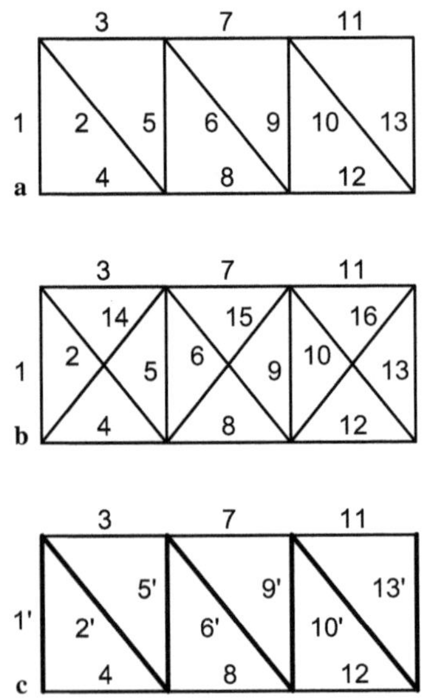

Figure 9.3 Different possible designs for a structure: (a) Case A – original base structure (13-bar plane truss), (b) Case B – original structure with redundant components (16-bar plane truss), and (c) Case C – original structure with reinforced components (13-bar plane truss)

Let us consider three different types of failure. Type 1 failure corresponds to slight damage, where the structure is no longer as robust as it was at the beginning but it can still be expected to perform the function it was built for. Type 2 failure corresponds to severe damage, where the structure is no longer stable but it is still standing. Finally, type 3 failure corresponds to complete structural failure, or collapse. Now we have four states to describe the structure, but only two (failed or not failed) to describe each component of the structure. We can track the state of the structure by tracking the states of its components. Also, we can compare the reliabilities of the three different structures over time, taking into account that different numbers of component failures will correspond to each type of structural failure depending on the structure. For example, a failure of one component in the case A and C bridges could lead to a type 2 failure (severe damage), while it will only lead to a type 1 failure (slight damage) in the case B bridge. In other words, for case B it will take at least two components to fail in the same section of the bridge before the structure experiences a type 2 failure.

In order to develop a numerical example, we assumed that the failure-time distributions associated with each individual truss are known. Table 9.1 shows these distributions. As explained before, this is a reasonable assumption since this information can be obtained either from historical data or from accelerated-life tests.

For cases A and C, only one minimal path must be considered since the structure will be severely damaged (the kind of "failure" we are interested in) whenever one of its components fails. However, for case B a total of 110 minimal paths were identified. The structure will not experience a type 2 failure if, and only if, all components in any of those minimal paths are still operative [8]. To numerically solve this case study we used the SURESIM software application [11], which implements the algorithms described in our methodology. We ran the experiments on a standard PC, Intel Pentium 4 CPU 2.8 GHz and 2 GB RAM. Each case was run for one million iterations, each iteration representing a structural life-cycle for a total of 1E6 observations. The total computational time employed for running all iterations was

Table 9.1 Failure-time distributions at component level for each truss

Component	Distribution	Shape	Scale	Component	Distribution	Shape	Scale
1	Weibull	4	22	9	Weibull	4	22
1′	Weibull	6	28	9′	Weibull	6	28
2	Weibull	6	18	10	Weibull	6	18
2′	Weibull	6	28	10′	Weibull	6	28
3	Weibull	5	30	11	Weibull	5	30
4	Weibull	5	30	12	Weibull	5	30
5	Weibull	4	22	13	Weibull	4	22
5′	Weibull	6	28	13′	Weibull	6	28
6	Weibull	6	18	14	Weibull	6	18
6′	Weibull	6	28	15	Weibull	6	18
7	Weibull	5	30	16	Weibull	6	18
8	Weibull	5	30	–	–	–	–

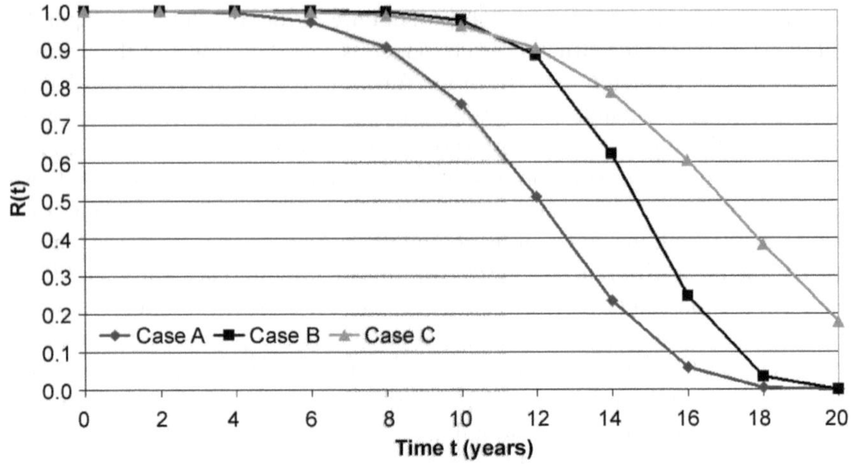

Figure 9.4 Survival functions for different alternative designs

Table 9.2 Estimated mean time to type 2 failure for each bridge (estimated values from simulation)

Case	Years
A	11.86
B	14.52
C	16.73

below 10 seconds for the two tests related to cases A and C – the ones with just one minimal path – and below 60 seconds for the test related to case B. Figure 9.4 shows, for a type 2 failure, the survival (reliability) functions obtained in each case – notice that similar curves could be obtained for other types of failures. This survival function shows the probability that each bridge will not have failed – according to the definition of a type 2 failure – after some time (expressed in years). As expected, both cases B and C represent more reliable structures than case A. In this example, case B (redundant components) shows itself to be a design at least as reliable as case C (reinforced components) for some time period (about 11 years), after which case C is the most reliable design. Notice that this conclusion holds only for the current values in Table 9.1. That is, should the shape and scale parameters change (e.g., by changing the quality of reinforced components), the survival functions could be different.

Table 9.2 shows the estimated structural mean time to a type 2 failure (severe damage) for each bridge design. Notice that case C is the one offering a larger value for this parameter.

Finally, Figure 9.5 shows failure criticality indices for case A; similar graphs could be obtained for cases B and C from the simulation output. Notice that the most critical components are trusses 2, 6, and 10. Since there is only one minimal path, this could have been predicted based on the distribution parameters assigned to each

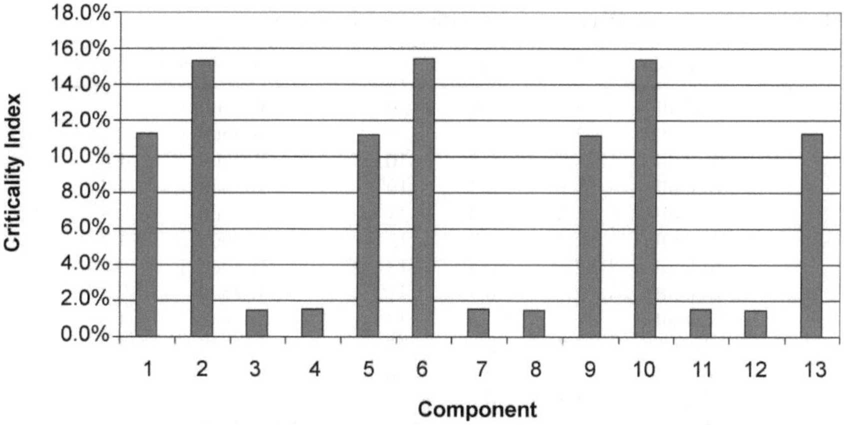

Figure 9.5 Failure criticality indices for case A

component. Components 1, 5, 9, and 13 also show high criticality indices. Knowing these indices could be very useful during the design phase, since they reveal those components that are responsible for most structural failures and, therefore, give clear hints on how to improve structural reliability either through direct reinforcement of those components or through adding redundancies.

9.9 Numerical Example 2: Structural Availability

For the purposes of illustrating our methodology, we will continue with a simplified maintainability analysis of the three bridge cases presented above. We have already introduced the benefits of being able to track a structure through time in DES in terms of measuring its reliability. With DES, one can also consider the effect of maintenance policies – modeled as random repair times for each component – and eventually track the structural availability function as well as the associated costs of those repairs. This could be a valuable extension of the example presented previously, because being able to consider the affects of maintenance policies could help in deciding between multiple designs for a structure.

Theoretically, this technique can be applied to any structure or system for which the component lifetimes and failure probabilities are known. It could be well suited for analyzing the reliability and maintenance costs of structures that are subjected to persistent natural degrading forces, such as wind turbines deployed in the ocean, bridges subjected to high winds, or perhaps even spacecraft that sustain a great deal of damage as they reenter the atmosphere. This method could also be especially valuable in the design phase of structures with moving parts that will undergo accelerated degradation, such as draw bridges, vehicles, rides at theme parks, or robotics used in manufacturing. For these structures, repairs should happen relatively fre-

Table 9.3 Repair-time distributions at component level for each truss

Component	Distribution	Shape	Scale	Component	Distribution	Shape	Scale
1	Weibull	2	0.5	9	Weibull	2	0.5
1′	Weibull	2	0.5	9′	Weibull	2	0.5
2	Weibull	1.8	0.5	10	Weibull	1.8	0.5
2′	Weibull	1.8	0.5	10′	Weibull	1.8	0.5
3	Weibull	1.8	0.3	11	Weibull	1.8	0.3
4	Weibull	1.8	0.3	12	Weibull	1.8	0.3
5	Weibull	2	0.5	13	Weibull	2	0.5
5′	Weibull	2	0.5	13′	Weibull	2	0.5
6	Weibull	1.8	0.5	14	Weibull	1.8	0.5
6′	Weibull	1.8	0.5	15	Weibull	1.8	0.5
7	Weibull	1.8	0.3	16	Weibull	1.8	0.5
8	Weibull	1.8	0.3	–	–	–	–

quently because they will need to operate at a higher level of reliability, especially where human lives could potentially be at risk.

Table 9.3 shows repair-time distributions for each of the trusses. As before, for illustration purposes it will be assumed that this data is known, e.g., that it has been obtained from historical observations. Again, our DES-based algorithms were used to analyze this new scenario. The goal was to obtain information about structural availability through time, i.e., about the probability that each possible structure will be operative – not suffering a type 2 or type 3 failure – at any given moment in the years to come. Figure 9.6 shows availability functions obtained for each alternative design. These functions consider a time interval of 100 years. Notice that this time there are not any significant differences between cases A and C. Since we are now considering repairs at component level, reinforcing some components (case C) will

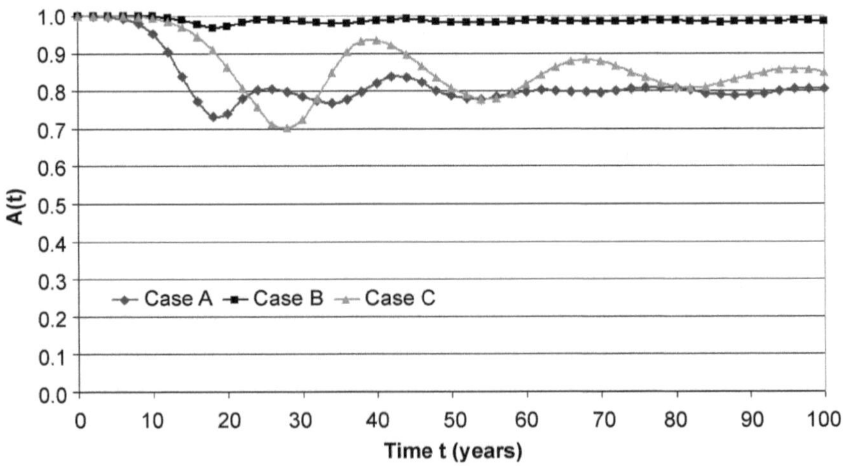

Figure 9.6 Availability functions for different alternative designs

basically shift the availability curve to the right, but not upwards. On the other hand, adding redundancies (case B) has shown to be more effective from an availability point of view. Since we are repairing components as they fail, and since repair times are much smaller than failure times, it is unlikely that two in the same section will be in a state of failure at the same time. Of course, costs associated with each strategy should also be considered in real-life whenever a decision on the final design must be made. Simulation can also be helpful in this task by providing estimates for the number of component repairs that will be necessary in each case.

9.10 Future Work: Adding Fuzzy Rule-based Systems

Based on what has been discussed so far, at any given time each structural component will have a certain level of operability. Recall that multiple states could be considered for components. As described before, this time-dependent component state can often be determined by using statistical distributions to model components' reliability and/or availability functions. Sometimes, though, this modeling process can be difficult to perform. Also, there might be situations in which it is not possible to accurately determine the current state of a component at a given moment but, instead, it is possible to perform visual or sensor-based inspections, which could then be analyzed by either human or system experts to obtain estimates about the component's state. Therefore, it seems reasonable to consider alternative strategies to model uncertainty at component-level. To that end, we propose the use of a fuzzy rule-based system (Figure 9.7). Some basic ideas behind this approach are given below, and a more detailed discussion of the concepts being involved can be found in [1].

Fuzzy sets allow the modeling of vagueness and uncertainty, which are very often present in real-life scenarios. A fuzzy set A defined on a set of elements U is represented by a membership function $\mu_A : U \rightarrow [0, 1]$, in such a way that for any element u in U the value $\mu_A(u)$ measures the degree of membership of u in the fuzzy set A. An example of such a membership function in the context of structural reliability can be found in [14]. In the structural reliability arena, a set of n observable proprieties, $u_i(t)$, $i = 1, 2, \ldots, n$, could be considered for each structural component at any given moment t. Each of these properties has an associated fuzzy set A_i, which usually consists of a list of desirable conditions to be satisfied by the component. Then, by defining $x_i(t) = \mu_{A_i}(u_i(t))$, the vector of inputs $(x_1(t), x_2(t), \ldots, x_n(t))$ is obtained. This vector describes how the associated component is performing with respect to the each of the n observable properties that are being considered. From this information, a corresponding output can be generated by using the so-called aggregation functions [1]. This output provides an index value that can be interpreted as a measure of the current component state, i.e., it can be interpreted as a measure of how far the component is from being in a failure state or, put in other words, how likely the component is of being in some operative state.

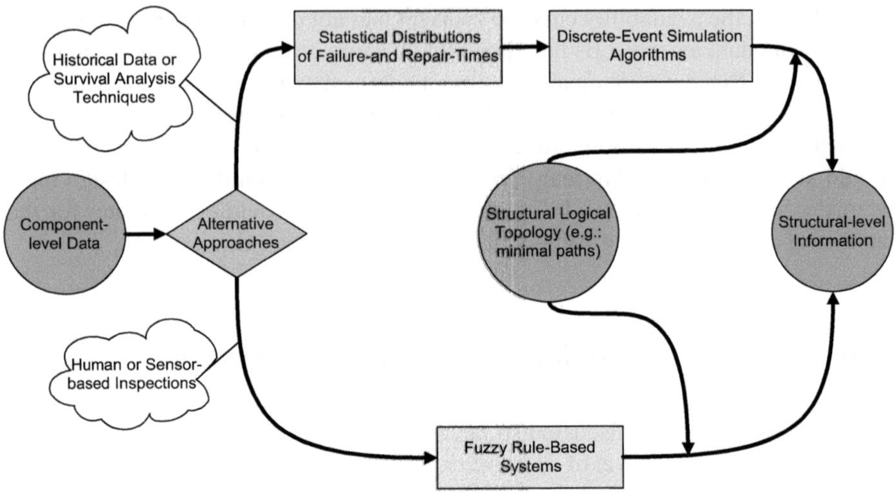

Figure 9.7 Alternative approaches to the structural reliability problem

The aforementioned aggregation functions represent a set of logical rules, which have the following form:

if $\{u_1 \in A_1\}$ and/or $\{u_2 \in A_2\}\ldots$ and/or $\{u_n \in A_n\}$ then *conclusion*

Fuzzy rule-based systems involve aggregation of various numerical scores, which correspond to degrees of satisfaction of antecedents associated with m rules. The initial form of the membership functions for fuzzy rules require a configuration process, since these rules employ some fuzzy expressions. The fuzzy rule-based system performs a fuzzy inference for calculating scores of judgment items [32]. Finally, notice that the number of fuzzy sets for each input item, the initial form of each membership function, and the initial score value in each rule must be set by discussion with building and civil engineering experts. As the main goal of our approach is to provide engineers with a practical and efficient tool to design more reliable structures, future work will be focused into implementing and testing this rule-based system approach into our SURESIM software [10].

9.11 Conclusions

In this chapter, the convenience of using probabilistic methods to estimate reliability and availability in time-dependent building and civil engineering structures has been discussed. Among the available methods, DES seems to be the most realistic choice, especially during the design stage, since it allows for comparison of different scenarios. DES offers clear advantages over other approaches, namely: (1) the opportunity of creating models which accurately reflect the structure's characteristics and behavior, including possible dependences among components' failure and repair times, and (2) the possibility of obtaining additional information about the system's internal functioning and about its critical components. Therefore, a simulation-based approach is recommended for practical purposes, since it can consider details such as multi-state structures, dependencies among failure- and repair-times, or non-perfect maintenance policies. The numerical examples discussed in this chapter provide some insight on how DES can be used to estimate structural R&A functions when analytical methods are not available, how it can contribute to detect critical components in a structure that should be reinforced or improved, and how to make better designing decisions that consider not only construction but also maintainability policies. Finally, we also discuss the potential applications of fuzzy rule-based systems as an alternative to the use of statistical distributions. One of the major advantages of the former approach is the possibility of incorporating the engineer's experience in order to improve the reliability of the structures, its design and its maintenance, so we consider it a valuable topic for future research in the structural reliability arena.

Acknowledgements This work has been partially supported by the IN3-UOC Knowledge Community Program (HAROSA) and by the Institute of Statistics and Mathematics Applied to the Building Construction (EPSEB – UPC).

References

1. Beliakov G, Pradera A, Calvo T (2007) Aggregation functions: a guide for practitioners. In: Studies in fuzziness and soft computing, Vol 221. Springer, Berlin
2. Billinton R, Wang P (1999) Teaching distribution systems reliability evaluation using Monte Carlo simulation. IEEE Trans Power Syst 14:397–403
3. Camarinopoulos L, Chatzoulis A, Frondistou-Yannas M, Kallidromitis V (1999) Assessment of the time-dependent structural reliability of buried water mains. Reliab Eng Syst Saf 65(1):41–53
4. Coit D (2000) System reliability prediction prioritization strategy. In: 2000 proceedings annual reliability and maintainability symposium, Los Angeles, CA. IEEE, Los Alamitos, CA, USA, pp 175–180
5. Ditlevsen O, Madsen H (2007) Structural reliability methods. John Wiley, Chichester, UK. Available at `http://www.web.mek.dtu.dk/staff/od/books.htm`
6. Fagan T, Wilson M (1968) Monte Carlo simulation of system reliability. In: Proceedings of the 23rd ACM national conference. ACM, New York, NY, USA, pp 289–293
7. Faulin J, Juan A, Serrat C, Bargueño V (2007) Using simulation to determine reliability and availability of telecommunication networks. Eur J Ind Eng 1(2):131–151
8. Faulin J, Juan A, Serrat C, Bargueño V (2008) Improving availability of time-dependent complex systems by using the SAEDES simulation algorithms. Reliab Eng Syst Saf 93(11):1761–1771
9. Frangopol D, Maute K (2003) Life-cycle reliability-based optimization of civil and aerospace structures. Comput Struct 81(7):397–410
10. Juan A, Faulin J, Serrat C, Sorroche M, Ferrer A (2008) A simulation-based algorithm to predict time-dependent structural reliability. In: Rabe M (ed) Advances in simulation for production and logistics applications. Fraunhofer IRB Verlag, Stuttgart, pp 555–564
11. Juan A, Faulin J, Sorroche M, Marques J (2007) J-SAEDES: A simulation software to improve reliability and availability of computer systems and networks. In: Proceedings of the 2007 winter simulation conference, Washington DC. IEEE Press, Piscataway, NJ, USA, pp 2285–2292

12. Juan A, Vila A (2002) SREMS: System reliability using Monte Carlo simulation with VBA and Excel. Qual Eng 15(2):333–340
13. Kamal H, Ayyub B (1999) Reliability assessment of structural systems using discrete-event simulation. In: 13th ASCE Engineering Mechanics Division specialty conference, Baltimore, MD. Available at `http://citeseer.ist.psu.edu/cache/papers/cs/13123/http:zSzzSzrongo.ce.jhu.eduzSzemd99zSsessionszSzpaperszSzkamal1.pdf/reliability-assessment-of-structural.pdf`
14. Kawamura K, Miyamoto A (2003) Condition state evaluation of existing reinforced concrete bridges using neuro-fuzzy hybrid system. Comput Struct 81:1931–1940
15. Laumakis P, Harlow G (2002) Structural reliability and Monte Carlo simulation. Int J Math Educ Sci Technol 33(3):377–387
16. Law A (2007) Simulation modeling and analysis. McGraw-Hill, New York, NY, USA
17. Lertwongkornkit P, Chung H, Manuel L (2001) The use of computer applications for teaching structural reliability. In: Proceedings of the 2001 ASEE Gulf-Southwest Section annual conference, Austin, TX. Available at `http://www.ce.utexas.edu/prof/Manuel/Papers/asee2001.PDF`
18. Li C (1995) Computation of the failure probability of deteriorating structural systems. Comput Struct 56(6):1073–1079
19. Mahadevan S, Raghothamachar P (2000) Adaptive simulation for system reliability analysis of large structures. Comput Struct 77:725–734
20. Marek P, Gustar M, Anagnos T (1996) Simulation based reliability assessment for structural engineers. CRC Press, Boca Raton, FL
21. Marquez A, Sanchez A, Iung B (2005) Monte Carlo-based assessment of system availability. a case study for co-generation plants. Reliab Eng Syst Saf 88(3):273–289
22. Meeker W, Escobar L (1998) Statistical methods for reliability data. John Wiley & Sons, New York, NY, USA
23. Melchers R (1999) Structural reliability: analysis and prediction. John Wiley & Sons, Chichester, UK
24. Nilson A, Darwin D, Dolan C (2003) Design of concrete structures. McGraw-Hill Science, New York, NY, USA
25. Park S, Choi S, Sikorsky C, Stubbs N (2004) Efficient method for calculation of system reliability of a complex structure. Int J Solid Struct 41:5035–5050
26. Petryna Y, Krätzig W (2005) Computational framework for long-term reliability analysis of RC structures. Comput Meth Appl Mech Eng 194(12-16):1619–1639
27. Piegat A (2005) A new definition of the fuzzy set. Int J Appl Math Comput Sci 15(1):125–140
28. Song J, Kang W (2009) System reliability and sensitivity under statistical dependence by matrix-based system reliability method. Struct Saf 31(2):148–156
29. Stewart M, Rosowsky D (1998) Time-dependent reliability of deteriorating reinforced concrete bridge decks. Struct Saf 20:91–109
30. Thoft-Christensen P, Murotsu Y (1986) Application of structural systems reliability theory. Springer, New York, NY, USA
31. Vukazich S, Marek P (2001) Structural design using simulation based reliability assessment. Acta Polytech 41(4–5):85–92
32. Zimmerman H (1996) Fuzzy sets theory and its applications. Kluwer, Boston, MA

Part III
Simulation Applications in Availability and Maintenance

Chapter 10
Maintenance Manpower Modeling: A Tool for Human Systems Integration Practitioners to Estimate Manpower, Personnel, and Training Requirements

Mala Gosakan and Susan Murray

Abstract This chapter discusses the maintenance manpower modeling capability in the Improved Performance Research Integration Tool (IMPRINT) that supports the Army's unit of action. IMPRINT has been developed by the US Army Research Laboratory (ARL) Human Research and Engineering Directorate (HRED) in order to support the Army's need to consider soldiers' capabilities during the early phases of the weapon system acquisition process. The purpose of IMPRINT modeling is to consider soldiers' performance as one element of the total system readiness equation. IMPRINT has been available since the mid 1990s, but the newest version includes significant advances.

10.1 Introduction

Even as the far-reaching implications of the next generation of weapons and information systems are being constantly redefined, one piece which has been and will continue to be central to the process is human involvement. The impacts of human performance on system performance are significant. Human systems integration (HSI) is primarily a concept to focus on the human element in the system design process [18]. The ability thus to include and consider human involvement early in the process of system development cycle will only ease mobilization, readiness, and sustainability of the newly developed system. The Department of Defense therefore has placed increased emphasis on applying HSI concepts to evaluate and improve the performance of complex systems [16].

M. Gosakan
Alion Science & Technology, MA&D Operation, 4949 Pearl East Circle, Suite 200, Boulder, CO 80301, USA (email: e-mail: mgosakan@alionscience.com)

S. Murray
Missouri University of Science and Technology, 1870 Miner Circle Rolla, MO 65409, USA (email: e-mail: murray@mst.edu)

P. Faulin, A. Juan, S. Martorell, and J.E. Ramírez-Márquez (eds), *Simulation Methods for Reliability and Availability of Complex Systems*.

The US Army was the first large organization to implement HSI approach and reap the benefits of it by creating the Manpower and Personnel Integration Management and Technical Program (MANPRINT) [24, 25]. As stated in the MANPRINT handbook, MANPRINT is a comprehensive management and technical program that focuses on the integration of human considerations (i.e., capabilities and limitations) into the system acquisition process. The goal of MANPRINT is to enhance soldier-system design, reduce life-cycle ownership costs, and optimize total system performance. To facilitate this, MANPRINT is divided into the following seven domains: manpower, personnel capabilities, training, human factors engineering, system safety, health hazards, and soldier survivability.

The manpower domain focuses on the number of people required and available to operate, maintain, sustain, and provide training for systems. The domain of personnel addresses the cognitive and physical characteristics and capabilities required to be able to train for, operate, maintain, and sustain materiel and information systems. The training domain is defined as the instruction, education, on-the-job, or self-development training required providing all personnel and units with essential job skills, and knowledge to effectively operate, deploy/employ, maintain, and support the system. One such software tool which aids HSI and MANPRINT practitioners in studying and assessing system performance as a function of human performance is IMPRINT. Since the focus of one of the capabilities, namely the maintenance manpower modeling capability of IMPRINT, discussed later in this chapter, aids in conducting quantitative trade-off analysis that applies to the first three domains, namely the manpower, personnel, and training (MPT), high-level definitions for these three domains was presented. For a more detailed description of all the seven domains please refer to the MANPRINT Handbook [24].

The following section entails the history and capabilities of one particular MPT tool, IMPRINT.

10.2 IMPRINT – an Human Systems Integration and MANPRINT Tool

IMPRINT is a simulation and modeling tool that provides means for estimation MPT requirements and to identify constraints for new weapon systems early in the acquisition process. The IMPRINT tool grew out of common US Air Force, Navy, and Army MPT concerns identified in the mid-1970s [8, 13–15, 20–22]. It is government-owned software and consists of a set of automated aids to assist analysts in conducting human performance analyses [6, 7]. IMPRINT has been available as a government product free of charge since the mid 1990s to the following organizations: US government agencies, US private industry with US government contract, and US colleges and universities working in HSI. It is supported by commercial-quality users' documentation, a training course, and a technical support organization [4]. Upgrades and enhancements to IMPRINT have been driven by user requirements, human modeling research, and changes in the state of the art in computer simulation [1, 2, 10, 11]. IMPRINT provides a powerful and flexible environment in which to develop human performance models, and has unique capabilities for assessing the impact of stressors (e.g., noise, heat, sleep deprivation, protective gear) on performance [5]. One of the most powerful and unique capabilities in IMPRINT is the method through which soldier characteristics and environmental stressors can be used to impact task performance [9]. This is achieved through an embedded simulation engine, based upon the commercial Micro Saint Sharp (`http://www.alionscience.com/index.cfm?fuseaction=Products.view&productid=35`) discrete event simulation tool [3, 12] and supplemented by human performance algorithms. The application includes a graphical user interface (GUI) shell that elicits information from the user needed to assess human performance issues associated with the operations and maintenance tasks of a weapon system. The simulation and analysis capabilities in IMPRINT along with the embedded data and GUI have been demonstrated to enable human factors professionals to impact system design and acquisition decisions based on early estimation of soldiers' abilities to operate, maintain, and support the system [19, 27].

A main component of IMPRINT is the capability to develop detailed models of maintenance manpower and manhour requirements as a function of the operational scenario and the system's component-level reliability. The maintenance module is updated in keeping with the emerging Army doctrine. The maintenance module was granted accreditation in June 2005 by the Army Data Council of Colonels. By its accreditation, IMPRINT was certified for use as a tool for materiel developers to use to support the Army manpower requirements criteria maintenance data standard methodology process by evaluating and estimating direct productive annual maintenance manhours under various scenarios. As a corollary action, IMPRINT may be used to conduct sensitivity analyses on parameters of interests (e.g., human performance effects, operational scenarios, and system reliability and maintainability). The remainder of this chapter discusses the maintenance module, its importance to the Army and future direction.

10.3 Understanding the Maintenance Module

The IMPRINT maintenance module consists of three elements: the GUI shell, the data set, and a static model. The GUI provides a way for the user to describe the inputs to the model, and the data set is used to store the input data as well as the results of the analysis. The static model is a task network model. The simulation model is created when the static model is parameterized from the input data. Sections 10.3.1 and 10.3.2 discuss the two areas that the user has to input data through the GUI; the system and the operational scenario in which the system is being operated. Section 10.4 discusses the structure of the static network model and describes its purpose.

10.3.1 System Data

The system to be defined is the particular system for which the manpower assessment is being studied, for example, the M1 Abrams Tank. A system is made of subsystems. Subsystems are made of components and components are made up of repair tasks. The repair task is the level at which all the system-level data is defined. As shown in Figure 10.1 the system being studied is a tank; armament is one of the subsystems that make up the tank, armament-other is a component of the armament subsystem and one of the repair tasks that are performed on this component is Adjust & Repair.

As shown in Figure 10.2, the repair tasks have the following attributes.

- *Repair task.* This describes the type of repair task that is needed. The complete list of maintenance task types in the logistics system analysis (LSA) standard consists of 33 separate task types. This field is populated with these 33 task types, some of which are Adjust & Repair, Inspect, Remove & Replace, Test & Check, and Troubleshoot.
- *Maintenance type.* There are two types of maintenance actions, preventive and corrective. Preventive maintenance is scheduled at fixed intervals. Corrective maintenance is required when a component fails because of usage or combat damage.
- *Organization level.* This data element identifies the maintenance organization that will perform the maintenance action. There are three possible maintenance echelons available in the IMPRINT in addition to contact team and crew-level

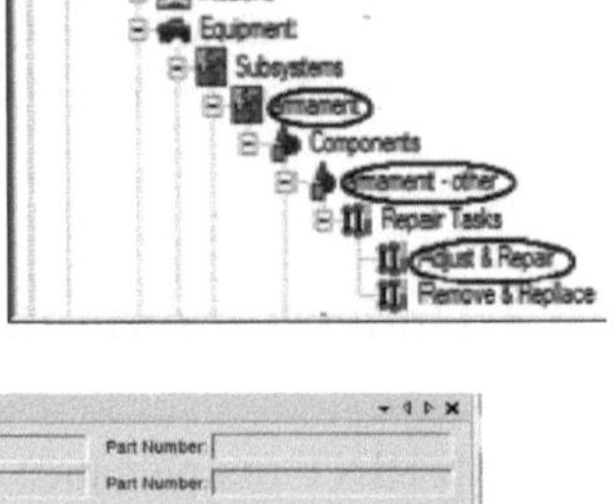

Figure 10.1 System decomposition

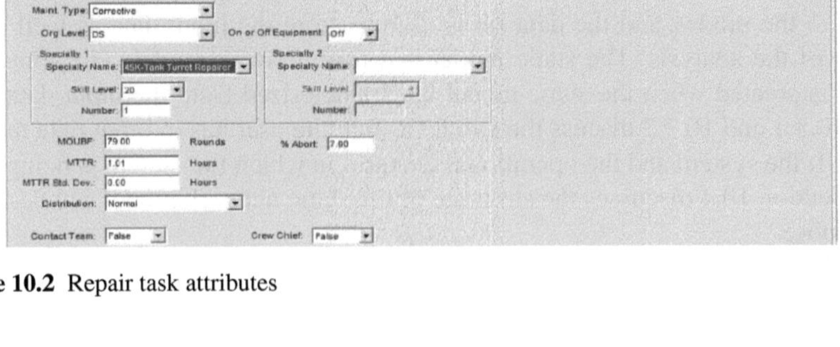

Figure 10.2 Repair task attributes

maintenance. Although the labels can be modified, the default labels for these maintenance levels are:

- organizational (Org);
- direct support (DS);
- general support (GS).

- *On- or off-equipment.* This field represents whether the repair is done on the equipment or off the equipment. All Org-level maintenance is assumed to be performed on-equipment. All GS-level maintenance is assumed to be performed off-equipment. DS maintenance can be modeled as either off-equipment or on-equipment. On-equipment maintenance makes the system unavailable during the time that maintenance is being performed. An example of an on equipment task is changing a tire or a filter. Off-equipment maintenance are repairs that are performed once a part has been removed from the system. The system itself remains available for missions. An example of an off-equipment task is fixing a hole in a tire after it has already replaced the tire with a spare.
- *Manpower requirements.* The next six columns in the data spreadsheet are used to define the military occupational specialties (MOS1, MOS2) that are required to perform the maintenance, the skill levels (10, 20, 30, 40, or 50 as defined by the duty positions for each MOS), and the number of maintainers needed (#MOS1, #MOS2). Up to two different MOSs can be selected.
- *Reliability.* The frequency of the maintenance action is expressed as mean operational units between failure. This is the number of operational units between failure, or the number of operational units between the need for this maintenance action. The units could be rounds fired, distance traveled, or the amount of time that the system has been operating. The actual time when the need for this action will occur in the simulation is drawn from an exponential distribution specified by this mean value. Although the Weibull distribution is the most widely used distributions in reliability engineering, the model draws from an exponential distribution as the IMRPINT users with real system data will be unable to provide the parameters needed for a three-parameter Weibull distribution. Also a two-parameter Weibull distribution best approximates to an exponential and hence the model chooses from an exponential distribution.
- *Maintainability.* The maintainability of each component is expressed as the mean time to repair (MTTR). This is expressed as a mean, standard deviation, and distribution type (the current choices for the distributions being normal, gamma, and lognormal) that describes the average time it takes to perform this maintenance action. These values are used to generate a simulated time for this maintenance action, and will be recalculated for each occurrence of the action.
- *Criticality.* The criticality of each maintenance action is expressed as the likelihood that the occurrence of a maintenance action will cause the entire system to interrupt or abort its current mission in order to have the maintenance done immediately. This is labeled as Abort % on the input menu.

The next two repair task attributes, contact team and crew chief, are the most recent additions to the maintenance model [11]. Specifically, the emerging doctrine for the

Army's unit of action indicates a heavier reliance on maintenance being performed by the crew chief and by mobile contact teams.

- *Contact team.* This field contains an indication of whether this maintenance action could be performed by a contact team. This does not necessarily mean that a contact team will perform this maintenance action. It also depends on whether a contact team has been defined, whether it has been enabled for the current run and whether there are enough contact team maintainers to perform this action.
- *Crew chief.* This field contains an indication of whether the operational crew is qualified and equipped to perform this maintenance action. If the maintenance action is needed, and the simulation model predicts that any required spares are available, and the user has entered a yes in this column and if the user has marked operators to be crew maintainers, then the maintenance task will be performed by the crew.

10.3.2 Scenario Data

A scenario is built in which the system (or a multiple number of the same system) will operate. Scenarios can be defined to run of a period of n days. Scenarios are described using the following attributes:

- *Segment* (a scenario comprises one or more segments). The operations tempo (OPTEMPO) of the mission is set by the analyst. The user describes the OPTEMPO by defining the parameters as shown on the Segment Info tab in Figure 10.3. The parameters are segment start time, duration, whether the segment repeats and if yes, how often it repeats, the minimum and maximum systems requested, and cancellation time. The other properties attached to the segment

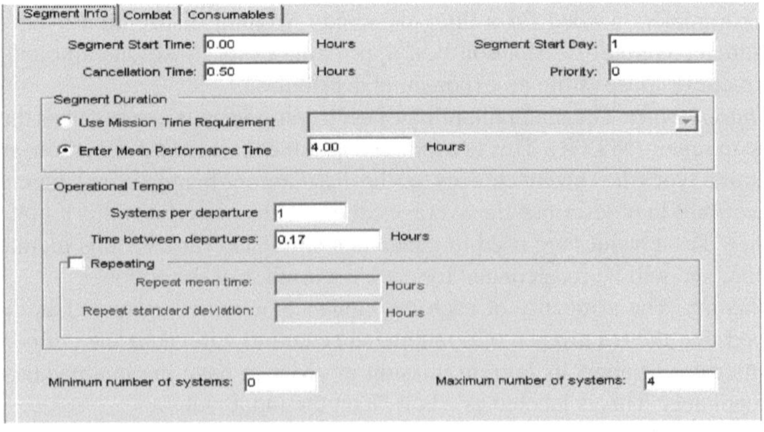

Figure 10.3 Segment attributes

are the combat damage data and the consumables usage data. The model currently has quite a simplistic representation of the combat damage effects on a segment. Based upon the probability of hit the system may encounter damage. Once a hit is determined, the amount of time required to repair the system is obtained, or if it was a kill how much time to replace the system is used. When a system is assessed any combat damage, the effect of combat damage is dealt with at the system level as opposed to component level. The consumables usage data is dependent on the types of subsystems defined for the system. Since the system accrues usage based on the distance traveled, the rounds fired, or the fuel consumed, these are the attributes attached to the consumables usage data.

- *Fuel Supply & Ammo Supply*. IMPRINT will generate reports for Fuel & Ammo Supply that estimate the number of transporters and the associated manpower required to supply the necessary fuel and/or ammo needed for the scenario. These estimates are based on the data entered for the capacity of the transporter, the load time, the specialties needed, and maximum number of daily trips.
- *Travel Time*. The travel times represent the amount of delay time to move the component to and between the different maintenance levels. When this is greater than zero, operational readiness will be affected.
- *Spare Parts*. In order not to burden the user with too much data entry, the spare part information is inputted at the subsystem level. The data entered are the likelihood that a part is available and also the wait time if the part is not readily available.
- *Maintenance Crew*. The data in these fields represents the number and types of maintainers on each shift available at each Org level to perform maintenance.
- *Contact Team*. The user can identify the number of contact teams, the number of maintainers within each team, and the maximum number of repair actions that can be in each team's queue at one time. All of these parameters combine to enable IMPRINT to model the impact of contact teams on operational readiness.

10.4 Maintenance Modeling Architecture

A high-level overview of the model is illustrated in Figure 10.4. During the simulation systems (i.e., tanks) are sent to perform missions. At the completion of the mission (or sooner if critical failures occur), the failures associated with component-level reliabilities and combat damage are tabulated and sent to the appropriate maintenance organization in the model. Once all required maintenance is performed on each system it is returned to the system available pool, and is made available for upcoming scheduled missions.

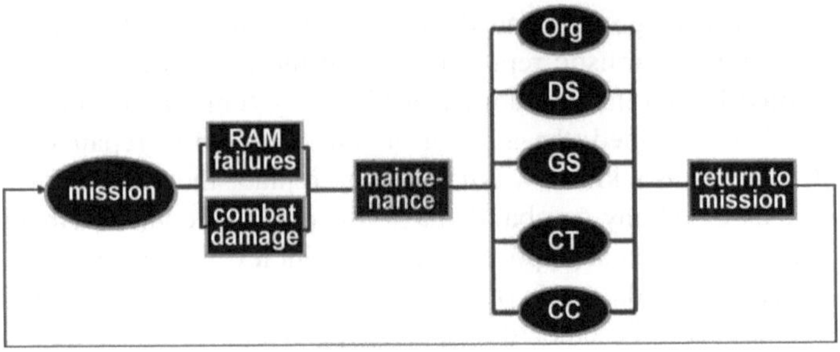

Figure 10.4 Model overview

10.4.1 The Static Model – the Brain Behind It All

The static model is a task network model that was built using Micro Saint Sharp. This model can be thought of as having three separate, but interrelated parts. The first part, shown in Figure 10.5, controls the flow of systems into mission segments. In this part of the model, the entities flowing through the network represent individual systems (e.g., an M1 Tank). This part of the model controls the accrual of usage to each individual component of each system (based on the distance traveled, rounds fired, and time operated). It also predicts any combat damage. Before sending a system out to perform a mission segment, IMPRINT looks ahead to see whether the mission segment will be aborted due to a failure of a critical component. If it determines that the mission segment will be aborted, it is careful to accrue only the completed proportion of usage to all components in that particular system. When the system returns from a mission segment, each non-abort component in each system is checked to determine whether the accrued usage is greater than the failure clock. It is important to note the amount of fidelity that is represented in the model. IMPRINT tracks separate failure clocks for each maintenance action (i.e., combination of repair task and component) on each system. This is a powerful and unique feature of IMPRINT.

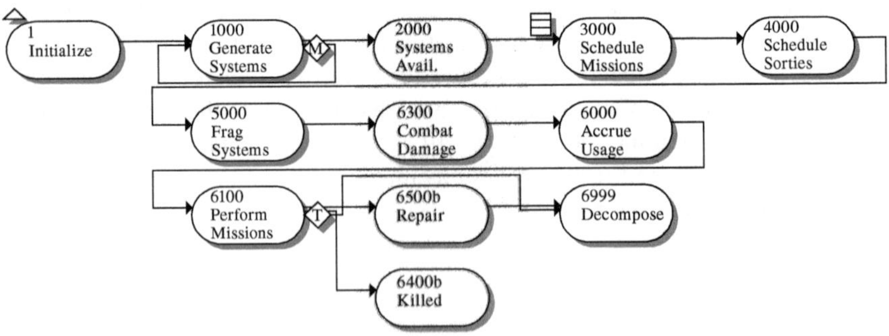

Figure 10.5 Flow of systems into mission segments

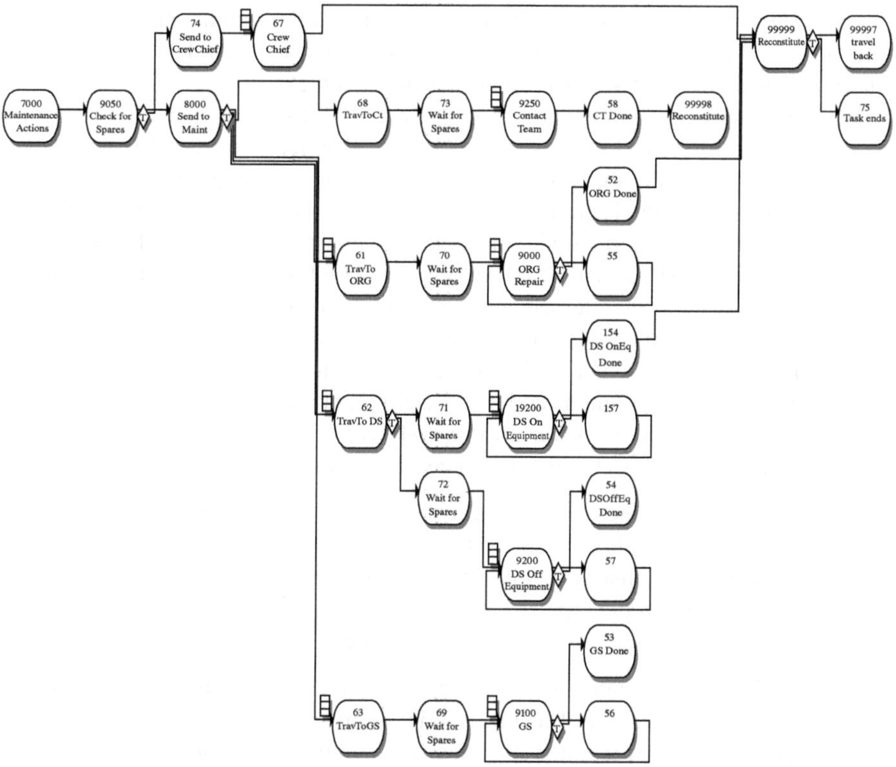

Figure 10.6 Flow of systems into mission segments

For any system that now has components in need of maintenance, the parent system is removed from the systems available pool, and the maintenance actions are sent to the second part of the model, depicted in Figure 10.6.

In Part 2 of the model, the maintenance actions are performed by the appropriate organizations. In this portion of the model, the entities flowing through the network represent maintenance actions as shown in Figure 10.6. Maintenance actions are queued up in front of their respective Org levels. If the maintenance action is remove and replace and the maintenance task is marked as a crew chief task, if operators are identified as crew maintainers, then the spare parts parameters for the parent subsystem are examined to see if the spare is actually needed, and if so, whether it is available. If it is not available, the system repair is not routed to the crew chief for maintenance but is routed to its default maintenance organization and is delayed for the appropriate time needed to procure the spare.

If a maintenance action has been marked for contact team maintenance, then the contact team capacity is assessed to determine whether there is sufficient room in the contact team queue for the new maintenance action. If sufficient capacity is not available, as specified on the contact team GUI, then the maintenance action is routed to the selected organization level.

The maintenance actions for a system are managed through the process in a logical flow, and the queues at each Org level are sorted by complex strategies that maximize availability in an operationally realistic context.

The total predicted maintenance time for each system is estimated by summing the MTTR for all the tasks of a specific system. The maintenance actions are then placed in an initial order that gives priority to the system with the shortest estimated total maintenance time. Then, the manpower requirements of the maintenance actions in the queues are compared to the available manpower pool by Military Occupational Specialty (MOS) and skill level for each Org level. The maximum numbers of repair tasks that can be released are then sent into the maintenance echelon where maintenance is performed. This strategy is careful to keep maintenance actions from being holed. This means that if a maintenance action takes fewer maintainers than one that is above it in the queue and insufficient maintainers are available to process the high priority action, the lower priority task will be released.

Critical assumptions to the maintenance process include:

- Crew chief maintenance can be performed in parallel with any other Org level
- Jobs flow from the contact team, to the Org level, to DS
- The contact team consists of soldiers that can perform all maintenance that the user has selected (in the repair task spreadsheet)
- The crew chief can do one task at a time for each system.

One final issue associated with this process is that all maintenance actions that are not complete at the end of a shift will be interrupted until enough maintainers are available on the next manned shift in order to complete the action. Maintenance actions that are interrupted are always given a higher priority than actions that have not yet begun. Note that the crew chief and contact team maintenance are not subject to a shift length limitation.

When all maintenance actions for a particular system are complete, the system is reconstituted and sent back to the system available pool. It is then available to be assigned to any upcoming mission segments.

Part 3 of the model runs in parallel to the first two parts. In this part, as shown in Figure 10.7 the entities flowing through the network represent mission segments.

The purpose of this portion of the model is to schedule mission segments and to determine whether they should be released or canceled. Mission segments are released if there are enough systems in the available pool to meet the minimum

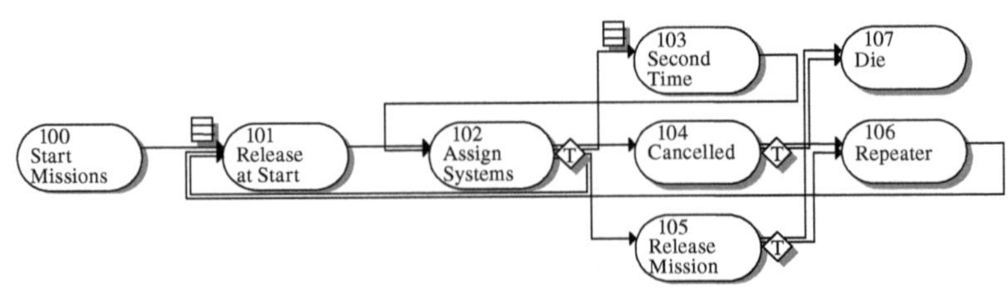

Figure 10.7 Scheduling of mission segments

number required at the mission start time. If the mission segment is not filled to its minimum at that time, the model continues to try to gather enough available systems by the mission cancellation time. If enough systems are not available at cancellation time, the mission is canceled and all systems are returned to the available pool.

The scheduler uses the mission segment priority to determine which mission segment systems will be assigned to if more than one segment is scheduled to leave at the same time. If this happens, then the model will attempt to fill each mission segment's minimum, before filling the mission segment to the maximum.

10.4.2 A Simple Example – Putting It All Together

The objective of the job assigned to the analyst is to study the effect of manpower allocation at the various Org levels on operational readiness rate (ORR) for the M1 Abrams Tank. The first step the analyst would go through is to populate the system data as explained in Section 10.3.1 or as explained later in Section 10.6.1. Next the analyst in working with subject matter experts would set up a scenario in which the M1 Tank would operate under.

An example of such a scenario would be seven M1 Tanks (available pool of systems for that scenario) on a seven-day mission pulse running for 365 days. The seven-day mission would translate to seven different segments, segment 1 through segment 7 with the modeler having the capability to create mission profiles for each of the segments. Segment 1, for example would have the following attributes: segment start time and day of 00:00 on Day 1, with duration of 12:00 hours, repeating every 132 hours and with a cancellation time of 0.50 hours. Variability to the repeat time can be introduced by adding a standard deviation and the adjusted repeat time would then be set by pulling from a normal distribution.

Further in this example, the systems requested corresponding to this segment is set as follows: six maximum systems and a minimum of two systems request with a grouping of systems requested at two systems spaced 10 minutes apart. At the very beginning of the simulation when the first segment request comes up on Day 1 at 00:00 hours, since all the requested maximum systems (six in this case) are available the first group of systems go out at clock time of 0, then after 10 minutes (as set in time between departure group), the next two systems (number of systems grouped together) are sent out and after 10 more minutes the final two systems are sent out.

The system available pool (in this example set at seven, see above) has to be at least equal to or greater than the minimum systems (set at two in this example) requested for the segment to start. If during the course of the 365-day run, if there happens to be one system available at the time of the segment request (as the remaining systems are down for maintenance), then the scheduler checks after 0.50 hours to see if there are at least a minimum of two systems available for the segment request to be met. If the number of available systems at that time is equal to or greater than the minimum systems then the systems are sent on mission accruing usage.

Each of these systems accrues 12 hours of usage. Should there be impact of combat damage assessed on the system and or any component subject to a critical failure in the middle of the mission, the mission time would then be adjusted accordingly.

Once the system and scenario data are populated, the analyst would run the simulation in an unconstrained mode. Unconstrained mode refers to an unlimited number of bodies – manpower available to perform maintenance actions. The analyst would then look at the reports, particularly the headcount frequency report. This report provides a measure of specialty utilization, or more specifically, it illustrates the frequency with which different numbers of people in each specialty were used and overall Org-level types. This should give the analyst an idea of how to populate the manning at each shift for the scenario. The analyst can then re-run the simulation in constrained mode (is limited by the number of bodies available at each shift) based on the adjustment made and note the ORR and the total direct maintenance man hours (DMMH). ORR is calculated as segments accomplished divided by segments requested. The total DMMH is sum of DMMH across all the organization levels. The analyst can then vary the number of bodies available accordingly and note how the ORR varies based on this. While the above discussion is intended to give the reader an idea of the kind of studies that can be undertaken using the maintenance module, it does not necessarily address all the factors that affect operational readiness. For a more detailed discussion of the various factors that affect ORR please refer to [26].

10.5 Results

The results of the maintenance simulation are gathered as the model runs. These results are then accessible through the Maintenance Results option on the IMPRINT Reports pull-down menu. Results include maintenance manhour requirements, tabulated by individual maintenance action, and summed for each MOS and Org level. Additionally, results are compiled by subsystem (e.g., engine, tracks) so that the high-driver subsystem, that is, subsystems which required the most maintenance, can be discriminated. Finally, several measures of availability and readiness are reported, enabling the user to trade off component reliability, maintainability, and Org manning against operational readiness, for a selected operational tempo. The types of questions the maintenance model helps answers are:

- Has the required operational readiness been achieved?
- How does applying performance moderators affect operational readiness?
- How many people of each specialty are needed to meet the system availability requirement?
- Which pieces of equipment (i.e., subsystems) are the high drivers for maintenance?
- How should each Org level be manned?
- How sensitive are the maintenance manpower requirements to the failure rates of individual components?

10.6 Additional Powerful Features

To augment the high-fidelity maintenance modeling capabilities which were described above, the maintenance module in IMPRINT has a few other powerful features to provide IMPRINT users with more modeling power. The discussion in this section focuses on three particular features. They are alternate data importing capabilities, effects of performance moderators on repair task times, and visualization capability.

10.6.1 System Data Importing Capabilities

The analyst has more than one option to populate the component data. One method to enter the data is manual entry. It is unlikely that an analyst would resort to this approach. If the system is in use, the analyst typically would use data from an existing database such as the Army's sample data collection. But if the system currently is being developed, the system manufacturer would provide the system design documentation (i.e., parts inventory) to the system program manager. The analyst will be then able to get this data from the program manager. Since this is a time-consuming process, IMPRINT currently has two methods through which data can be imported. One is the capability to accept the LSA 001 report, and second is an Excel template format where the user can pre-populate all of the component data which then be read into IMPRINT.

10.6.2 Performance Moderator Effects on Repair Times

The ability to perform repair actions under ideal conditions may differ drastically from ability to perform under stressful conditions. The environment in which military operations (tasks and missions) are conducted can be very stressful. IMPRINT has unique capabilities for assessing the impact of stressors such as noise, temperature, sleep deprivation, and protective gear on performance [5]. One of the most powerful and unique capabilities in IMPRINT is the method through which soldier characteristics and environmental stressors can be used to impact task performance [9].

It is important to note that not all repair tasks are affected in the same way. To model this effect, IMPRINT uses a task category weighting scheme. Currently, in IMPRINT nine categories or taxons used to describe a task [17]. Taxons can be described as the type of effort needed to perform the repair task. Examples of task types include motor, visual, and cognitive. In the maintenance module each of the 33 repair tasks is pre-mapped to these taxons. To see the impact of taxons on the task times, the user needs to check the PTS Adjustments box in the Executions Settings

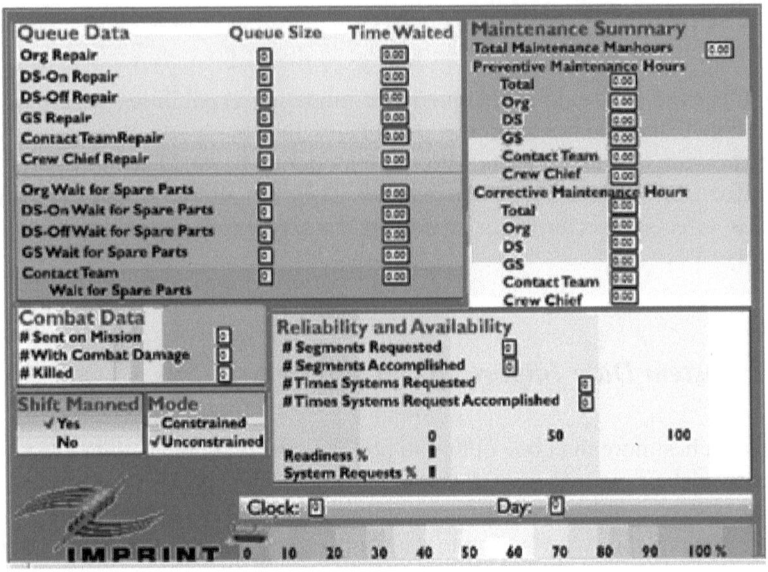

Figure 10.8 Animation screen during execution

menu before they run the model. This powerful feature equips the analyst to see the effect of environmental conditions on operational readiness.

10.6.3 Visualization

To provide the users with insight into the model's execution a visualization capability is provided as shown in Figure 10.8. This screen enables the users to see the impact of manning levels on queue sizes in front of each maintenance echelons. The impact of spare availability on ORR and total preventive and/or corrective manhours being spent by each Org levels is also depicted. Operational readiness as the mission progresses can also be assessed. This capability equips the user with a visual that will help them identify and diagnose problems in the maintenance concept as they work towards meeting a goal such as a target readiness rate.

10.7 Summary

Whether an analyst's goal is to study the impacts of emerging Army two-level maintenance concept on operational readiness or to be able see the effects of reliability, maintainability, and operational requirements on operational readiness or to be able estimate manpower assessments, the maintenance module has been developed and updated to reflect the changing Army transformation. The IMPRINT maintenance module is a tool that allows HSI practitioners to make an impact in the design of next-generation weapon systems.

References

1. Adkins R, Dahl SG (1993) Final report for HARDMAN III, Version 4.0. Report E-482U, prepared for US Army Research Laboratory. Micro Analysis & Design, Boulder, CO
2. Adkins R, Dahl SG (1992) Front-end analysis for HARDMAN III in Windows. Report E-21512U, Prepared for Army Research Institute for the Behavioral and Social Sciences. Micro Analysis & Design, Boulder, CO
3. Alion Science and Technology (2008) Micro Saint Sharp version 3.0 User Manual. Boulder, CO
4. Alion Science and Technology and ARL (2009) IMPRINT Pro V3.0 User Guide
5. Allender L *et al.* (1999) Evaluation of human performance under diverse conditions via modeling technology. In: Improved performance research integration tool (IMPRINT), user's guide (Appendix A). US ARL, Aberdeen Proving Ground, MD
6. Allender L *et al.* (1995) Verification, validation, and accreditation of a soldier-system modeling tool. In: Proceedings of the 39th human factors and ergonomics society meeting, October 9–13, San Diego, CA. Human Factors and Ergonomics Society, Santa Monica, CA
7. `http://www.arl.army.mil/IMPRINT`
8. Archer R *et al.* (1987) Product 5: manpower determination aid. Final concept paper for US ARI. Micro Analysis & Design, Boulder, CO
9. Archer S, Adkins R (1999) IMPRINT user's guide prepared for US Army Research Laboratory. Human Research and Engineering Directorate, Boulder, CO
10. Archer SG, Allender L (2001) New Capabilities in the Army's Human Performance Modeling Tool, Proceedings of the Military, Government, and Aerospace Simulation Conference. editor Michael Chinni, Seattle, WA, pp 22–27
11. Archer SG, Gosakan M *et al.* (2005) New capabilities of the army's maintenance manpower modeling tool. J Int Test Eval Assoc 26(1):19–26
12. Bloechle W, Schunk D (2003) Micro Saint Sharp Simulation. In: Proceedings of the 2003 Winter Simulation Conference, New Orleans, LA
13. Dahl SG (1993) A study of unit measures of effectiveness to support unit MANPRINT. Final report prepared for Ft. Huachuca Field Unit. US Army Research Laboratory, Boulder, CO
14. Dahl SG (1992) Integrating manpower, personnel and training factors into technology selection and design. In: Proceedings of the International Ergonomics Society. Micro Analysis & Design, Boulder, CO
15. Dahl *et al.* (1990) Final report for concepts on MPT estimation (Development of MANPRINT methods, Report E-17611U. Prepared for US Army Research Institute for the Behavioral and Social Sciences. Micro Analysis & Design, Boulder, CO
16. Defense acquisition guidebook (DAG). Chapter 6, Human systems integration. `https://acc.dau.mil/CommunityBrowser.aspx?id=314774&lang=en-US`
17. Fleishman EA, Quaintance MK (1984) Taxonomies of human performance: the description of human tasks. Academic Press, Orlando, FL
18. Booher HR (2003) Introduction: human systems integration. In: Handbook of human systems integration. Wiley, Hoboken, NJ
19. Hoagland DG *et al.* (2000) Representing goal-oriented human performance in constructive simulations: validation of a model performing complex time-critical-target missions. SIW conference. Simulation Interoperability Standards Organization, San Diego, CA, Paper Number 01S-SIW-137

20. Kaplan JD *et al.* (1989) MANPRINT methods. In: Aiding the development of manned system performance criteria. Technical report 852, US Army Research Institute for the Behavioral and Social Sciences, Alexandria, VA
21. Laughery KR *et al.* (2005) Modeling human performance in complex systems. In: Salvendy G (ed) Handbook of industrial engineering, 4th edn. Wiley, New York
22. Laughery KR *et al.* (1988) A manpower determination aid based upon system performance requirements. In: Proceedings of the Human Factors Society 32nd annual meeting. Human Factors and Ergonomics Society, Santa Monica, CA, pp 1060–1064
23. Lockett JF, Archer SG (2009) Impact of digital human modeling on military human-systems integration and impact of the military on digital human modeling. In: Duffy VG (ed) Handbook of digital human modeling – research for applied ergonomics and human factors engineering. CRC Press, Boca Raton, FL
24. Manpower and personnel integration MANPRINT handbook, Office of the Deputy Chief of Staff G1. MANPRINT Directorate, Washington
25. `http://www.manprint.army.mil/manprint/docs/MEMOS/skelton/manprintforthearmy.html`
26. Simpson J *et al.* (2006) IMPRINT output analysis final report. Technical report prepared by FSU-FAMU College of Engineering Simulation Modeling Group for MA&D and ARL-HRED, April. Tallahassee, FL
27. Wojciechowski JQ *et al.* (1999) Modeling human command and control performance sensor to shooter. Proceedings of human performance, situation awareness, and automation conference, Savannah, GA

Chapter 11
Application of Monte Carlo Simulation for the Estimation of Production Availability in Offshore Installations

Kwang Pil Chang, Daejun Chang, and Enrico Zio

Abstract The purpose of this chapter is to show the practical application of the Monte Carlo simulation method in the evaluation of the production availability of offshore facilities, accounting for realistic aspects of system behavior. A Monte Carlo simulation model is developed for a case study to demonstrate the effect of maintenance strategies on the production availability, e.g., by comparing the system performance under different preventive maintenance tasks.

11.1 Introduction

11.1.1 Offshore Installations

Offshore installations are central elements in the supply chain of offshore oil and gas. As shown in Figure 11.1, the supply chain consists of four stages: well drilling, production & storage, transport, and supply to consumers. Offshore installations are the facilities in the first two upstream stages, including drilling rigs and ships for the well-drilling stage, and fixed-platform floating production, storage, and offloading units (FPSOs), and floating storage and offloading units (FSOs) for the production and storage stage. As the new supply chain of liquefied natural gas (LNG) is under

---

K.P. Chang
Hyundai Industrial Research Institute, Hyundai Heavy Industries, Ulsan, Korea

D. Chang
Dept. of Ocean Systems Engineering, Korea Advanced Institute of Science and Technology, Daejeon, Korea (Former, Hyundai Industrial Research Institute, Hyundai Heavy Industries, Ulsan, Korea)

E. Zio
Energy Department, Politecnico di Milano, Milan, Italy

P. Faulin, A. Juan, S. Martorell, and J.E. Ramírez-Márquez (eds), *Simulation Methods for Reliability and Availability of Complex Systems*. © Springer 2010

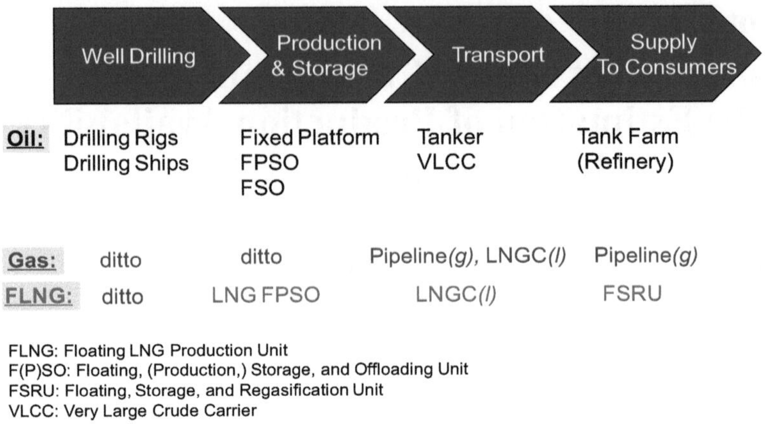

Figure 11.1 Supply chain of offshore oil and gas with offshore installations in well-drilling and production & storage stages

development, LNG FPSOs seem to arise as a noticeable member of the offshore installation.

11.1.2 Reliability Engineering Features of Offshore Installations

Offshore installations differ from other industrial facilities in that they are unique, safety-critical, independent, and subject to varying operating conditions. Each of these features is worth a more elaborate explanation. An offshore installation is fixed in a site and designed to perform its specific function, e.g., to process a well fluid to gas and crude streams. The environmental conditions and the target fluids are different for every installation.

Safety-criticality is another intrinsic feature of offshore installations as they process flammable gas and crude in a congested space. Fire and explosion are the most challenging hazards. Moreover, offshore installations have various additional hazards including collision with supply and stand-by vessels, helicopter crash, and dropped objects during crane operations.

Each offshore installation is designed to operate by itself, in an independent manner and be protected in case of emergency. Under normal operating circumstances the utilities of electric power, cooling water, heating medium, instrument air, etc. should be generated within the offshore installation itself. Even the operator resides within the installation. The presence of onsite maintenance engineers and spare-parts storage enhances the recovery of system performance after its components or equipments fall into failure. When a catastrophic accident takes place, the installation should activate its safety systems to mitigate the severity and evolution of the accident. In an emergency, the installation should provide for the accommodated personnel to escape safely.

The operating conditions of an offshore installation usually vary over its life cycle. Take as an example a floating production installation: it receives the feed well fluid from the subsea wellhead, which is connected to the underground wells; as the production continues, the properties of the well fluid keep changing; typically, the well pressure decreases, the oil portion decreases with the gas and water portion increasing. This means that the floating installation should handle a feed with properties which change over the long-term period.

These features typical of offshore installations represent significant challenges of reliability engineering in terms of:

- assigning proper reliability levels for various safety systems, i.e., the commonly called safety integrity levels (SILs);
- verifying the SILs taking into account the realistic reliability information of the components of the safety systems;
- optimizing maintenance including the preventive maintenance intervals and the stock of spare parts for the corrective maintenance;
- estimating the realistic production level or production availability, considering the failure and repair behaviors of the process components and equipment;
- optimizing process configuration to maximize the life cycle cost considering the capital expenditure and operating expenditure with production availability and accidental operation interruption taken into account.

Some of these challenges still require research developments. Reliability engineering and risk analysis come into full effect for the analysis and evaluation of the detailed design, where many of the details are frozen; on the other hand, some "smart" approach of reliability engineering is needed in the early stage of the conceptual design to optimize the backbone of the offshore installation design under development.

11.1.3 Production Availability for Offshore Installations

Production availability is defined as the ratio of the actual production to the planned one over a specified period of time [1]. It is an important indicator of the performance of offshore installations since it describes how much the system is capable of meeting the delivery demand.

The offshore industry requires a rigorous estimation of the production availability not just for knowing the evolution of the average production of the facility, but also for optimizing the components and systems maintenance strategies, e.g., in terms of their maintenance intervals and spare-parts holdings. In this sense, the analysis of production availability serves as integration work and cornerstone of all other efforts of reliability engineering for the offshore installation since it extensively considers all the aspects of design and operation.

Indeed, production availability is affected by the frequencies of corrective and preventive maintenance tasks. Furthermore, the spare-parts holding requirements must comply with the limits of space and weight at the offshore facility. Also, the items with long lead times for replacement can produce a serious negative effect on the production availabilities.

The following are typical results which are expected from a production availability analysis:

- to quantify the oil/gas production efficiency for the system over the considered field life;
- to identify the critical items which have a dominant effect on the production shortfall;
- to verify the intervention and maintenance strategies planned during production;
- to determine the (minimum) spare-parts holdings.

To produce these results under realistic system conditions, Monte Carlo simulation is increasingly being used to estimate the production availabilities of offshore installations [2]. The main attractiveness of such simulation is that it allows accounting for realistic maintenance strategies and operational scenarios, and for this reason it is being widely used in various fields of application [3–5].

This chapter focuses on the problem of estimating the production availability of offshore installations. The purpose of the chapter is to show the application of the Monte Carlo simulation method for production availability estimation, which enables accounting for realistic aspects of system behavior. A case study is considered which exploits a Monte Carlo simulation model to investigate the effects of alternative maintenance strategies on the production availability of the offshore facility.

The next section summarizes the classical Monte Carlo simulation approach for system reliability and availability analysis [6, 7]. Section 11.3 presents a pilot case model to show the application of the Monte Carlo simulation method for production availability estimation on a case study. Section 11.4 provides a short list of commercial tools available for the evaluation of production availability in offshore production facilities. Section 11.5 concludes the chapter with some remarks on the work presented.

11.2 Availability Estimation by Monte Carlo Simulation

In practice, the evaluation of system availability by Monte Carlo Simulation is done by performing a virtual observation of a large number of identical stochastic systems, each one behaving differently due to the stochastic character of the system behavior, and recording the instances in which they are found failed [6, 7]. To do this, the stochastic process of transition among the system states is modeled and a large number of realizations are generated by sampling the times and arrival states of the occurring transitions. Figure 11.2 shows a number of such realizations on the plane system configuration *vs.* time: in such a plane, the realizations take the form of random walks made of straight segments parallel to the time axis in-between tran-

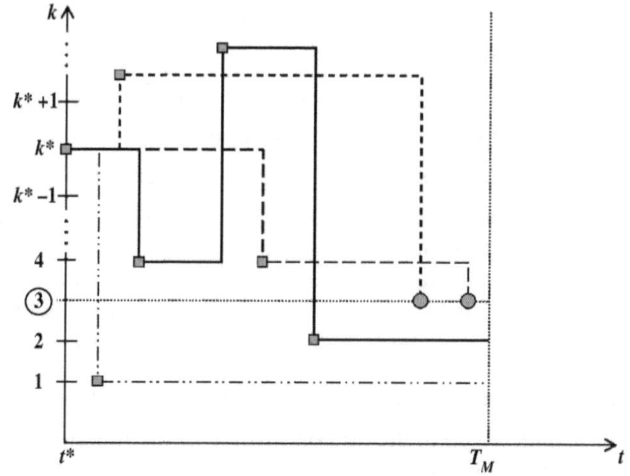

Figure 11.2 System random walks in the system configuration *vs.* time plane. System configuration 3 is circled as a faulty configuration. The *squares* identify points of transition (t, k); the *bullets* identify faulty states. The *dashed lines* identify realizations leading to system failure before the mission time T_{M}

sitions when the system is in a given configuration, and vertical stochastic jumps to new system configurations at the stochastic times when transitions occur [8].

For the purpose of estimation of system availability, a subset Γ of the system configurations is identified as the set of faulty states. Whenever the system enters one such configuration, its failure is recorded together with its time of occurrence and all the successive times during which the system remains down, before being repaired. With reference to a given time t of interest, an estimate of the system instantaneous unavailability at time t, i.e., of the probability that the system is down at such time, can be obtained by the frequency of instances in which the system is found failed at t, computed by dividing the number of system random walk realizations which record a failed state at t by the total number of random walk realizations simulated.

The Monte Carlo simulation of one single system random walk (also called history or trial) entails the repeated sampling from the probabilistic transport kernel defining the process of occurrence of the next system transition, i.e., the sampling of the time t and the new configuration k reached by the system as a consequence of the transition, starting from the current system configuration k' at t'. In this chapter, this is done by the so-called direct Monte Carlo simulation approach [7].

In the direct Monte Carlo simulation approach, the system transitions are generated by sampling directly the times of all possible transitions of all individual components of the system and then arranging the transitions along a timeline in increasing order, in accordance to their times of occurrence. The component which actually performs the transition is the one corresponding to the first transition in the timeline. Obviously, this timeline is updated after each transition occurs, to include the new possible transitions that the transient component can perform from its new state. In other words, during a history starting from a given system configuration k' at t', we sample the times of transition $t^i_{j'_i \to m_i}$, $m_i = 1, 2, \ldots, N_{S_i}$, of each component i, $i = 1, 2, \ldots, N_c$, leaving its current state $t^i_{j'_i \to m_i}$ and arriving at the state m_i from the corresponding transition time probability distributions $f_T^{i, j'_i \to m_i}(t|t')$. The time instants $t^i_{j'_i \to m_i}$ thereby obtained are then arranged in ascending order along a timeline from $t_{\min}$ to $t_{\max} \leqslant T_{\mathrm{M}}$. The clock time of the trial is then moved to the first occurring transition time $t_{\min} = t^*$ in correspondence of which the system configuration is changed, i.e., the component i^* undergoing the transition is moved to its new state m^*_i. At this point, the new times of transition $t^{i^*}_{m_{i^*} \to l^{i^*}_{i^*}}$, $l_{i^*} = 1, 2, \ldots, N_{S_{i^*}}$ of component i^* out of its current state m^*_i are sampled from the corresponding transition time probability distributions, $f_T^{i^*}$, $m^{l_{i^*}}_{i^*} \to l_{i^*} (t|t^*)$, and placed in the proper position of the timeline. The clock time and the system are then moved to the next first occurring transition time and corresponding new configuration, respectively, and the procedure repeats until the next first occurring transition time falls beyond the mission time, i.e., $t_{\min} > T_{\mathrm{M}}$.

For illustration purposes, consider for example the system in Figure 11.3, consisting of components A and B in active parallel followed by component C in series. Components A and B have two distinct modes of operation and a failure state whereas component C has three modes of operation and a failure state. For example, if A and B were pumps, the two modes of operation could represent the 50% and 100% flow modes; if C were a valve, the three modes of operation could represent the "fully open," "half-open," and "closed" modes.

For simplicity of the illustration, but with no loss of generality, let us assume that the components times of transition between states are exponentially distributed and denote by $\lambda^i_{j_i \to m_i}$ the rate of transition of component i going from its state j_i to the state m_i. Table 11.1 gives the transition rates matrices in symbolic form for components A, B, and C of the example (with the rate of self-transition $\lambda^i_{j_i \to j_i} = 0$ by definition).

The components are initially ($t = 0$) in their nominal states, which we label with the index 1 (e.g., pumps A and B at 50% flow and valve C fully open) whereas the failure states are labeled with the index 3 for the components A and B and with the index 4 for component C.

The logic of operation is such that there is one minimal cut set of order 1, corresponding to component C in state 4, and one minimal cut set of order 2, corresponding to both components A and B being in their respective failed states 3.

Starting at $t = 0$ with the system in nominal configuration (1, 1, 1) one would sample the times of all the possible components transitions by the inverse transform

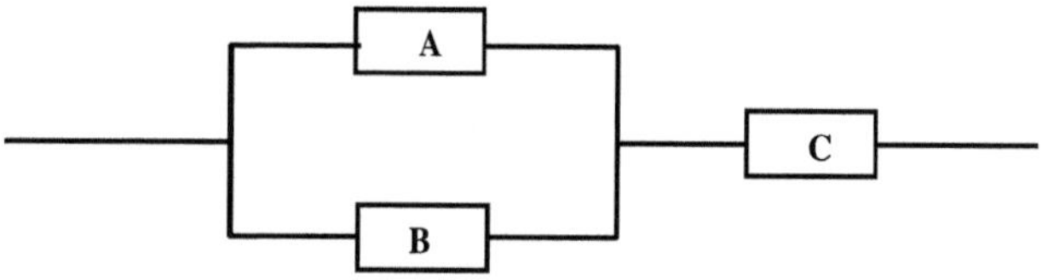

Figure 11.3 A simple series–parallel logic

Table 11.1 Component transition rates

Initial	Arrival 1	2	3
1	0	$\lambda_{1\to 2}^{A(B)}$	$\lambda_{1\to 3}^{A(B)}$
2	$\lambda_{2\to 1}^{A(B)}$	0	$\lambda_{2\to 3}^{A(B)}$
3	$\lambda_{3\to 1}^{A(B)}$	$\lambda_{3\to 2}^{A(B)}$	0

Initial	Arrival 1	2	3	4
1	0	$\lambda_{1\to 2}^{C}$	$\lambda_{1\to 3}^{C}$	$\lambda_{1\to 4}^{C}$
2	$\lambda_{2\to 1}^{C}$	0	$\lambda_{2\to 3}^{C}$	$\lambda_{2\to 4}^{C}$
3	$\lambda_{3\to 1}^{C}$	$\lambda_{3\to 2}^{C}$	0	$\lambda_{3\to 4}^{C}$
4	$\lambda_{4\to 1}^{C}$	$\lambda_{4\to 2}^{C}$	$\lambda_{4\to 3}^{C}$	0

method [9], which in the case of exponentially distributed transition times gives

$$t_{1\to m_i}^{i} = t_0 - \frac{1}{\lambda_{1\to m_i}^{i}} \ln\left(1 - R_{t,1\to m_i}^{i}\right) \tag{11.1}$$

$$i = \mathrm{A}, \mathrm{B}, \mathrm{C}$$
$$m_i = 2, 3 \quad \text{for } i = \mathrm{A}, \mathrm{B}$$
$$m_i = 2, 3, 4 \quad \text{for } i = \mathrm{C}$$

where, $R_{t,1\to m_i}^{i} \sim U[0, 1]$.

These transition times would then be ordered in ascending order from $t_{\min}$ to $t_{\max} \leqslant T_{\mathrm{M}}$.

Let us assume that $t_{\min}$ corresponds to the transition of component A to state 3 of failure, i.e., $t_{\min} = t_{1\to 3}^{\mathrm{A}}$ (Figure 11.4). The other sampled transition time relating to component A, namely $t_{1\to 2}^{\mathrm{A}}$, is canceled from the timeline and the current time is moved to $t_1 = t_{\min}$ in correspondence with which the system configuration changes to (3, 1, 1) still operational, due to the occurred transition. The new transition times

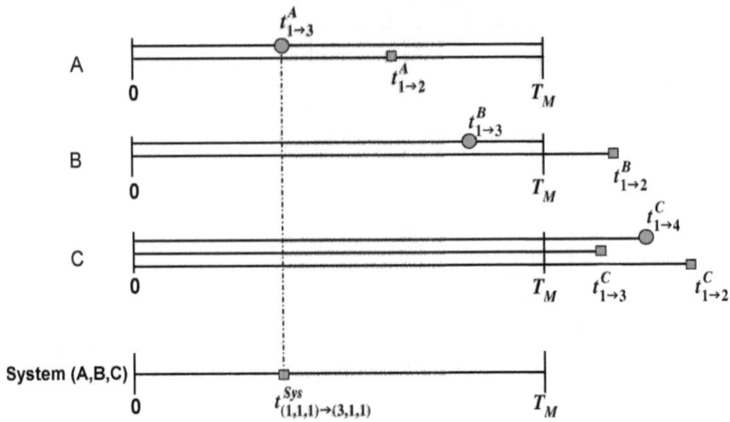

Figure 11.4 Direct simulation method. The squares identify component transitions; the bullets identify fault states

of component A are then sampled:

$$\begin{aligned} t^{\mathrm{A}}_{3\to m_{\mathrm{A}}} &= t_1 - \frac{1}{\lambda^{\mathrm{A}}_{3\to m_{\mathrm{A}}}} \ln\left(1 - R^{\mathrm{A}}_{t,3\to m_{\mathrm{A}}}\right) \\ m_{\mathrm{A}} &= 1, 2 \\ R^{\mathrm{A}}_{t,3\to m_{\mathrm{A}}} &\sim U[0, 1) \end{aligned} \tag{11.2}$$

and placed at the proper position in the timeline of the succession of transitions. The simulation then proceeds to the successive times in the list, in correspondence of which a system transition occurs. After each transition, the timeline is updated by canceling the times of the transitions relating to the component which has undergone the last transition and by inserting the newly sampled times of the transitions of the same component from its new state.

The trial simulation of the system random walk proceeds through the various transitions from one system configuration to another, until the mission time T_{M}. When the system enters a failed configuration (*, *, 4) or (3, 3, *), where the * denotes any state of the component, its time of occurrence is recorded together with all the successive times in which the system remains down, until it is repaired. More specifically, from the point of view of the practical implementation into computer code, the system mission time is subdivided in N_t intervals of length Δt and to each time interval an unavailability counter $C^{\mathrm{A}}(t)$ is associated to record the fact that the system is down at time t: at the time τ when the system enters a fault state a one is collected into all the unavailability counters $C^{\mathrm{A}}(t)$ associated to times successive to the failure occurrence time up to the time of repair. After simulating a large number of random walk trials M, an estimate of the system instantaneous unavailability at time t can be obtained by simply dividing by M and by the time interval Δt the accumulated contents of the counters $C^{\mathrm{A}}(t)$, $t \in [0, T_{\mathrm{M}}]$.

11.3 A Pilot Case Study: Production Availability Estimation

The procedure of production availability analysis by Monte Carlo simulation is illustrated in Figure 11.5. The availability is calculated by a Monte Carlo model for simulating the complicated interactions occurring among the components of the system, including time-based events and life-cycle logistic, operation, and reconfiguration constraints.

The first step for the calculation of the availability is to define the functional flow diagram of the system. Next, it is necessary to identify the potential failure modes of each component of the system and its production loss level with the failure event. The failure model due to the failure events is developed by a FMECA-like study. After constructing the failure model, the data and operational information should be collected for input to the simulation. Operation scenarios such as flaring policy, planned shutdown for inspection and failure management strategies are usually specified as a minimum. The failure management strategies are mainly focused on the planning of the preventive maintenance tasks. The feasible preventive maintenance task types and schedules can be determined based on RCM task decision logic or component suppliers" maintenance guidance. A simulation model is pre-

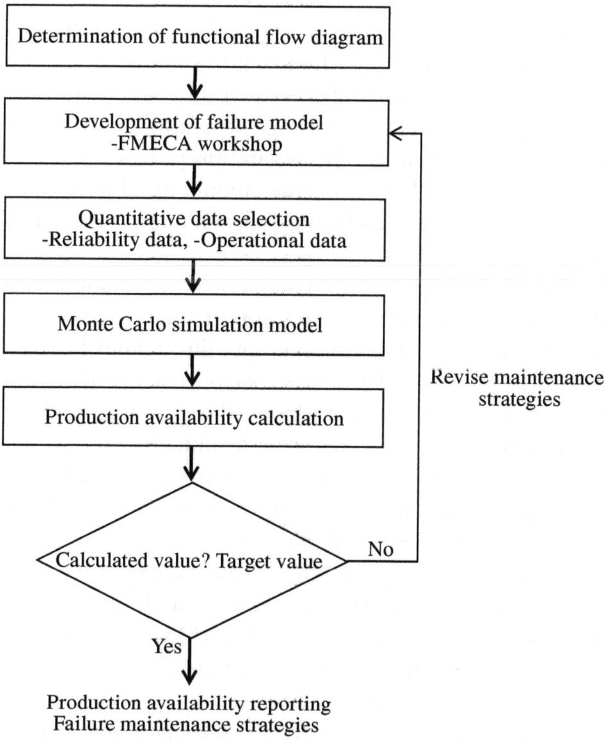

Figure 11.5 A procedure of the production availability analysis

pared based on the functional diagram and it imports the system configuration with the detailed information of the components, including their failure rates and repair times, and the system operational information. The simulation of the system life is repeated a specified number of times M. Each trial of the Monte Carlo simulation consists in generating a random walk of the system from one configuration to another at successive times. Let A_i be the production availability in the i-th system random walk, $i = 1, 2, \ldots, M$. The system availability A is then estimated as the sample mean of the individual random walks [10]:

$$A = \frac{\sum_{i=1}^{M} A_i}{M} \tag{11.3}$$

Finally, the estimated production availability is compared with the target value and if it does not satisfy the production requirements, the simulation system must be re-assessed.

11.3.1 System Functional Description

A prototypical offshore production process plant is taken as the pilot system for production availability assessment by Monte Carlo simulation (Figure 11.6).

The three-phase fluid produced in the production well enters a main separation system which works a single-train, three-stage separation process. The well fluid is separated into oil, water, and gas by the separation process. The well produces at its maximum 30,000 m^3/d of oil, which is the amount of oil which the separator can handle. The separated oil is exported by the export pumping unit, also with capacity of 30,000 m^3/d of oil.

Off-gas from the separator is routed to the main compressor unit, with two compressors running and one standby a *2oo3* voting. Each compressor can process a maximum of 3.0 MMscm/d. The nominal gas throughput for the system is assumed to be 6.0 MMscm/d, and the system performance will be evaluated at this rate. Gas dehydration is required for the lift gas, the export gas and the fuel gas. The dehydration is performed by a $1 \times 100\%$ glycol contactor on the total gas flowrate, based on gas saturated with water at conditions downstream of the compressor. The total maximum gas processing throughput is assumed to be 6.0 MMscm/d, limited by the main compression and dehydration trains.

To ensure the nominal level of production of the well, the lift gas is supplied from the discharge of the compression, after dehydration, and routed to the lift gas risers under flow control on each riser. An amount of 1.0 MMscm/d is compressed by the compressor for lift gas and injected back into the production well.

Water is injected into the producing reservoirs to enhance oil production and recovery. The water separated in the separator and treated seawater is injected in the field. The capacity of water injection system is assumed to be 5,000 m^3/d.

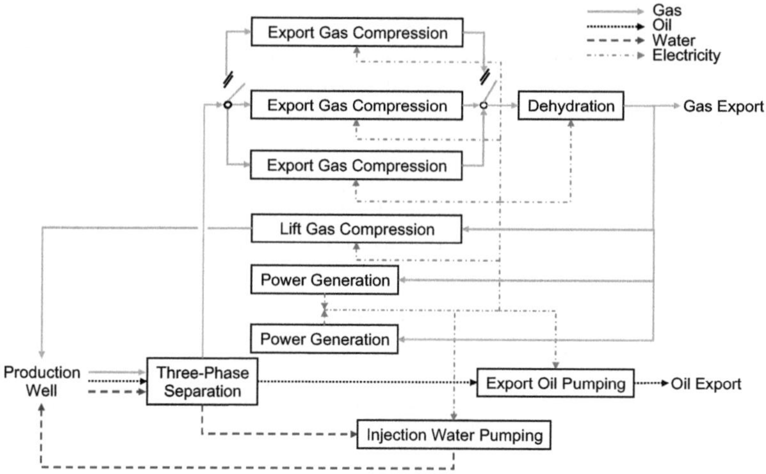

Figure 11.6 Functional block diagram of the offshore production process plant

The 25 MW power requirements on the production system will be met by 2 × 17 MW gas turbine-driven power generation units.

11.3.2 Component Failures and Repair Rates

For simplicity, the study considers in details the stochastic failure and maintenance behaviors of only the *2oo3* compressor system (one in standby) for the gas export and the *2oo2* power generation system; the other components have only two states "functioning" and "failed."

The transition rates of the components with only two transition states are given in Table 11.2.

Table 11.2 Transition rates of the components

Component	MTTF (per 10^6 h)	MTTR (h)
Dehydration	280	96
Lift gas compressor	246	91
Export oil pump	221	150
Injection water pump	146	127
Three-phase separator	61.6	5.8
Export gas compressor	246	91
Power generator	500	50

The compressor and power generation systems are subjected to stochastic behaviors due to their voting configuration. The failures and repair events for both compressor and power generation systems are described in Section 11.3.4 in detail.

The required actual performance data or test data of the components are typically collected from the component supplier companies.

If it is impossible to collect the data directly from the suppliers, then generic data may be used as an alternative to estimate the component failure rates. Some generic reliability databases used for production availability analysis are:

- OREDA (Offshore Reliability Data).
- NPRD (Non-electronic Parts Reliability Data).
- EIREDA (European Industry Reliability Data Bank).

In many cases, the generic data are adjusted or reviewed with experts for production availability analysis.

11.3.3 Production Reconfiguration

The failure of the components and systems are assumed to have the following effects on the production level:

- Failure of any component immediately causes the production level to decrease by one step.
- Failure of the lift gas compression or water injection pump reduces the oil production by 10,000 m^3/day (30% of total oil production rate) and the gas production by 2.0 MMscm/day.
- Failure of both the lift gas compression and injection water pumping reduces the oil production by 20,000 m^3/day and the gas production by 4.0 MMscm/day.
- Failure of two export gas compressors or one generator forces the compression flow rate to decrease from 6.0 MMscm/day to 3.0 MMscm, facing the oil production rate to reduce accordingly from 30,000 m^3/day to 15,000 m^3/day.
- Failure of the dehydration unit, all three export gas compressors, or both power generators results in total system shutdown.

The strategy of production reconfiguration against the failure of the components in the system is illustrated in Table 11.3.

11.3.4 Maintenance Strategies

11.3.4.1 Corrective Maintenance

Once failures occur in the system, it is assumed that a corrective maintenance is immediately implemented by only a single team apt to repair the failures. In the

Table 11.3 Summary of different production levels upon component failures

Production level (Capacity, %)	Failure events	Oil (km^3/d)	Gas (MMsm/d)	Water injection (km^3/d)
100	None	30	6	5
70	Lift gas compressor	20	4	4
70	Water injection pump	20	4	0
50	Two export gas compressors One power generator Two export gas compressor and one power generator together	15	3	5
50	Two export gas compressors and injection water pumping	15	3	0
30	Lift gas compressor and injection water pump	10	2	0
0	Dehydration unit All three export gas compressors Both power generators	0	0	0

case that two or more components are failed at the same time, the maintenance tasks are carried out according to the sequence of occurrence of the failure events.

The failure and repair events of the export gas compressor system and power generation are more complicated than those of the other components. Figure 11.7 shows the state diagram of the export compression system. As shown in Figure 11.7, common-cause failures which would result in total system shutdown are not considered in the study. The compressors in the export compression system are considered to be identical. The times of transition from a state to another are assumed to be exponentially distributed; this assumption describes the stochastic transition behavior of the components during their useful life, at constant transition rates, and is often made in practice when the data available are not sufficient to estimate more than the transition rates. Assumptions on the component stochastic behavior of transition other than the exponential (e.g., the Weibull distribution to describe aging processes) can be implemented in a straightforward manner within the Monte Carlo simulation scheme, by changing formula 11.1 of the inverse transform method for sampling the component transition times [9]. Obviously, in practice any assumption on the components stochastic behavior, i.e., on the distribution of the transition times, must be supported by statistical data to estimate the parameters of the stochastic model which arise.

The export compression system can be in four different states. State 0 corresponds to two active compressors running at 100% capacity. State 1 corresponds to one of the two active compressors being failed and the third (standby) compressor being switched on while the repair task is carried out; the switch is considered perfect and therefore state 1 produces the same capacity as state 0. State 2 represents operation with only one active compressor or one standby compressor (two failed compressors), i.e., 50% capacity; the export compression system can transfer

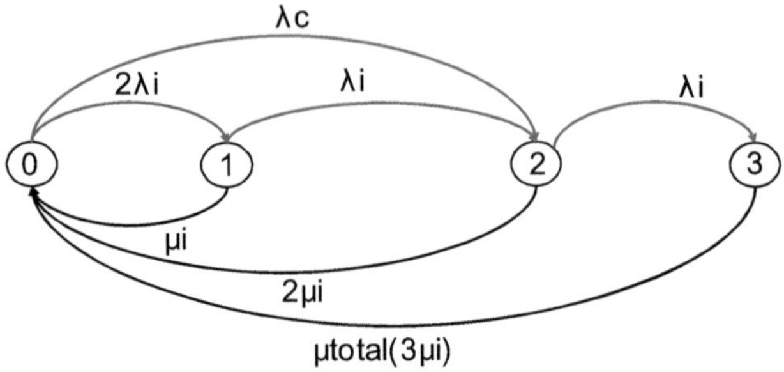

Figure 11.7 State diagram of export compression system

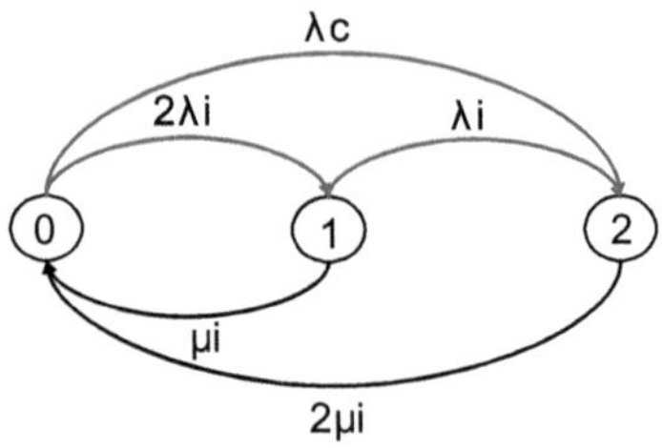

Figure 11.8 State diagram of power generation system

to state 2 by transition from either state 0 directly (due to common cause failure of two of the three compressors) or from state 1 (due to failure of an additional compressor). State 3 corresponds to total system shutdown, due to failure of all three compressors.

The same assumptions of the export compression system apply to the power generation system, although there are only three states given the parallel system logic. The state diagram is shown in Figure 11.8. Repairs allow returning to states of higher capacity from lower ones.

11.3.4.2 Preventive Maintenance

The following is assumed for the preventive maintenance tasks:

- Scheduled preventive maintenance is only implemented to the compressor system for the gas export and to the power generation system.
- Scheduled maintenance tasks of the compressors and the power generation system are carried out at the same time, to minimize downtime.
- Well should be shut down during preventive maintenance.

The scheduled maintenance intervals for both systems are given in Table 11.4.

Table 11.4 Scheduled maintenance interval for compressors and power generators

Period (month)	Task type	Downtime (h)
2	Detergent washing	6
4	Service/cleaning	24
12	Baroscopic inspection/generator check	72
60	Overhaul or replacement	120
48	Planned shutdown	240

11.3.5 Operational Information

In addition to the information provided in Sections 11.3.1 to 11.3.4, much additional operational information should be incorporated in the simulation model. The principal operation scenarios necessary to be considered during estimation of production availability for offshore facilities are:

- flaring policy;
- start-up time;
- planned downtime:
  - emergency shutdown test,
  - fire and gas detection system test,
  - total shutdown with inspection.

No flaring and no production delay at start-up time are assumed in the study. Every 4 years, the facility is totally shut down for 10 days due to planned inspection.

11.3.6 Monte Carlo Simulation Model

The system stochastic failure/repair/maintenance behavior has been modeled by Monte Carlo simulation and quantified by a dedicated computer code implemented by Visual BASIC programming.

11.3.6.1 Model Algorithm

Figure 11.9 illustrates the flowchart of the Monte Carlo simulator developed in the study. First of all, the program imports the system configuration with the detailed information of the components including the failure rates, repair times, preventive maintenance intervals, and required downtimes. Then, the simulator proceeds to determining the next transition times for all the components. These depend on the current states of the components. When the component is under corrective or preventive maintenance, its next transition occurs after completion of the maintenance

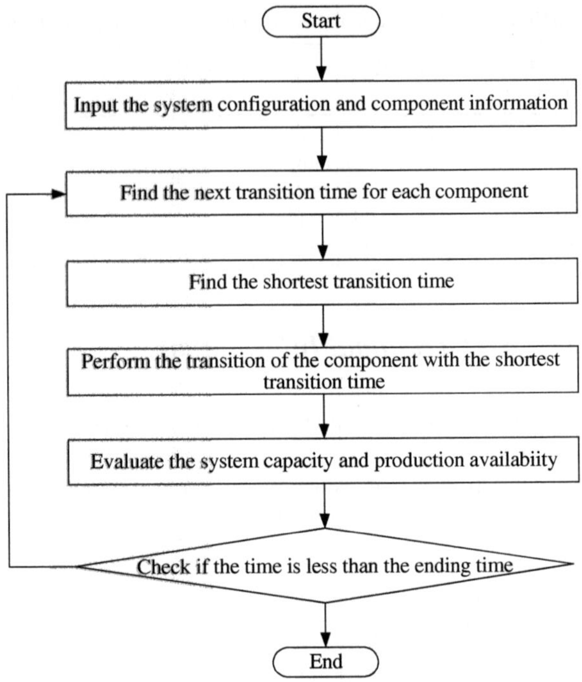

Figure 11.9 Flow chart for developed simulation program

action; this maintenance time is predetermined. When the component is in operation (not necessarily with 100% capacity), the next transition time is sampled by the direct Monte Carlo simulation method of Section 11.2 [7].

11.3.6.2 Numerical Results

Figure 11.10 shows the values of plant production availability over the mission time for 10,000 system life realizations (histories), each one representing a plausible evolution of the system performance over the 30-year analysis period. The sample mean of 93.4% gives an estimate of the system performance.

The key contributors to production losses are shown in Figure 11.11. Lift gas compressor, dehydration package, and export oil pump account for 82% of the production loss. The key contributors to production loss can be classified into two groups:

- *Type I.* Components having high failure rates with no redundancy: dehydration system, oil export pump.
- *Type II.* Components subject to frequent preventive tasks and no redundancy: lift gas compressor, impact of scheduled maintenance task ≫ impact of critical failures of component.

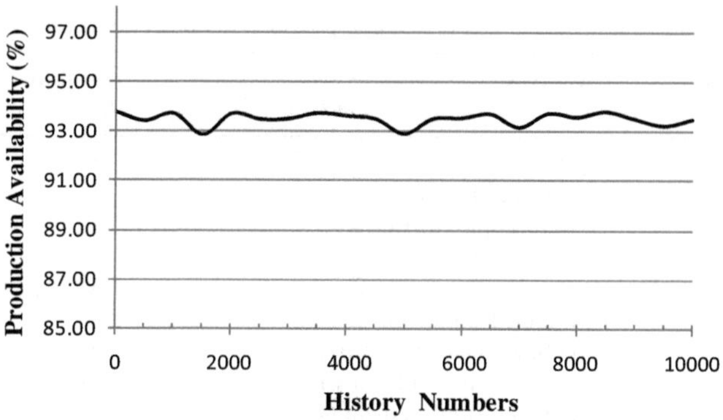

Figure 11.10 Production availability values of the 10,000 Monte Carlo Simulation histories

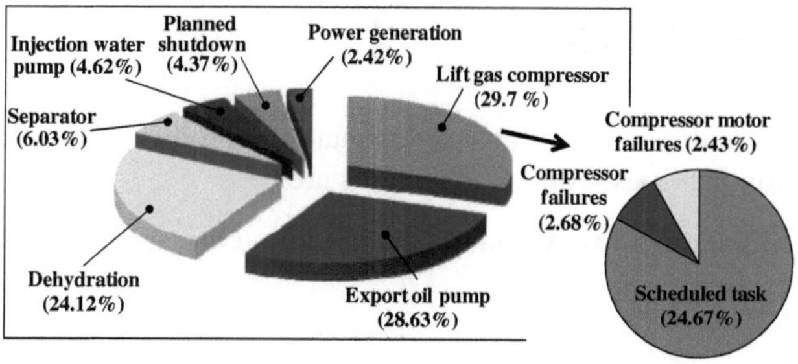

Figure 11.11 Key contributors to production losses

11.3.6.3 Effect of Preventive Maintenance

According to Figure 11.11, the preventive maintenance tasks of lift gas compressors and generators are identified as one of the key contributors to production losses.

Table 11.5 shows an example of the effect of preventive maintenance tasks on the production availability and how the information for identification of key contributors to production availability could be used to improve the system performance. The comparison between the results of the nominal case (case 1) described in Table 11.4 and the case with the reduction of frequencies for preventive maintenance tasks (case 2) is shown. The case 2 results are prepared based on the combination of each maintenance task identified in the nominal case with respect to maintenance job similarity. For example, the combined task of case 2 is defined to conduct each preventive maintenance task identified in the nominal case, detergent washing, service/cleaning, and inspection/generator check, at the same time every 12 months.

According to the Table 11.5, the more frequent preventive maintenance actions slightly decrease the production availability; this result is due to the assumption that components do not age (i.e., their failure behavior is characterized by constant failure rates), so that maintenance has the sole effect of rendering them unavailable while under maintenance.

Table 11.5 Schedule maintenance interval for compressors and power generators

	Period (month)	Task type	Downtime (h)	Availability (%)
Case 1	2	1. Detergent washing	6	93.4
	4	2. Service/cleaning	24	
	12	3. Baroscopic inspection/ generator check	72	
	60	4. Overhaul or replacement	120	
	48	5. Planned shutdown	240	
Case 2	12	Combined task (1 + 2 + 3)	100	94.1
	48	Planned total shutdown with overhaul or replacement (4 + 5)	360	

11.4 Commercial Tools

Commercial simulators are available to estimate the production availability of offshore production facilities. Some known tools are:

- MAROS (Maintainability Availability Reliability Operability Simulator);
- MIRIAM Regina;
- OPTAGON.

These commercial simulators are based on Monte Carlo simulation schemes with similar technical characteristics. For example, the flow algorithm is one feature common to all and the simulation model can consider a wide variety of complicated components, system behaviors, and operational and maintenance philosophies including production profile, start, and logistic delays. These realistic aspects of production are not readily implementable in analytical models.

MAROS applies a direct simulation algorithm structured on the sampling and scheduling of the next occurring event. The main input and output are summarized in Table 11.6 (`http://www.jardinetechnology.com/products/maros.htm`).

The OPTAGON package is a tool for production availability developed by BG Technology (`http://www.advanticagroup.com/`). OPTAGON uses reliability block diagrams with partial operating modeling to represent the functionality of a system in terms of its components, similarly to MAROS. The probability distributions used in OPTAGON are exponential, Weibull, normal, and lognormal or user-defined. The main output of the simulation by OPTAGON are shortfall, unavailability, system failure rate, and costs such as cost of shortfall, capital and operating costs, maintenance costs, and spares holding costs.

MIRIAM Regina is also commonly used to evaluate the operational performance of continuous process plants in terms of equipment availability, production capability, and maintenance resource requirements (`http://www.miriam.as/`). The main difference from other commercial tools is the modeling based on the flow algorithm which can handle multiple flows and records production availability for

Table 11.6 Main input and simulation output of MAROS

Model input	Simulation output
Economics	Production analysis
– unit costs, product pricing	– availability
– CAPEX	– production efficiency
Production	– equipment criticality
– reservoir decline	– contract/production shortfalls
– plant phase-in/out	
Operations	Net product value (NPV) cash flows
– item reliability	
– redundancy	Maintenance analysis
Maintenance	– manpower expenditure
– resources, priority of repair	– mobilization frequency
– work shifts, campaign/opportune	– planned maintenance scheduling
– logistics	– spare/manpower utilization
Transportation	
– round-trip delays	
– weather factors	
– standby/service vessel	

several boundary points. The probability distribution types available in MIRIAM Regina are as follows: constant, uniform, triangular, exponential, gamma, and log-normal.

11.5 Conclusions

In this chapter, the problem of estimating the production availability in offshore installations has been tackled by standard Monte Carlo simulation. Reference has been made to a case study for which a Monte Carlo simulation model has been developed, capable of accounting for a number of realistic operation and maintenance procedures. The illustrative example has served the purpose to show the applicability and added value of a Monte Carlo simulation analysis of production availability. The simulation environment allows closely following the realistic behavior of the system without encountering the difficulties which typically affect analytical modeling approaches. Yet, it seems important to remark that the actual exploitation of the detailed modeling power offered by the Monte Carlo simulation method still rests on the availability of reliable data for the estimation of the parameters of the model.

Acknowledgements The authors wish to express their gratitude to the anonymous reviewers for their thorough revision which has led to significant improvements to the presentation of the work performed.

References

1. NORSOK Standard (Z-016) (1998) Regularity management & reliability technology. Norwegian Technology Standards Institution, Oslo, Norway
2. Zio E, Baraldi P, Patelli E (2006) Assessment of the availability of an offshore installation by Monte Carlo simulation. Int J Pressure Vessels Pip 83:312–320
3. Juan A, Faulin J, Serrat C, Bargueño V (2008) Improving availability of time-dependent complex systems by using the SAEDES simulation algorithms. Reliab Eng Syst Saf 93(11):1761–1771
4. Juan A, Faulin J, Serrat C, Sorroche M, Ferrer A (2008) A simulation-based algorithm to predict time-dependent structural reliability. In: Rabe M (eds) Advances in simulation for production and logistics applications. Fraunhofer IRB Verlag, Stuttgart, pp 555–564 (ISBN: 978-3-8167-7798-4)
5. Juan A, Faulin J, Sorroche M, Marques J (2007) J-SAEDES: A simulation software to improve reliability and availability of computer systems and networks. In: Proceedings of the 2007 winter simulation conference, Washington DC, December 9–12, pp 2285–2292
6. Dubi A (1999) Monte Carlo applications in systems engineering. Wiley, Hoboken, NJ, USA
7. Marseguerra M, Zio E (2002) Basics of the Monte Carlo method with application to system reliability. LiLoLe-Verlag, Hagen, Germany
8. Zio E (2009) Computational methods for reliability and risk analysis. World Scientific Publishing, Singapore
9. Labeau PE, Zio E (2002) Procedures of Monte Carlo transport simulation for applications in system engineering. Reliab Eng Syst Saf 77:217–228
10. Rausand M, Hoyland A (2004) System reliability theory: models, statistical methods, and application, 2nd edn. Wiley-Interscience, Hoboken, NJ, USA

Chapter 12
Simulation of Maintained Multicomponent Systems for Dependability Assessment

V. Zille, C. Bérenguer, A. Grall, and A. Despujols

Abstract In this chapter, we propose a modeling approach of both degradation and failure processes and the maintenance strategy applied on a multicomponent system. In particular, we describe the method implementation using stochastic synchronized Petri nets and Monte Carlo simulation. The structured and modular model developed allows consideration of dependences between system components due either to failures or to operating and environmental conditions. Maintenance activity effectiveness is also modeled to represent the ability of preventive actions to detect component degradation, and the ability of both preventive and corrective actions to modify and keep under control degradation mechanism evolution in order to avoid occurrence of a failure. The results obtained from part of a nuclear power plant are presented to underline specificities of the method.

12.1 Maintenance Modeling for Availability Assessment

Maintenance tasks are performed to prevent failure-mode occurrences or to repair failed components. It is a fundamental aspect of industrial system dependability. Therefore, the large impact of the maintenance process on system behavior should be fully taken into account in any reliability and availability analysis (Zio 2009).

It is difficult to assess the results of the application over several years of a complex maintenance program, resulting for example from implementation of the widely used reliability-centered maintenance (RCM) method (Rausand 1998). These difficulties are due to:

- the complexity of the systems, consisting of several dependent components, with several degradation mechanisms and several failure modes possibly in competition to produce a system failure;

V. Zille · C. Bérenguer · A. Grall
University of Technology of Troyes, France

A. Despujols
EDF R&D, France

P. Faulin, A. Juan, S. Martorell, and J.E. Ramírez-Márquez (eds), *Simulation Methods for Reliability and Availability of Complex Systems*. © Springer 2010

- the complexity of maintenance programs – large diversity of maintenance tasks and complexity of program structure.

For this reason, the numerous performance and cost models developed for maintenance strategies (Cho and Parlar 1991; Valdez-Flores and Feldman 1989), cannot be applied. Thus, it is desirable to develop methods to assess the effects of maintenance actions and to quantify the resulting system availability (Martorell *et al.* 1999).

In the RCM method, different maintenance tasks are defined with their own characteristics of duration, costs, and effects on component degradation and failure processes (Rausand 1998). Among them, we consider:

- corrective maintenance repairs, undertaken after a failure;
- preventive scheduled replacement using a new component, according to the maintenance program;
- preventive condition-based repair, performed according to the component state.

Within condition-based maintenance, component degradation states can be observed through detection tasks such as overhauls, external inspections, and tests (Wang 2002). All these monitoring actions differ in terms of cost, unavailability induced by component maintenance, and efficiency of detection (Barros *et al.* 2006; Grall *et al.* 2002). Depending on the component state observation, a preventive repair may be activated.

Overhauls consist of a long and detailed observation of the component to evaluate its degradation state. Their realization implies both a scheduled unavailability of the component and a high cost but it is highly efficient in terms of detection.

External inspections are less expensive than overhauls and consist of observing the component without stopping it. However, these two advantages imply a larger distance from the degradation and any associated error risks of non-detection or false alarm need to be considered. Typically, this kind of task can easily be used to observe some potential degradation symptoms, that is, measurable observations which characterize one or more degradation mechanism evolution. Thus, some error of appreciation can exist when decisions of preventive repair are taken (treatment of the wrong degradation mechanism while another one is still evolving with an increasing probability of failure).

Tests are performed on stand-by components to detect any potential failure before component activation. They can have an impact on the component degradation since they imply a subsequent activation.

To obtain a detailed representation of how various maintenance tasks applied within a complex maintenance program can impact a multicomponent system, it is important to take into account the entire causal chain presented in Figure 12.1 (Zille *et al.* 2008).

The aspects and relations described in Figure 12.1 can be modeled and simulated to describe individual system component behavior. These behaviors are consequences of different degradation mechanism evolutions that impact on components and may lead to some failure-mode occurrences. Thus, it is necessary to describe these evolutions and the way maintenance tasks can detect them (directly or through

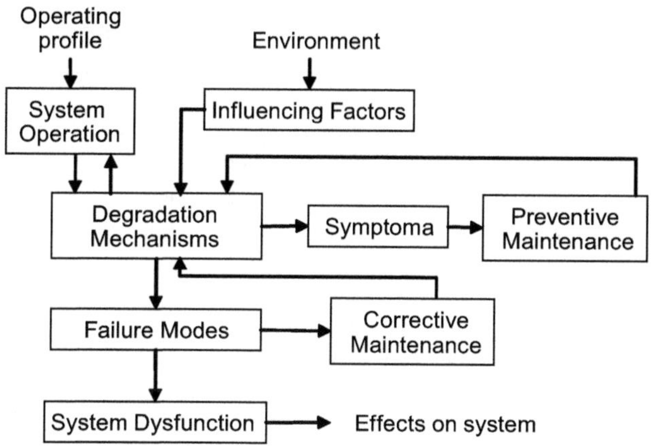

Figure 12.1 The causal chain describing component behavior and its impact on system availability

symptom detection) and repair, if necessary, in order to prevent or correct the effects on the system.

Thus, the behavior of the system composed by the above-described components has to be represented. The objective is to detail how the system can become unavailable, in a scheduled or in an unscheduled way. This is done by taking into consideration the dependences between components (Dekker 1996), and the modeling of:

- the occurrences of component failures;
- the impact of component failures on system functioning;
- the effects of maintenance tasks on components.

12.2 A Generic Approach to Model Complex Maintained Systems

Industrial complex systems contain numerous components. The availability of each component is submitted to failure-mode occurrences which may lead to the dysfunction of the system (Cho and Parlar 1991). Thus, to evaluate system availability, it seems convenient to represent the behavior of both the system and its components.

Therefore, four models can be developed and integrated together within a two-level model which takes into account both the degradation and failure phenomena and the maintenance process applications on components and the system (Bérenguer *et al.* 2004). In the proposed approach, this is done through the global framework presented in Figure 12.2 with the gray elements that refer to the system level and the white elements that refer to the component level. Within this overall structure, we distinguish the elements of the causal chain described in Figure 12.1.

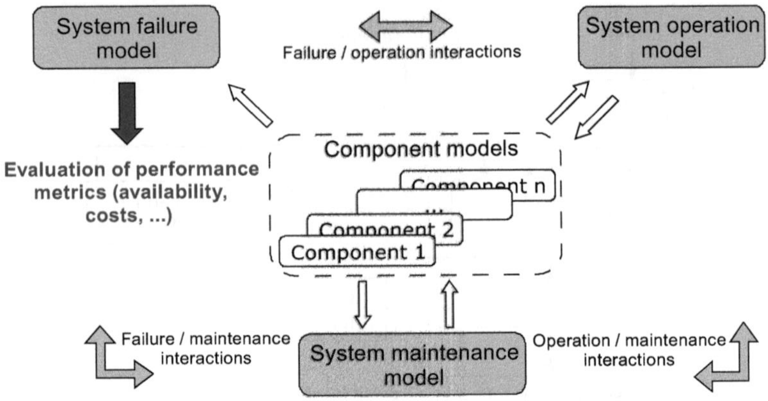

Figure 12.2 Overall structure for maintained complex system modeling

The three system-level models and the component-level model interact together in order to fully represent the system behavior, its unavailability and expenditure, according to the behavior of its components and the maintenance tasks carried out.

The nominal behavior and the operating rules of the system are defined in the system operation model, which interacts with the component model and evolves according to the operating profile and to the needs of the system (activating of a required component, stopping of a superfluous component, etc.).

The component level consists of a basic model developed for each component of the system by using a generic model taking into account both the physical states (sound state, degraded, hidden, or obvious failure) and the functional states (in maintenance, in stand-by, operating) of a component. It describes the degradation process and all the maintenance tasks that impact upon the component availability.

In addition, the system maintenance strategy applied is defined in the system maintenance model, whereas individual maintenance procedures are considered only at the component modeling level.

Finally, the system failure model describes all the degradation/failure scenarios of the system. It gives the global performance indicators of the maintained system. In particular, this model allows system unavailability evaluation, either due to a failure or to maintenance actions.

The proposed framework is hierarchical since the system behavior description, by means of the three models of the system level, is based on the component behavior evolution, described by the different models of the component level. Moreover, the overall model describes both probabilistic phenomena and processes and deterministic actions. Thus, a hybrid implementation is needed to simulate the model. These observations lead one to consider Petri nets as an appropriate implementation tool and more precisely, the stochastic synchronized Petri nets (SSPN), since SSPN use classical properties of Petri nets to treat the sequential and parallel processes, with stochastic and deterministic behaviors and flows of information called "messages" which are very useful in the proposed approach to describe the relations between the different models and levels within the global framework.

12.3 Use of Petri Nets for Maintained System Modeling

The proposed generic methodology has been developed using SSPN coupled with Monte Carlo simulation to assess industrial system performance (Bérenguer *et al.* 2004; Lindeman 1998; Dubi 2000).

For system dependability studies, SSPN offer a powerful modeling tool that allows for the description of:

- random phenomena, such as failure occurrence;
- deterministic phenomena, such as maintenance action realization;
- discrete phenomena, such as event occurrence;
- continuous phenomena, such as degradation mechanism evolution.

Several Petri net elements are built to model all the different aspects that are under consideration in Figures 12.1 and 12.2. System dependability studies can then be carried out by instantiating the generic elements. This allows a very large number of systems and strategies to be considered.

12.3.1 Petri Nets Basics

The Petri net is a directed graph modeling approach consisting of places, transitions, and directed arcs, as in Figure 12.3 (Alla and David 1998). Nets are formed by a five-tuple $N = (P, T, A, W, M_0)$ where P is a finite set of places, T is a finite set of transitions, A is a set of arcs, W is a weight function, and M_0 is an initial marking vector. Arcs run between places and transitions, from an input place to an output place. Places may contain any non-negative number of tokens. In this case, places are said to be marked.

A transition of a Petri net may *fire* whenever there is a token at the end of all input arcs; when it fires, it consumes these tokens, and sends the tokens to the end of all the output arcs. In other words:

- Firing a transition t in a marking M consumes $W(s, t)$ tokens from each of its input places s, and produces $W(t, s)$ tokens in each of its output places s.

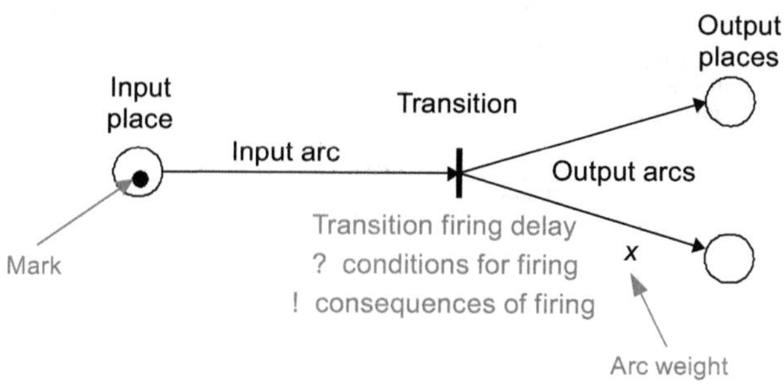

Figure 12.3 Petri net concepts

- The transition is *enabled* and may *fire* in M if there are enough tokens in its input places for the consumption to be possible and if conditions for firing are validated.
- The transition firing may lead to the update of messages or consequences (for example the value of a variable).

12.3.2 Component Modeling

Within the basic component-level model, a Petri net is built for each degradation mechanism to represent its evolution through several degradation levels and the respective risk of failure-mode occurrence, see Figure 12.4. It is a phase-type model (Pérez-Ocon and Montoro-Cazorla 2006), which can give a fine and detailed description of a large part of degradation evolutions with classical modeling tools. In particular, it is possible to represent mechanisms that evolve, e.g., according to life-time distribution (Wang 2002), or according to random shocks (Bogdanoff and Kozin 1985), as well as failures that occur in a random way at any time of the component life or, on the contrary, after a given lifetime (Marseguerra and Zio 2000).

Figure 12.4 describes the evolution of degradation mechanism 1 and the existing relations with maintenance and failure-mode occurrence. The black elements refer to the degradation, the dark gray elements refer to failure modes and the light gray elements refer to the impact of maintenance on the degradation level.

Transitions between two successive levels of degradation are fired according to probability laws taking into account the various influencing factors that have an impact on the mechanism evolution such as environmental conditions, failure of another component, etc. ... And the token goes from one place to another to describe the behavior of the considered component. Failure modes can occur at every

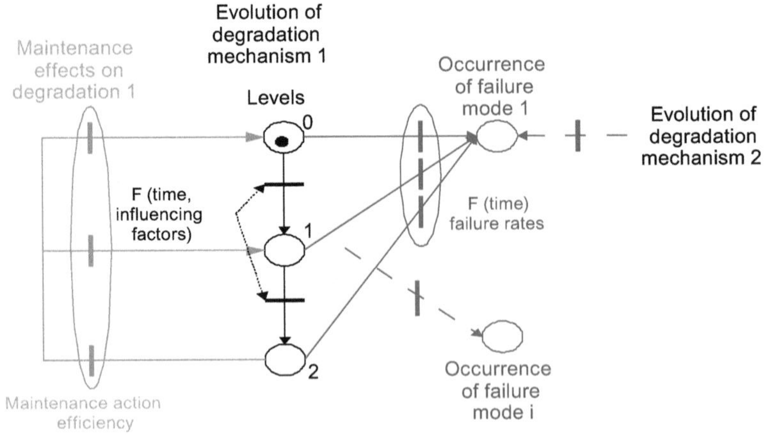

Figure 12.4 Representation of a component degradation and failure processes by using Petri nets

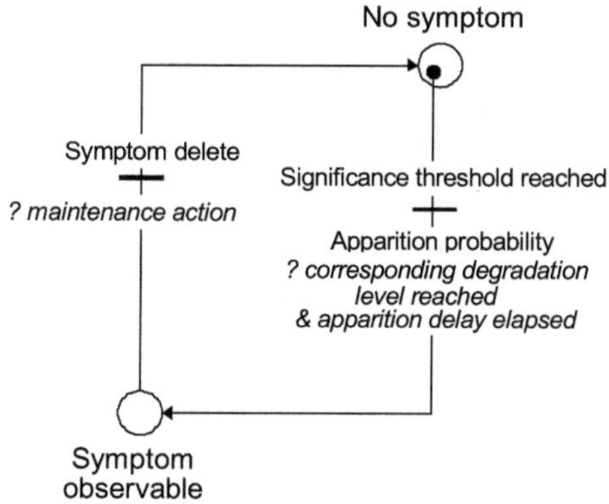

Figure 12.5 Petri net modeling of symptom appearance

degradation level, with a corresponding failure rate, represented by the firing of the corresponding transition, increasing with the degradation level. The return to a lower degradation level is due to maintenance task performance and depends on its effectiveness (Brown and Proschan 1983).

Figure 12.4 also represents the fact that a failure mode can appear due to various degradation mechanisms, as well as the fact that a degradation mechanism can cause more than one failure mode.

In addition, symptoms, that is, observations that appear and characterize degradation mechanism evolution, are represented. This allows for the description of condition-based maintenance tasks such as external inspections, which give information about the component degradation level and make it possible to decide to carry out a preventive repair (Jardine *et al.* 1999). Figure 12.5 shows the Petri net modeling of symptom appearance: when a symptom reaches a given threshold, it becomes a sign of a degradation evolution. Its detection, during an external inspection, may avoid the degradation observation to make a decision about repairing the component if necessary. Obviously, symptom appearance is linked to the evolution of the degradation. A symptom testifies to a degradation level and is deleted after a repair.

By representing failure occurrence, degradation evolution, and symptom appearance, all the RCM maintenance tasks shown in Table 12.1 can be considered: predetermined maintenance tasks (scheduled replacement), condition-based maintenance tasks (external inspection, condition monitoring, test, overhaul), and corrective maintenance (repair). Their associated effects on the various behavior phenomena are modeled, as well as their performance corresponding to the maintenance program defined.

Since all tasks in Table 12.1 have their own characteristics, it is important to create an appropriate description of each one. Thus, specific Petri net models are proposed. As an example, Figures 12.6 and 12.7 describe the representation of overhauls and preventive repair of a component.

According to the preventive maintenance program, when the time period is elapsed, an overhaul is performed on the component to detect its degradation state.

Table 12.1 RCM method maintenance tasks characteristics

Task	Activation	Effects
Corrective maintenance		
Repair	Failure-mode occurrence	Unavailability Failure repair
Systematic or predetermined preventive maintenance		
Scheduled replacement	Time period elapsed	Unavailability
External inspection	Time period elapsed	No unavailability Symptom observation
Overhaul	Time period elapsed	Unavailability Degradation observation
Test	Time period elasped Performed on stand-by components	Unavailability Failure observation
Condition-based preventive maintenance		
Preventive repair	Symptom detected OR Degradation > threshold	Unavailability Degradation repair

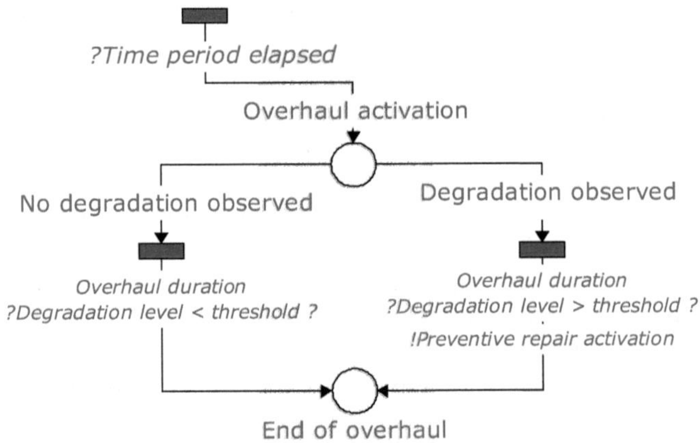

Figure 12.6 Petri net modeling of overhaul realization

Thus, a token is created to enter the net when the time period for overhaul elapses and to describe the realization of the maintenance task.

The decision of performing a preventive repair is based on the degradation level observed. In the overall model, Petri nets are interacting together due to information transfer (value of a Boolean variable, firing condition based on the number of tokens in a place, etc.) (Simeu-Abazi and Sassine 1999). In particular, transitions of the various nets dedicated to maintenance actions depend on information from the degradation mechanism evolution.

Then, depending on the degradation level observed during the overhaul, a preventive repair can be decided upon and performed. Such a decision is modeled through the variable "preventive repair action" which takes the value "true". As a consequence, a token is created to enter the net which modeled the corresponding pre-

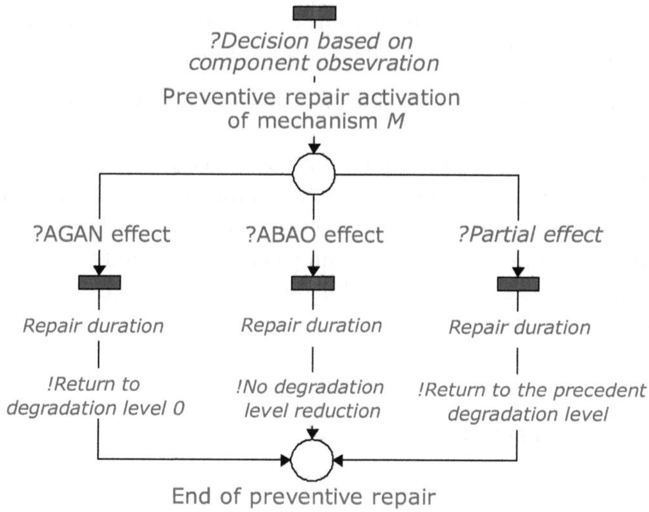

Figure 12.7 Petri net modeling of preventive repair realization

ventive repair action. Finally, preventive repair makes the considered degradation mechanism return to a lower level.

Regarding their efficiency, corrective and preventive repair actions are considered either as good as new, as bad as old, or partial (Brown and Proschan 1983).

The proposed way of modeling a maintained component gives a detailed representation of how various maintenance tasks applied within a complex maintenance program can impact on the degradation and failure processes of the components. It defines the way that each component of the system can enter the unavailability state, either for maintenance or for failure.

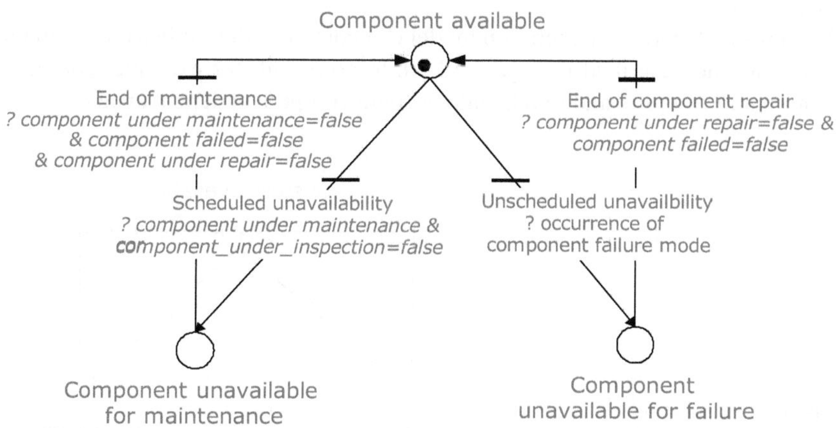

Figure 12.8 Petri net modeling of component availability

Based on the information coming from the specific Petri nets, the component state of availability can then be described, as in Figure 12.8:

- The component becomes unavailable in an unscheduled way when a failure mode occurs.
- The component becomes unavailable in a scheduled way when a maintenance task that engenders the component unavailability is performed, that is all the different preventive repair and detection tasks except the external inspections.
- The component becomes available again when all the maintenance tasks are finished. In the specific case of unscheduled unavailability, these actions only consist in corrective repair.

12.3.3 System Modeling

Within the global structure described in Figure 12.2, the component-level model gives information on component states (failure, unavailability for maintenance) and maintenance costs, to the three other system-level models.

Then, at the system level, the representation of the system dysfunction in the system failure model is based on this input data. More precisely, classical dependability analyses such as failure and event trees are carried out to define the scenarios that lead to system unavailability for failure or for maintenance (Rausand and Hoyland 2004). Boolean expressions are defined to transcribe all the unavailability scenarios such as conditions for transition firing validation.

For example, in Figure 12.9, the condition "system unavailable for failure" consists in a Boolean variable that holds true if one of the system failure scenarios is verified. Each scenario is defined as a possible combination of events, such as component failures, that lead to the occurrence of the system failure. It is therefore similar to the minimal cut sets obtained from failure tree analysis (Malhotra and Trivesi 1995).

In addition, the system operation model can contain different Petri net elements, such as the one described in Figure 12.10, to represent, for example, component activation and stopping and to take into account component dependences.

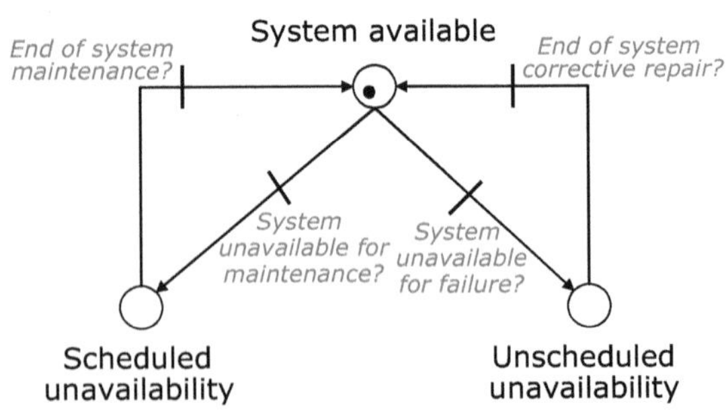

Figure 12.9 Petri net representing the system failure model

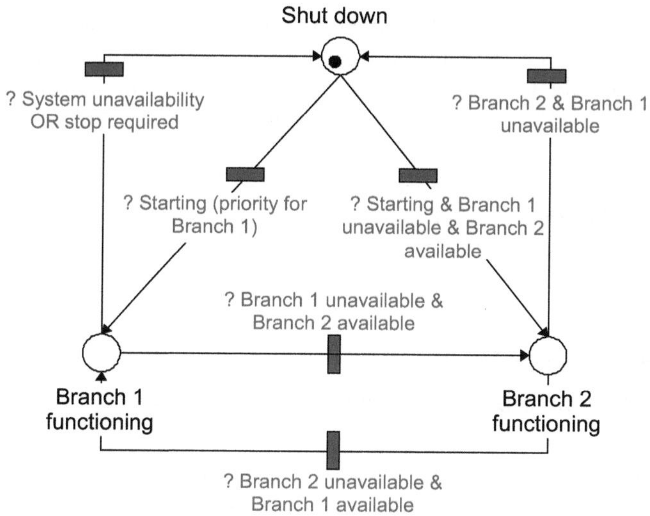

Figure 12.10 System model, Petri net representation of switch-over between two parallel branches

Figure 12.10 refers to a two-branch system and presents the operating rules of switch-over from one branch to the other. The model evolves according to the relative component branch states. It also activates necessary components after a switch-over. Other elements can represent the activation of stand-by components in the case of failure, or a system-scheduled stand-by period.

Finally, the system maintenance model essentially consists in maintenance rules which send data to the component-level model, for example to force the maintenance of a component coupled together with a component already in maintenance. By so doing, it is possible to take into account component dependences for maintenance grouping (Thomas 1986), such as opportunistic maintenance.

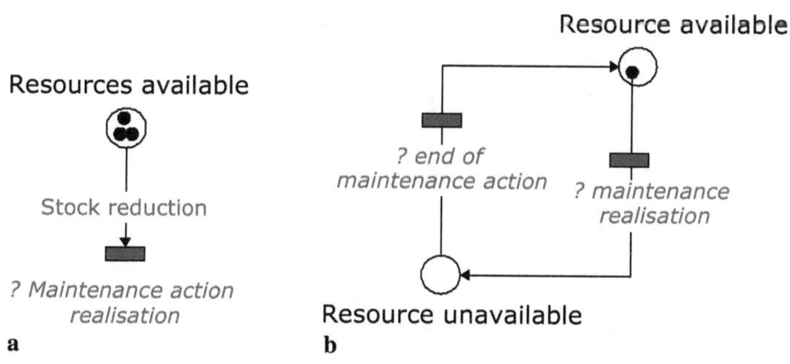

Figure 12.11 Two Petri net representations for maintenance resources use: (a) resources that are consumed, such as spare parts, and (b) unavailability of resources such as specific tools or equipment

The other specific aspect of maintenance resources can also be modeled, as described in Figure 12.11. It can describe situations of resource sharing of limited equipments which can lead to maintenance task postponement or cancellation and have consequences on the system dependability (Dimesh Kumar *et al.* 2000).

Since the three models of the system level are interacting with the component level, the global framework can consider complex systems, made of several dependant components (Ozekici 1996).

12.4 Model Simulation and Dependability Performance Assessment

For system dependability studies, SSPN offer a powerful and versatile modeling tool which can be used jointly with Monte Carlo simulation (Dubi 2000). The SSPN use classical properties of Petri nets to treat sequential and parallel processes, with stochastic and deterministic behaviors and flows of information called "messages" which are very useful in the proposed approach to characterize interactions between the four models.

As described in Figure 12.12, inverse transform Monte Carlo simulation is applied to compute the delay d between transition enabling and firing for all the different Petri net transitions based on their associated distribution laws (Lindeman 1998). By so doing, each transition firing time is sampled until the end of the mission time considered. The entire sequence of transition firing times reproduces one of the possible system behaviors. This simulation process is repeated considerable number of times in order to give the estimation of quantities of information useful for the system performance assessment.

During the simulation of each history, quantities of interest are recorded in appropriate counters (Marseguerra and Zio 2000).

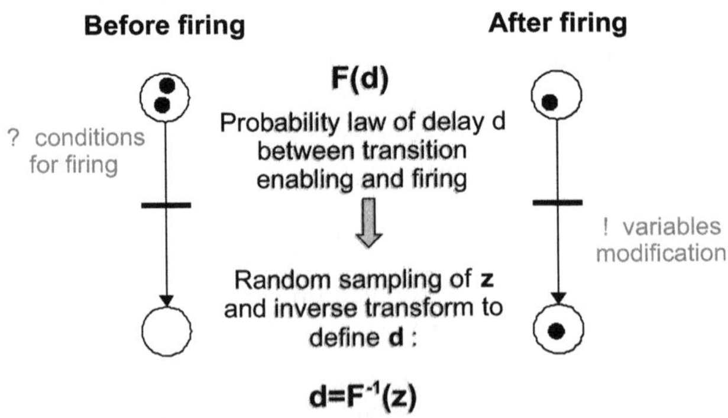

Figure 12.12 Petri net simulation by Monte Carlo method

We implement the proposed approach by using the software MOCA-RP (Dutuit *et al.* 1997), so as to render possible consideration of:

- the time each Petri net place is marked, to give the time the system and the components are in the different states of functioning, failure, availability, scheduled unavailability, unscheduled unavailability, etc.;
- the time each Petri net transition is fired, to give the number of occurred events such as failure, maintenance tasks, etc.;
- the number of tokens in each place at the end of the simulation, to count the spent resources for example.

At the end of the simulation of all the histories, the contents of the counters gives the statistical estimates of the associated quantities of interest over the simulation trials. In particular, the Monte Carlo simulation of the model gives:

- the estimated number of maintenance tasks of each type performed on each component;
- the estimated time the system is unavailable for maintenance;
- the estimated time the system is unavailable for failure;
- the estimated number of system failures;
- the estimated time the different components are in the functioning or unavailable state, degraded or failed state.

Finally, from this information, system dependability performance indicators can be assessed such as the system unavailability or the maintenance costs for example (Leroy and Signoret 1989).

12.5 Performance Assessment of a Turbo-lubricating System

Through the various possible applications, studies performed in collaboration with EDF on real complex systems have given the percentage of time a turbo-pump lubricating system is unavailable for a given maintenance strategy.

12.5.1 Presentation of the Case Study

We provide here results obtained on a simplified turbo-pump lubricating system (described in Figure 12.13). Simulations have been made to study the effects of parameter variations, such as maintenance tasks period, on the system behavior.

The system described in Figure 12.14 is a complex system composed of different types of components. Each one is characterized by different behavior phenomena, that lead to different possible maintenance tasks (Zille *et al.* 2008).

Expert interrogations and data collected analysis define:

- for each component, as described in Tables 12.2 and 12.3 for pumps 03PO and 05PO:

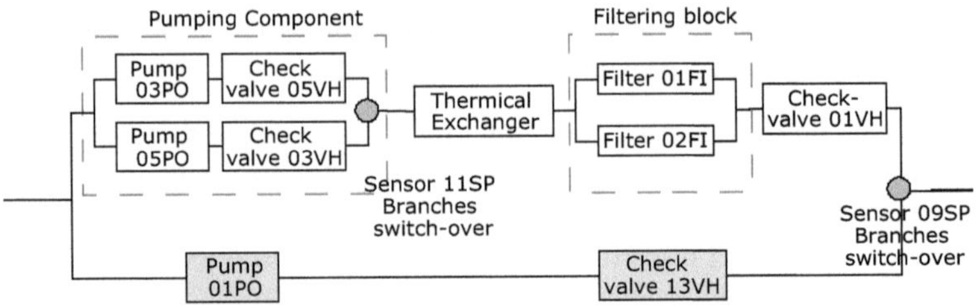

Figure 12.13 Part of a turbo-lubricating turbo-pump system

Table 12.2 Maintenance task parameters for pumps 03PO and 05PO

Preventive maintenance: detection	Duration (days)	Cost (k€)	False-alarm error risks	Non-detection error risks
Overhauls	3	40	No	No
Inspections	0.1	0.2	0.001	0.002

Preventive and corrective repair	Duration (days)	Cost (k€)	Repair type
Preventive repair	3	40	As good as new
Corrective repair	10	95	As good as new

  - the degradation mechanisms, with relative number of evolution levels, and probabilistic laws of transition from one level to the next,
  - the failure modes, with the failure rates associated with the different degradation levels,
  - the symptoms, and how they can be detected corresponding to the degradation mechanisms,
  - the maintenance tasks possibly performed, with their effects, duration, costs, resources,
  - the relations between the different aspects, as shown in Figure 12.1;
- the system failure scenarios, and the way it can become unavailable for maintenance;
- the system operation rules such as the activation of components, the scheduled stopping of the system, the switch-over rules for parallel structures;
- the system maintenance rules and the maintenance grouping procedures.

By so doing, all the different elements of the overall modeling structure can be compiled in order to be simulated.

We can also note that the system studied can be divided into parts to take advantage of the incremental construction of the model. In particular, a first study can be devoted to the pumping-component structure (Zille *et al.* 2008), and then extended to the rest of the system by simply building the required generic models for components and adapting the three system-level models.

Table 12.3 Modeling parameters for Pumps 03PO and 05PO

Degradation	Evolution to successive level	Failure modes: Unscheduled shutdown	Failure modes: Impossible starting	Symptoms: Vibrations	Symptoms: Temperature	Influencing factors and conditions of evolution
–	–	Unscheduled shutdown	Impossible starting	Vibrations	Temperature	–
Mechanism B : Oxidation						
Level 0	Weib(4, 200)	Exp(10^{-4})	–	–	–	Evolution when component is functioning, depending on number of duty-cycle
Level 1	Weib(2, 100)	Exp(0.04)	–	Detection	–	
Level 2	–	Exp(0.02)	–	Detection	Detection	
Mechanism B : Oxidation						
Level 0	Weib(7, 250)	Exp(10^{-30})	Exp(10^{-5})	–	–	Evolution when component is in stand-by, depending on environmental conditions
Level 1	Weib(2, 100)	Exp(0.002)	Exp(0.005)	–	Detection	
Level 2	–	Exp(0.004)	Exp(0.02)	–	Detection	

Weib (x, y): states for a Weibull law with shape parameter x and scale parameter y; Exp(z): states for an exponential law with intensity parameter z; –: relation is not considered.

12.5.2 Assessment of the Maintained System Unavailability

In this section, we are interested in minimizing the system unavailability for maintenance, that is the time the system is stopped in order to perform some preventive maintenance actions (systematic replacement, overhauls, tests, preventive repairs).

We assume that until now the system considered is only maintained through corrective repairs of failed components after the system failure. To decrease the number of failures, one can prevent their occurrences by performing preventive maintenance tasks. However, their realization may induce some system scheduled unavailability which differs according to the various possible options.

To identify the best maintenance program among the propositions resulting from the RCM method application, we assess from the previously described approach the performance of each of the following strategies.

- In strategy S0, no preventive maintenance is performed, the system is only maintained through corrective repairs after its failure.
- In strategy S1, the system is entirely maintained by scheduled replacements of its components, without observing their degradation state.
- In strategy S2, components of the pumping-component structure defined in Figure 12.13 are maintained through condition-based maintenance and the others remain maintained by scheduled replacements. Condition-based preventive repairs are based on overhauls which observe component degradation levels and decide the need of preventive repair if a threshold is reached.
- In strategy S3, all the system components are maintained through condition-based maintenance. Overhauls are performed on the components of the pumping-component structure. On the others, external inspections are performed on functioning components and tests are made on those on stand-by. During inspection, symptoms such as vibration or temperature are observed to obtain information about the component degradation level; a test reveals a failure mode that has occurred during the stand-by period.

For each strategy, the optimal case, corresponding to the minimal system unavailability for maintenance, is identified. Unavailability for system failure is not considered in the present comparison. In particular, Figure 12.14 presents the results obtained for the variation of the pumping-component overhaul periodicity in case of strategy S2. The objective is here to identify the optimal pump overhaul periodicity. Thus, in Figure 12.15, the minimal system unavailability for maintenance associated to strategies S0 to S3 are compared, and the associated number of systems failures are presented.

In Figure 12.15, it appears that a lower system scheduled unavailability time can induce a greater number of system failures. These events can engender system unscheduled unavailability whose associated cost is often really higher than that of scheduled unavailability.

The antagonistic criteria of cost and unavailability make the optimization of the maintenance process difficult. That is why it is useful to base the optimization on a global dependability criteria or on a multi-objective criteria.

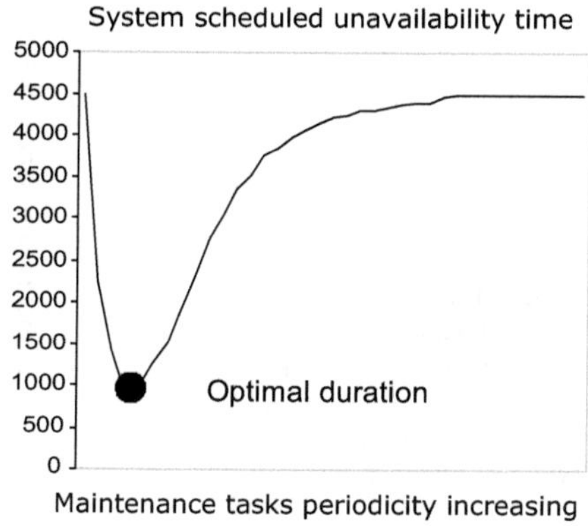

Figure 12.14 Variation of pumping-component overhaul periodicity to identify the minimal unavailability for maintenance of strategy S2

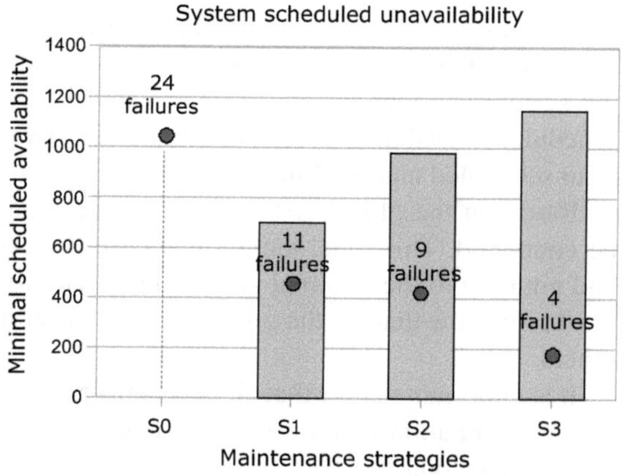

Figure 12.15 Comparison of the minimal unavailability for maintenance for strategies S0 to S3 and associated number of system failures

12.5.3 Other Dependability Analysis

The overall model presented allows for maintained multicomponent system unavailability assessment. It also gives the evaluation of the associated maintenance costs. Thus, system global dependability analysis can be performed by taking into account both the maintenance costs, depending on the number of tasks performed and the relative resources used, and the system availability and unavailability during its mission time.

This can be done through a multi-objective framework that consider simultaneously antagonistic criteria such as cost and availability (Martorell *et al.* 2005). Another possible method is to define a global dependability indicator (Simeu-Abazi and Sassine 1999). In the present study, we define by Equation 12.1 a global maintenance cost model:

$$\text{Cost(Strategy)} = \lim_{\text{TMiss}\to\infty} \frac{\sum_i n_i c_i + t_{\text{su}} c_{\text{su}} + t_{\text{uu}} c_{\text{uu}}}{\text{TMiss}} \qquad (12.1)$$

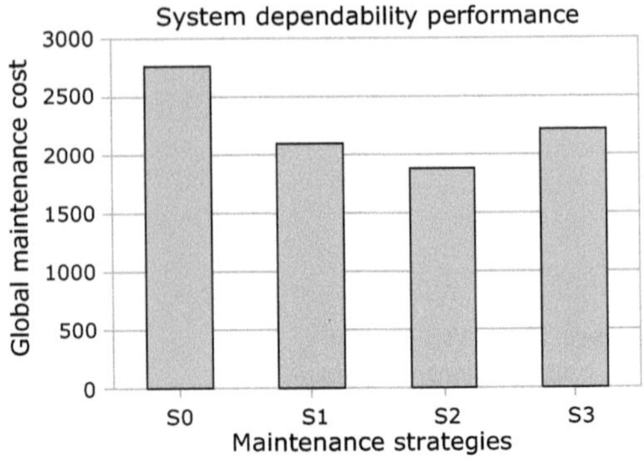

Figure 12.16 Comparison of maintenance strategies S0 to S3 based on the global dependability criterion defined as the optimal global maintenance cost

where Tmiss = the mission time throughout which the system operates, n_i = number of maintenance tasks i performed; c_i = cost of maintenance task i; t_{su} = time the system is under scheduled unavailability; t_{uu} = time the system is under unscheduled unavailability; c_{su} = cost rate of scheduled unavailability; c_{uu} = cost rate of unscheduled unavailability.

Based on the global cost criteria, the optimal case for strategies S0 to S3 can be compared. This time, the optimal case corresponds to the minimal global cost and not only to the minimal system unavailability for maintenance. Results in Figure 12.16 show that for the given parameters, strategy S2 should be preferred to the others.

It is important to note that all the results presented depend on the parameters used for the simulation and are not a formal and absolute comparison of the different maintenance policies.

12.6 Conclusion

In this chapter, a modeling approach of complex maintained systems has been proposed. A two-level modeling framework accurately describes the entire causal chain that can lead to system dysfunction. In particular, the way each component can be degraded and failed is modeled through Petri nets and the Monte Carlo simulation of their behaviors allows for the system availability assessment. The structured and modular model takes into consideration dependences between system components due either to failures or to operating and environmental conditions. Moreover, the detailed maintenance process representation makes it possible to assess the maintained system performance not only in terms of availability, but also in terms of maintenance costs, given the number of tasks performed and their costs. It can therefore be used as a decision-making aid tool to work out preventive maintenance programs on complex systems such as energy power plants.

References

Alla H, David R (1998) Continuous and hybrid Petri nets. J Circuits Syst Comput 8:159–188
Barros A, Bérenguer C, Grall A (2006) A maintenance policy for two-unit parallel systems based on imperfect monitoring information. Reliab Eng Syst Saf 91(2):131–136
Bérenguer C, Châtelet E, Langeron Y et al. (2004) Modeling and simulation of maintenance strategies using stochastic Petri nets. In: MMR 2004 proceedings, Santa Fe
Bogdanoff JL, Kozin F (1985) Probabilistic models of cumulative damage. John Wiley & Sons, New York
Brown M, Proschan F (1983) Imperfect repair. J Appl Probab 20:851–859
Cho DI, Parlar M (1991) A survey of maintenance models for multi-unit systems. Eur J Oper Res 51(1):1–23
Dekker R (1996) Applications of maintenance optimization models: a review and analysis. Reliab Eng Syst Saf 51(3):229–240
Dimesh Kumar U, Crocker J, Knezevic J et al. (2000) Reliability, maintenance and logistic support – a life cycle approach. Kluwer Academic Publishers
Dubi A (2000) Monte Carlo applications in systems engineering. John Wiley, New York
Dutuit Y (1999) Petri nets for reliability (in the field of engineering and dependability). LiLoLe Verlag, Hagen
Dutuit Y, Châtelet E, Signoret JP et al. (1997) Dependability modeling and evaluation by using stochastic Petri nets: application to two test cases. Reliab Eng Syst Saf 55:117–124
Grall A, Dieulle L, Bérenguer C et al. (2002) Continuous-time predictive-maintenance scheduling for a deteriorating system. IEEE Trans Reliab 51:141–150
Jardine AKS, Joseph T, Banjevic D (1999) Optimizing condition-based maintenance decisions for equipment subject to vibration monitoring. J Qual Maint Eng 5:192–202
Leroy A, Signoret JP (1989) Use of Petri nets in availability studies. In: Reliability 89 proceedings, Brighton
Lindeman C (1998) Performance modeling with deterministic and stochastic Petri nets. John Wiley, New York
Malhotra M, Trivesi KS (1995) Dependability modeling using Petri nets. IEEE Trans Reliab 44(3):428–440
Marseguerra M, Zio E (2000) Optimizing maintenance and repair policies via a combination of genetic algorithms and Monte Carlo simulation. Reliab Eng Syst Saf 68(1):69–83
Martorell S, Sanchez A, Serradell V (1999) Age-dependent reliability model considering effects of maintenance and working conditions. Reliab Eng Syst Saf 64(1):19–31
Martorell S, Villanueva JF, Carlos S et al. (2005) RAMS+C informed decision-making with application to multi-objective optimization of technical specifications and maintenance using genetic algorithms. Reliab Eng Syst Saf 87(1):65–75
Ozekici S (1996) Reliability and maintenance of complex systems. Springer, Berlin
Pérez-Ocón R, Montor-Cazorla D (2006) A multiple warm standby system with operational and repair times following phase-type distributions. Eur J Oper Res 169(1):78–188
Rausand M (1998) Reliability centered maintenance. Reliab Eng Syst Saf 60:121–132
Rausand M, Hoyland A (2004) System reliability theory – models, statistical methods and applications. Wiley, New York
Simeu-Abazi Z, Sassine C (1999) Maintenance integration in manufacturing systems by using stochastic Petri nets. Int J Prod Res 37(17):3927–3940
Thomas LC (1986) A survey of maintenance and replacement models for maintainability and reliability of multi-item systems. Reliab Eng 16:297–309
Valdez-Flores C, Feldman RM (1989) A survey of preventive maintenance models for stochastically deteriorating single-unit systems. Naval Res Logist Quart 36:419–446
Wang H (2002) A survey of maintenance policies of deteriorating systems. European Journal of Oper Res 139(3):469–489

Zille V, Bérenguer C, Grall A et al. (2008) Multi-component systems modeling for quantifying complex maintenance strategies. In: ESREL 2008 proceedings, Valencia

Zio E (2009) Reliability engineering: old problems and new challenges. Reliab Eng Syst Saf 94(2):125–141

Chapter 13
Availability Estimation via Simulation for Optical Wireless Communication

Farukh Nadeem and Erich Leitgeb

Abstract The physical systems due to inherent component variation and change in the surrounding environment are not completely failure free. There always exists the probability of failure that may cause unwanted and sometimes unexpected system behavior. It poses the requirement of detailed analysis of issues like availability, reliability, maintainability, and failure of a system. The availability of the system can be estimated though the analysis of system outcomes in the surrounding environment. In this chapter, the availability estimation has been performed for an optical wireless communication system through Monte Carlo simulation under different weather influences like fog, rain, and snow. The simulation has been supported by data measured for number of years. The measurement results have been compared with different theoretical models.

13.1 Introduction

The rising need for high-bandwidth transmission capability links, along with security and ease of installation, has led to increased interest in free-space optical (FSO) communication technology. It provides the highest data rates due to their high carrier frequency in the range of 300 THz. FSO is license free, secure, easily deployable, and offers low bit error rate links. These characteristics motivate the use of FSO as a solution to last-mile access bottlenecks. Wireless optical communication can find applications for delay-free web browsing, data library access, electronic commerce, streaming audio and video, video on demand, video teleconferencing, real-time medical imaging transfer, enterprise networking, work-sharing capabilities and high-speed interplanetary internet links (Acampora 2002).

In any communication system, transmission is influenced by the propagation channel. The propagation channel for FSO is the atmosphere. Despite great potential of FSO communication for its usage in the next generation of access net-

Institute of Broadband Communication, Technical University Graz, Austria

works, its widespread deployment has been hampered by reliability and availability issues related to atmospheric variations. Research studies have shown that optical signals suffer huge attenuations, i.e., weakening of the signal in moderate continental fog environments in winter, and even much higher attenuation in dense maritime fog environments in the summer months. Furthermore, in different fog conditions, weather effects like rain and snow prevent FSO from achieving the carrier class availability of 99.999% by inflicting significant attenuation losses to the transmitted optical signal. The physical parameters like visibility, rain rate, and snow rate determine fog, rain, and snow attenuation and subsequently availability of the optical wireless link. The existing theoretical models help to determine the attenuation in terms of these parameters. However, the random occurrence of these parameters makes it difficult to analyze the availability influenced by these parameters. The availability estimation has been performed in this chapter through simulation. It has been reported in Naylor *et al.* (1966) that simulation can help to study the effects of certain environmental changes on the operation of a system by making alterations in the model of the system and observing the effects of these alterations on the system behavior. The Monte Carlo method is the most powerful and commonly used technique for analyzing the complex problems (Reuven 1981). Many scientific and engineering disciplines have devoted considerable effort to develop Monte Carlo methods to solve these problems (Docuet *et al.* 2001). The performance measure of availability has been estimated for different weather conditions using Monte Carlo simulation while keeping the bit error ratio (BER) below a certain value to provide quality reception. The BER is the number of erroneous bits received divided by the total number of bits transmitted. A similar approach for link availability estimation can be found in Shengming *et al.* (2001, 2005).

13.2 Availability

The availability of a system is simply the time percentage a system remains fully operational. Generally, availability and reliability are confused with each other. The definitions of reliability and availability are given to clarify the difference.

- System *reliability* $R(t)$ is the probability that the system works correctly in the period of time t under defined environmental conditions.
- System *availability* $A(t)$ is the probability that the system works correctly at the time point t.

For example, ping is a computer network tool used to test whether a particular computer is reachable. If we use a ping test to measure the availability of a wireless link and we get acknowledgment of 800 out of 1000 ping tests, we simply say that the availability of the wireless link is 80%.

13.3 Availability Estimation

Equation 13.1 helps only if we have such measured data. The alternate solution is to use surrounding environment models that predict the availability under different conditions. Using our example of wireless optical communication link, the surrounding environment is the atmosphere:

$$A = \frac{T_{\mathrm{up}}}{T_{\mathrm{up}} + T_{\mathrm{down}}} .100\,\% \tag{13.1}$$

13.3.1 Fog Models

Among different atmospheric effects, fog is the most crucial and detrimental to wireless optical communication links. Basically three models proposed by Kruse, Kim, and Al Naboulsi (Kruse *et al.* 1962; Kim *et al.* 2001; Al Naboulsi *et al.* 2004; Bouchet *et al.* 2005) are used to predict the fog attenuation due to visibility. The specific attenuation in dB/km (decibels/kilometer) of a wireless optical communication link for the models proposed by Kim and Kruse is given by

$$a_{\mathrm{spec}} = \frac{10 \log V\%}{V(\mathrm{km})} \left(\frac{\lambda}{\lambda_0}\right)^{-q} \quad (\mathrm{dB/km}) \tag{13.2}$$

Here $V(\mathrm{km})$ stands for visibility in kilometers, $V\%$ stands for transmission of air drops to percentage of clear sky, λ in nm (nanometers) stands for wavelength, and λ_0 is the visibility reference (550 nm). For the model proposed by Kruse *et al.* (1962),

$$q = \begin{cases} 1.6 & \text{if } V > 50\,\mathrm{km} \\ 1.3 & \text{if } 6\,\mathrm{km} > V > 50\,\mathrm{km} \\ 0.585V^{1/3} & \text{if } V < 6\,\mathrm{km} \end{cases} \tag{13.3}$$

Equation 13.3 implies that for any meteorological condition, there will be less attenuation for higher wavelengths. The attenuation of 1550 nm is expected to be less than attenuation of shorter wavelengths. Kim rejected such wavelength-dependent attenuation for low visibility in dense fog. The variable q in Equation 13.2 for the Kim model (Kim *et al.* 2001) is given by

$$q = \begin{cases} 1.6 & \text{if } V > 50\,\mathrm{km} \\ 1.3 & \text{if } 6\,\mathrm{km} < V < 50\,\mathrm{km} \\ 0.16V + 0.34 & \text{if } 1\,\mathrm{km} < V < 6\,\mathrm{km} \\ V - 0.5 & \text{if } V < 0.5\,\mathrm{km} \end{cases} \tag{13.4}$$

The models proposed by Al Naboulsi (France Telecom models) in Al Naboulsi *et al.* (2004) and Bouchet *et al.* (2005) have provided relations to predict fog attenuation. They characterize advection and radiation fog separately. Advection fog is formed by the movements of wet and warm air masses above the colder maritime and terrestrial surfaces. Al Naboulsi provides the advection fog attenuation coefficients as (Al Naboulsi *et al.* 2004; Bouchet *et al.* 2005)

$$\gamma_{\text{ADV}}(\lambda) = \frac{0.11478\lambda + 3.8367}{V} \tag{13.5}$$

Radiation fog is related to the ground cooling by radiation. Al Naboulsi provides the radiation fog attenuation coefficients as (Al Naboulsi *et al.* 2004; Bouchet *et al.* 2005)

$$\gamma_{\text{RAD}}(\lambda) = \frac{0.18126\lambda^2 + 0.13709\lambda + 3.7502}{V} \tag{13.6}$$

The specific attenuation for both types of fog is given by Al Naboulsi as (Al Naboulsi *et al.* 2004; Bouchet *et al.* 2005)

$$a_{\text{spec}}\left(\frac{\text{dB}}{\text{km}}\right) = \frac{10}{\ln(10)}\gamma(\lambda) \tag{13.7}$$

The models proposed by Al Naboulsi give linear wavelength dependence of attenuation in the case of advection fog and quadratic wavelength dependence of attenuation in the case of radiation fog. Al Naboulsi *et al.* (2004) explained that the atmospheric transmission computer codes such as FASCODE (fast atmospheric signature codes), LOWTRAN, and MODTRAN, use the modified gamma distribution in order to model the effect of two types of fog (advection and radiation) on the atmospheric transmission. This model shows more wavelength dependence of attenuation for the radiation fog case. These models predict the attenuation in wireless optical communication link in terms of visibility. All these models can use visibility to find the attenuation. We can simulate the behavior of a wireless optical communication link for low visibility as shown in Figure 13.1.

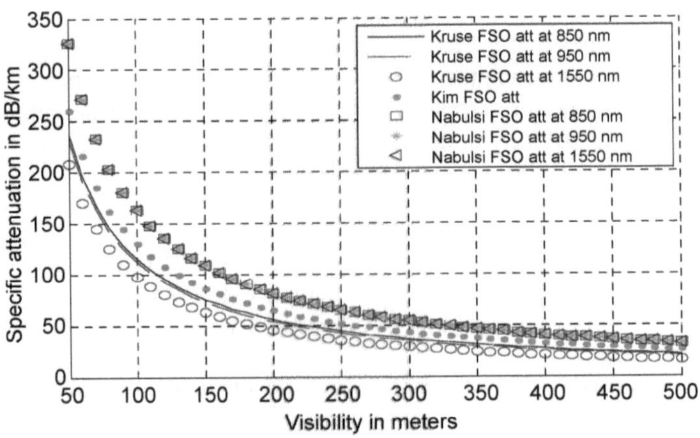

Figure 13.1 Specific attenuation behavior of a wireless optical communication link as predicated by different models

In all these models, visibility has been used for prediction of attenuation. The visibility is random occurring variable that requires Monte Carlo simulation for prediction of attenuation. The random variation in visibility does not allow to predict attenuation without simulating all probable random values taken by visibility. This attenuation can be used to estimate the availability depending upon the link budget. An alternate approach can use Mie scattering theory (Mie 1908) for precise and exact prediction of attenuation.

13.3.2 Rain Model

Another atmospheric factor influencing the optical wireless link is rain. The optical signal passes through the atmosphere and is randomly attenuated by fog and rain. The main attenuation factor for optical wireless link is fog. However, rain also imposes certain attenuation.

When the size of water droplets of rain increases, they become large enough to cause reflection and refraction processes. Most raindrops are in this category. These droplets cause wavelength-independent scattering (Carbonneau and Wisley 1998). It was found that attenuation linearly increases with rainfall rate, and the mean of the raindrop sizes increases with the rainfall rate and is in the order of a few millimeters (Achour 2002). The specific attenuation of a wireless optical link for rain rate of R mm/h is given by Carbonneau and Wisley (1998):

$$a_{\text{spec}} = 1.076R^{0.67} \tag{13.8}$$

This model can be used to simulate the behavior of a wireless optical communication link for different rain rates. Figure 13.2 shows this behavior for rain rate up to 155 mm/h.

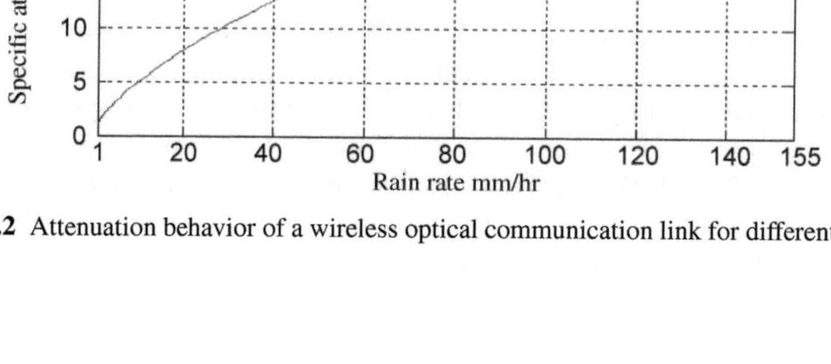

Figure 13.2 Attenuation behavior of a wireless optical communication link for different rain rates

The random occurrence of rain rate can change the attenuation. The rain rate has been taken as a random variable for Monte Carlo simulation. The predicted attenuation is used to estimate the availability depending upon link budget.

13.3.3 Snow Model

Similarly, other factors affecting wireless optical communication link can be used to evaluate the link behavior. One of the important attenuating factors for optical wireless communication is snow. The attenuation effects of snow can be found in terms of a randomly varying physical parameter of snow rate. This requires predicting the attenuation in terms of snow rate by using Monte Carlo simulation. This attenuation can further be used to simulate the availability by considering attenuation and link budget. The FSO attenuation due to snow has been classified into dry and wet snow attenuations. If S is the snow rate in mm/h, then specific attenuation in dB/km is given by (Sheikh Muhammad *et al.* 2005)

$$a_{\text{snow}} = a.S^b \tag{13.9}$$

If λ is the wavelength, a and b are as follows for dry snow:

$$a = 5.42 \times 10^{-5}\lambda + 5.4958776$$
$$b = 1.38$$

The same parameters for wet snow are given as follows:

$$a = 1.023 \times 10^{-4}\lambda + 3.7855466$$
$$b = 0.72$$

Figure 13.3 shows the specific attenuation of an FSO link with wavelength 850 nm for dry and wet snow. In this simulation, the specific attenuation has been predicted due to dry and wet snow at different snow rates.

13.3.4 Link Budget Consideration

The next step in this regard is to estimate the link availability using link budget, receiver sensitivity, and previously recorded weather parameters. As an example we consider the features of the GoC wireless optical communication system at Technical University Graz, Austria, mentioned in Table 13.1. This system is operated over a distance of 2.7 km. If the received signal strength is 3 dB above the receiver sensitivity, the BER reduces to 10^{-9} (Akbulut *et al.* 2005). If we reduce 3 dB from the fade margin, the specific margin to achieve 10^{-9} BER for a distance of 2.7 km be-

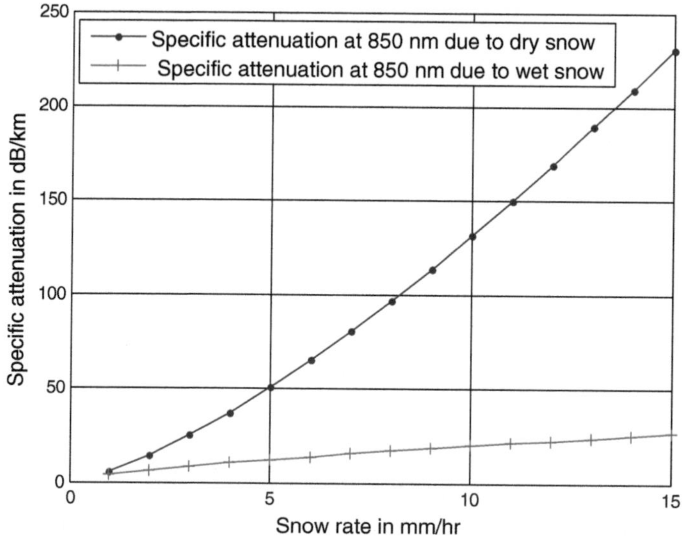

Figure 13.3 Specific attenuation of 850 nm wireless optical link for snow rate up to 15 mm/h

Table 13.1 Features of GoC wireless optical communication system

Parameters	Numerical values
T_X wavelength/ frequency	850 nm
T_X technology	VCSEL
T_X power	2 mW (+ 3 dBm)
T_X aperture diameter	4 × 25 mm lens
Beam divergence	2.5 mrad
R_X technology	Si-APD
R_X acceptance angle	2 mrad
R_X aperture	4 × 80 mm lens
R_X sensitivity	−41 dBm
Spec. margin	7 dB/km

comes 5.88 dB/km. It means whenever, the specific attenuation exceeds this threshold wireless optical communication is no more available to achieve 10^{-9} BER. Now we see how this information can be helpful to estimate availability via simulation for a measured fog event.

13.3.5 Measurement Setup and Availability Estimation via Simulation for Fog Events

The measurements campaign at Graz, Austria was carried out in the winter months of 2004–2005 and 2005–2006, and from January 2009. An infrared link at wave-

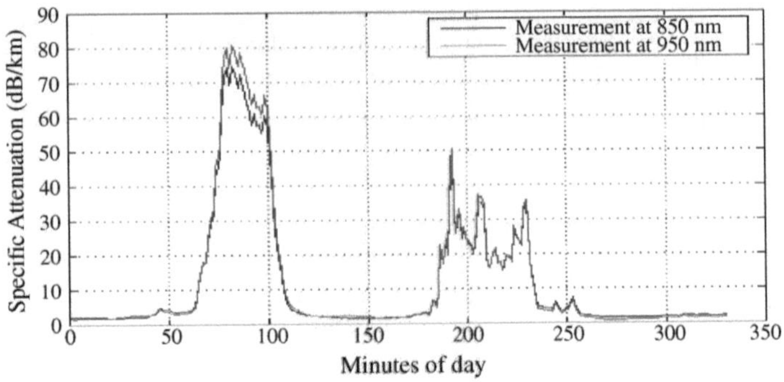

Figure 13.4 Specific attenuation measured for fog event of September 29, 2005

lengths of 850 nm and 950 nm was used for distances of 79.8 m and 650 m. The optical transmitter used has two independent LED-based light sources. One operates at 850 nm center wavelength with 50 nm spectral width at a full divergence of 2.4° which emits 8 mW average optical power; average emitted power in this case after the lens is about 3.5 mW. The second source operates at 950 nm center wavelength with 30 nm spectral widths at a beam divergence of 0.8° using four LEDs each emitting 1 mW to produce the same average power at the receiver. The data was collected and sampled at 1 s.

The following fog event was measured on September 29, 2005. The attenuation has been measured for both 850 nm and 950 nm wavelengths. Figure 13.4 shows the specific attenuation measured for both of the wavelengths. It can be observed that specific attenuation approaches as high as 70 dB/km and 80 dB/km for 850 nm and 950 nm wavelengths respectively.

The GoC wireless optical communication system uses 850 nm wavelength for communication. We use a specific margin of 5.88 dB/km as the limit to achieve availability with 10^{-9} BER. Figure 13.5 shows the availability simulation for the fog event of 19.09.2005 for 850 nm wavelength. The simulation has been performed using a measured value of attenuation and comparing it with the above-mentioned specific margin. The results for this fog event show that wireless optical communication link remained available for 230 minutes out of the total recorded 330 minutes of this fog event. Thus 69.69% availability can be achieved for this fog event. The availability value of 40 has been used to show the time instants when the link is available to achieve 10^{-9} BER, whereas a value of 10 has been used to show when the link is not available to achieve 10^{-9} BER.

Generally the visibility data can be used to predict the availability of a wireless optical communication link at any location. The models presented in Kruse *et al.* (1962), Kim *et al.* (2001), Al Naboulsi *et al.* (2004), and Bouchet *et al.* (2005) can be used to determine the specific attenuation at any location in terms of visibility. The specific attenuation can be used to determine availability by using the above-mentioned criterion. The choice of model for prediction of specific attenuation in terms of visibility can be based on a comparison of measured specific attenuation

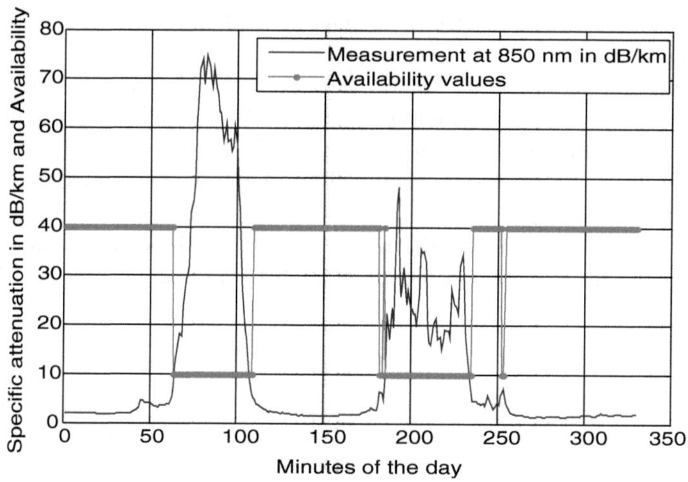

Figure 13.5 Wireless optical communication availability simulated for measured fog event

and predicted specific attenuation by different models. However, it requires simultaneous measurement of visibility as well as specific attenuation. The attenuation data was measured in La Turbie, France in 2004 under dense-fog advection conditions for this purpose. The measurement setup included a transmissiometer to measure visibility at 550 nm center wavelength, an infrared link for transmission measurement at 850 and 950 nm, and a personal-computer-based data logger to record the measured data. These measurements were used to show the comparison between measurements and fog attenuation predicted by different models (Figure 13.6) for the dense-fog advection case. It was concluded that it does not provide any reason to prefer any model over another (Sheikh Muhammad *et al.* 2007). Figure 13.6 shows

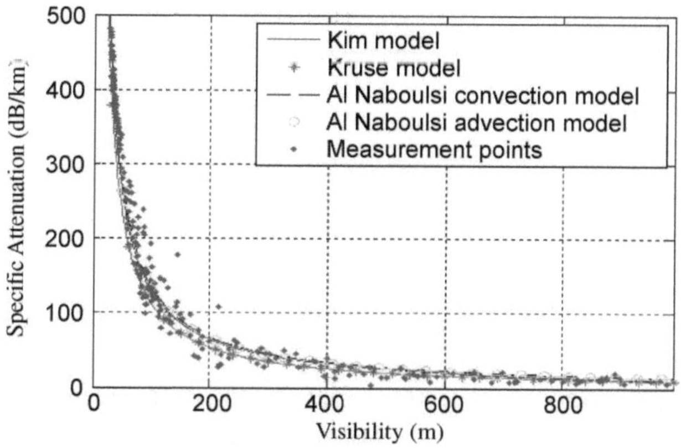

Figure 13.6 Measured specific attenuation for 950 nm and fog attenuation predicted by different models (Nadeem *et al.* 2008)

the comparison of measured specific attenuation and predicted specific attenuation by different models.

In Sheikh Muhammad *et al.* (2007), the magnified view up to 350 m visibility was presented. Here the magnified view up to 250 m visibility is presented in Figure 13.7. But this magnified view also does not help in favoring any model over others (Nadeem *et al.* 2008). A statistical analysis should be performed for the choice of specific model.

Another possibility can be to take the highest predicted specific attenuation. Then use the model with highest specific attenuation for prediction. Figure 13.8 shows

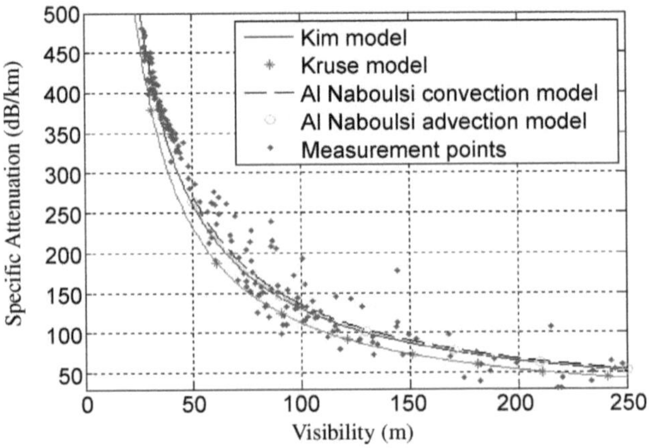

Figure 13.7 Magnified view comparing different models for measured attenuation data for 950 nm (Nadeem *et al.* 2008)

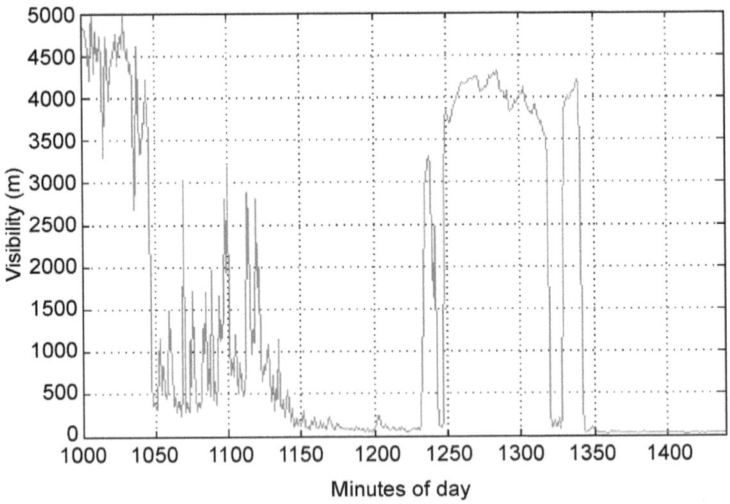

Figure 13.8 Visibility recorded on June 28, 2004

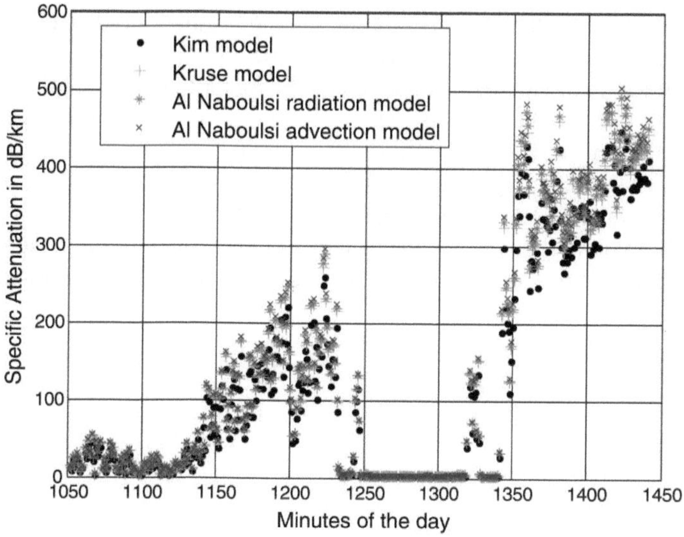

Figure 13.9 Specific attenuation predicted by different models for the recorded visibility

the visibility recorded on June 28, 2004 in La Turbie, France. Figure 13.9 shows the specific attenuation predicted by different models for 850 nm wavelength. The visibility data of June 28, 2004 has been used to simulate the specific attenuation predicted by different models.

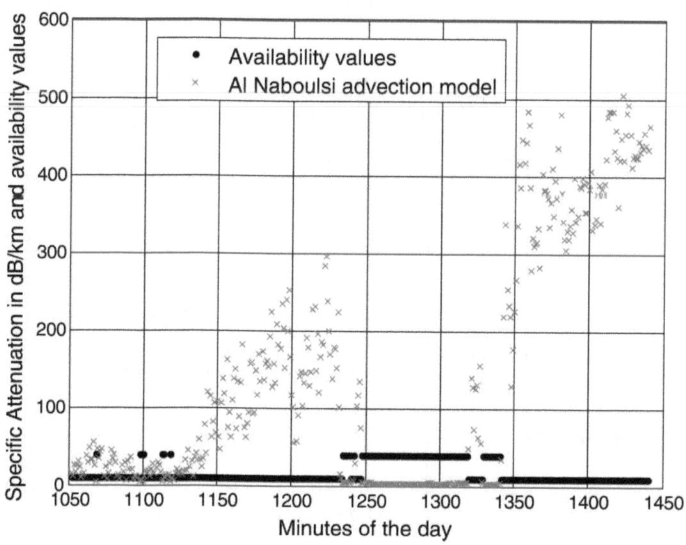

Figure 13.10 Availability estimation using Al Naboulsi specific attenuation prediction for the recorded visibility

Figure 13.9 shows that the specific attenuation values predicted by different models are close to one another. However, the specific attenuation predicted by Al Naboulsi advection model seems to be relatively higher than that predicted by other models. If we use the Al Naboulsi advection model for availability estimation, it can be said that the actual availability will be greater than or equal to availability predicted by this model. Figure 13.10 shows the availability estimated using the Al Naboulsi advection model. The estimated availability is 24.67%, which corresponds to a link being available for 96 minutes out of a total of 389 minutes to achieve 10^{-9} BER. The availability value of 40 has been used to show the time instants when

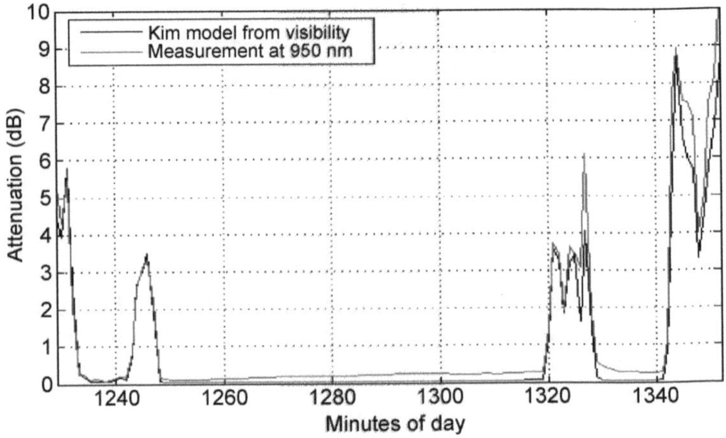

Figure 13.11 Attenuation measured for 950 nm wavelength and attenuation predicted by Kim model

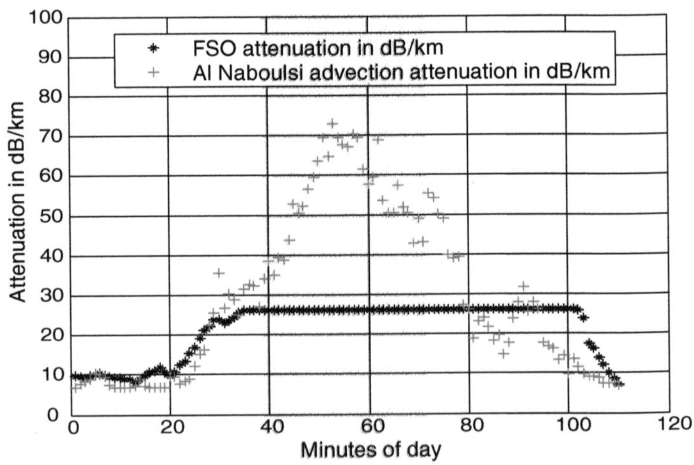

Figure 13.12 Comparison of measured specific attenuation of FSO link with prediction by Al Naboulsi advection model

link is available to achieve 10^{-9} BER, whereas a value of 10 has been used to show when link is not available to achieve 10^{-9} BER.

Figure 13.11 shows that measured values are in close approximation to the predicted attenuation by the model. Sometimes due to any measurement mismatch, measured and predicted specific attenuation can be different. Now we consider another case shown in Figure 13.12 where measured specific attenuation shows variation with the predicted values of attenuation for an FSO link. However, the availability measured by both measured and predicted specific attenuation is equal in this case.

13.3.5.1 Monte Carlo Simulation for Availability Estimation Under Fog Conditions

The above results use measured data. However, the randomly varying visibility motivates one to use it as a random variable and perform Monte Carlo simulation to predict attenuation for this random visibility. The Kruse model has been used to predict the attenuation from this random variable of visibility as the results of the Kruse model were close to the measured data. The random values of visibility between 400 m (extremely low visibility) and 10 km were generated using uniform distribution. The number of random values taken is 100 000. From these visibility values, the attenuation was evaluated using the Kruse model. These 100 000 attenuation values and link budget consideration were used to find the status of the

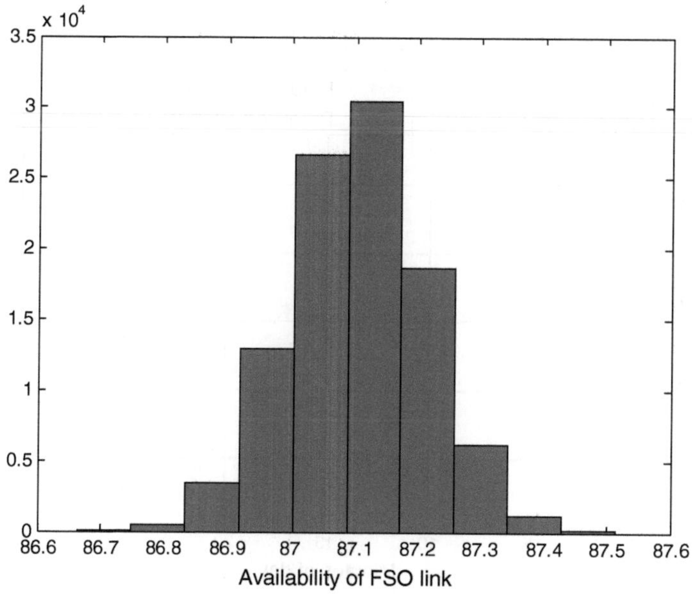

Figure 13.13 Histogram of FSO link availability for different visibility values

reception of the optical signal. These 100 000 optical signal reception status values were used to evaluate one availability value. The whole above process was repeated 100 000 times to find 100 000 availability values. The simulation was performed using Matlab. The results are presented in Figure 13.13. The results show that availability of the FSO link remains around 87% during different visibility values of fog conditions.

13.3.6 Measurement Setup and Availability Estimation via Simulation for Rain Events

An FSO link at 850 nm has been operated on a path length of about 850 m. The transmitted power is +16 dBm, the divergence angle is 9 mrad, and the optical receiver aperture is 515 cm^2. The recording fade margin is about 18 dB.

The meteorological conditions were recorded using a black-and-white video camera. Rain rate was measured using two tipping-bucket rain gauges with different collecting areas. Figure 13.14 shows the simulation of predicted attenuation compared to actual measured attenuation. The predicted attenuation has been simulated using the recorded visibility of the event and using the Al Naboulsi model.

The corresponding availabilities have also been simulated. Figure 13.15 shows comparison of availability simulated for measured attenuation data and availability predicted by the rain model using measured rain rate. The estimated availability for measured attenuation data is 52.38%, which corresponds to the link being available

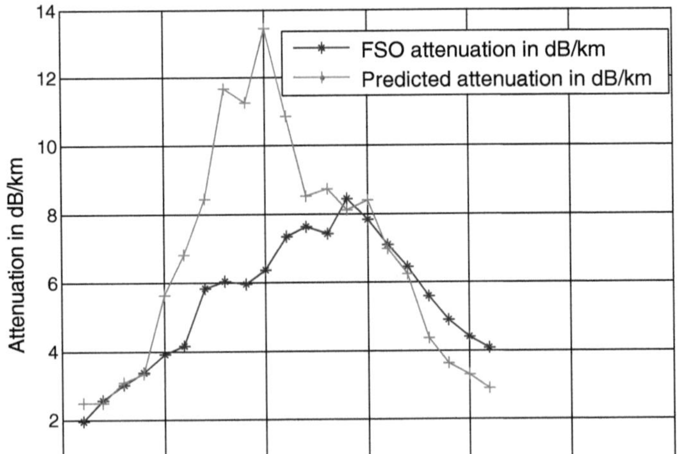

Figure 13.14 FSO measured and predicted attenuation in dB/km

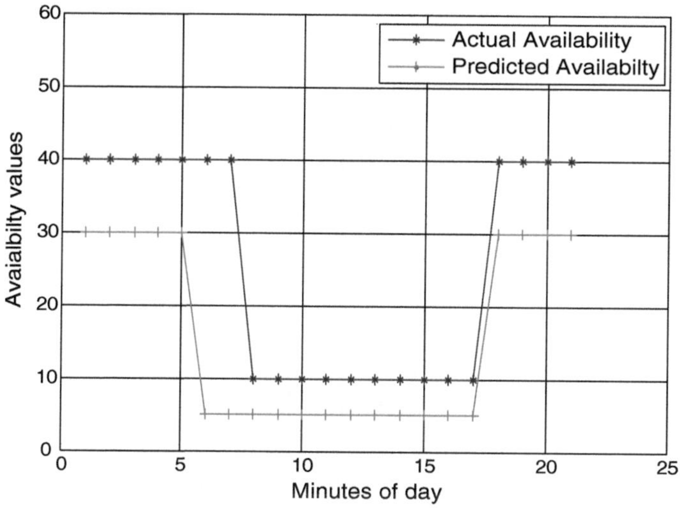

Figure 13.15 Comparison of availability simulated for measured attenuation data and availability predicted by rain model using measured rain rate

for 11 minutes out of a total of 21 minutes to achieve 10^{-9} BER. The availability value of 40 has been used to show the time instants when the link is available to achieve 10^{-9} BER, whereas a value of 10 has been used to show when link is not available to achieve 10^{-9} BER. The estimated availability for predicted attenuation using rain rate data is 42.86%, which corresponds to the link being available for 11 minutes out of a total 21 minutes to achieve 10^{-9} BER. The availability value of 30 has been used to show the time instants when the link is available to achieve 10^{-9} BER, whereas a value of 5 has been used to show when the link is not available to achieve 10^{-9} BER. This comparison shows that availability predicted by the rain model follows the trend of availability predicted by measured attenuation data. However, the model predicted availability is less and can help in more safe and careful estimation.

13.3.6.1 Monte Carlo Simulation for Availability Estimation Under Rain Conditions

The above results use measured data. However, the randomly varying rain rate motivates one to use it as a random variable and perform Monte Carlo simulation to predict attenuation for this random occurrence of rain rate. The random values of rain rate between 1 mm/h and 155 mm/h were generated using a uniform distribution. The total number of values taken is 100 000. From these rain rate values, the attenuation was evaluated using Equation 13.8. These 100 000 attenuation values and link budget consideration were used to find the status of reception of the optical signal. These 100 000 optical signal reception status values were used to evaluate one availability value. The whole above process was repeated 100 000 times to find

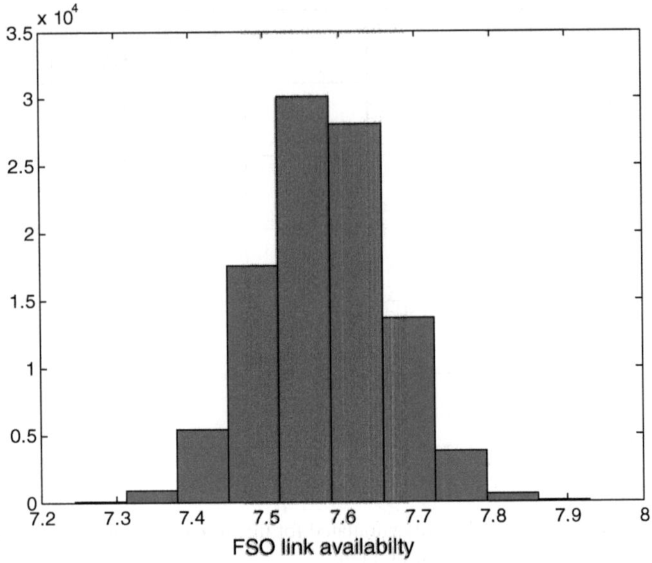

Figure 13.16 Histogram of FSO link availability for different rain rate values

100 000 availability values. The simulation was performed using Matlab. The results are presented in Figure 13.16. The results show that availability of the FSO link remains around 7.6% during different rain rate values.

13.3.7 Availability Estimation via Simulation for Snow Events

The specific attenuation due to snow was measured on November 28, 2005 for an FSO link. Figure 13.17 shows the specific attenuation measured.

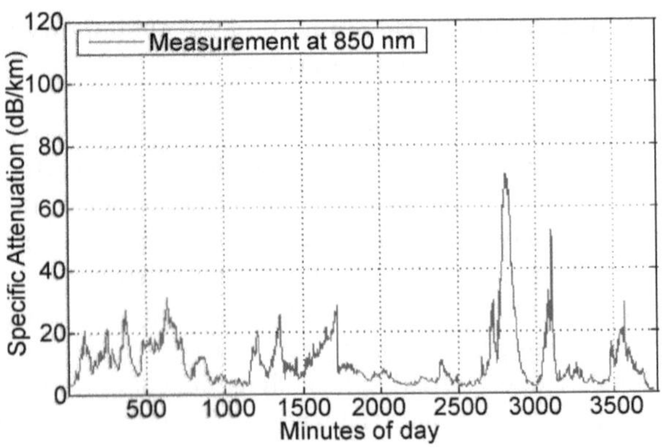

Figure 13.17 Specific attenuation measured for FSO with 850 nm wavelength for a snow event

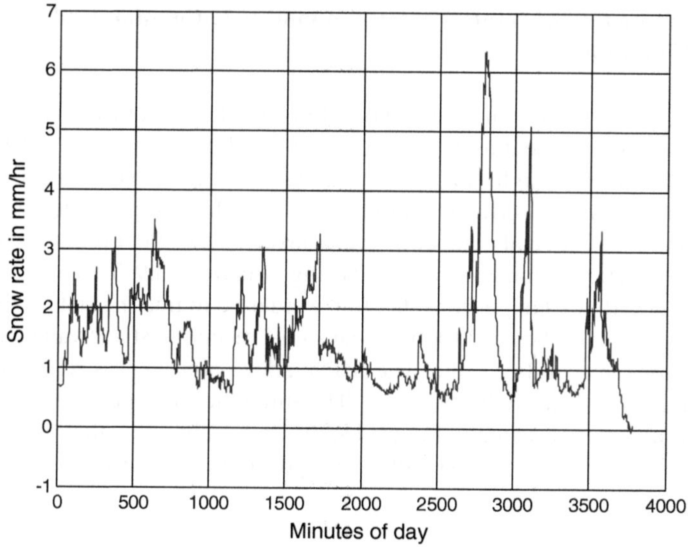

Figure 13.18 Snow rate simulated using a dry snow model

The corresponding snow rate has been simulated in Figure 13.18. As snow rate could not be measured, it has been simulated using a dry snow model.

Figure 13.19 shows the availability simulate using measured attenuation data. The estimated availability for measured attenuation data is 39.49%, which corresponds to the link being available for 1493 minutes out of a total of 3780 minutes to achieve 10^{-9} BER. The availability value of 40 has been used to show the time instants when the link is available to achieve 10^{-9} BER, whereas a value of 10 has been used to show when the link is not available to achieve 10^{-9} BER.

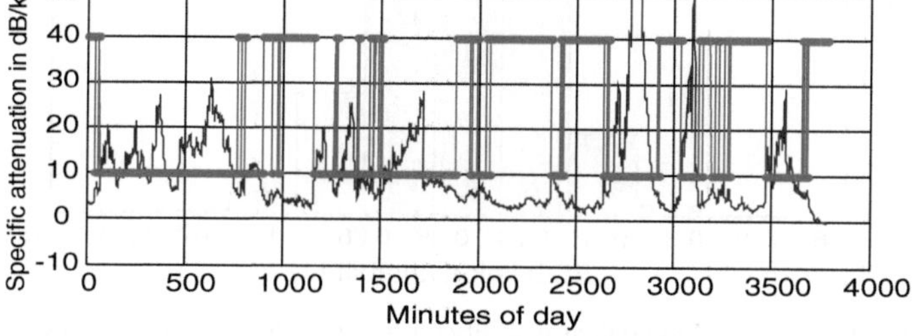

Figure 13.19 Availability simulated using measured attenuation data

13.3.7.1 Monte Carlo Simulation for Availability Estimation Under Dry Snow Conditions

The above simulations use measured data. However, the randomly varying dry snow rate motivates one to use it as a random variable and perform Monte Carlo simulation to predict attenuation for this random occurrence of dry snow rate. The random values of dry snow rate between 1 mm/h and 15 mm/h were generated using a uniform distribution. The total number of values taken is 100 000. From these dry snow rate values, the attenuation was evaluated using Equation 13.9. These 100 000 attenuation values and link budget consideration were used to find the status of reception of the optical signal. These 100 000 optical signal reception status values were used to evaluate one availability value. The whole above process was repeated 100 000 times to find 100 000 availability values. The simulation was performed using Matlab. The results are presented in Figure 13.20. The results show that the availability of the FSO link remains around 0.36% during different dry snow rate values.

13.3.8 Availability Estimation of Hybrid Networks: an Attempt to Improve Availability

Wireless optical communication has the tremendous potential to support the high data rates that will be demanded by future communication applications. However, high availability is the basic requirement of any communication link. We have observed that wireless optical communication link availability is 39.49%, 52.38% and

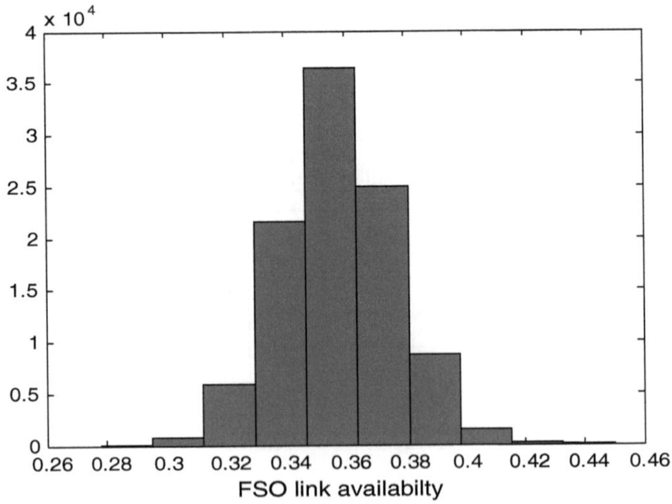

Figure 13.20 Histogram of FSO link availability for different dry snow rate values

Table 13.2 Features of 40 GHz backup link

System	Numerical values
T_X wavelength/ frequency	40 GHz
T_X technology	Semiconductor amplifier
T_X power	EIRP 16 dBW
T_X aperture diameter	Antenna gain 25 dB
Beam divergence	10 degrees
R_X technology	Semiconductor LNA
R_X acceptance angle	10 degrees
R_X sensitivity	Noise figure 6 dB
Spec. Margin	2.6 dB/km

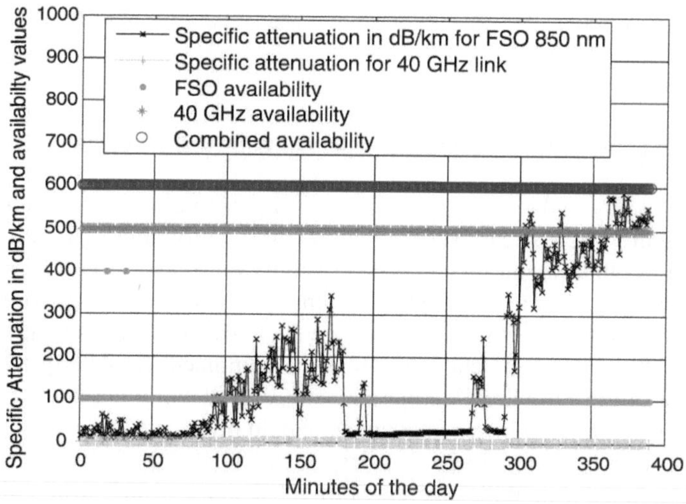

Figure 13.21 Comparison of the specific attenuation and availabilities of FSO and 40 GHz links and their combined availability for a fog event

24.67% availability for snow, rain, and fog events, respectively. This suggests using a backup link for improving reduced availability of wireless optical communication link. Keeping this aspect in view, a 40 GHz backup link was installed parallel to the FSO link mentioned in Table 13.1. Table 13.2 shows the features of the 40 GHz link.

The fog attenuation of the 40 GHz link has been simulated using Nadeem *et al.* (2008), Recommendation ITU-R P.840-3, and Eldrige (1966). The individual availability of the link and the combined availability of the hybrid network is shown in Figure 13.21 for a fog event. Due to high specific attenuation it has only 0.51% availability, whereas 100% availability of the 40 GHz link due to its negligible attenuation makes the combined availability 100%. Availability values of 600, 500, and 400 represent when the combined, 40 GHz, and FSO links are available, respectively, whereas availability values of 300, 200, and 100 represent when the com-

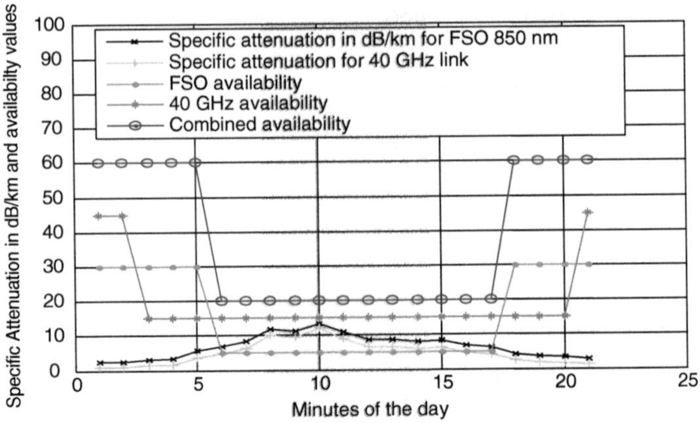

Figure 13.22 Comparison of the specific attenuation and availabilities of FSO and 40 GHz links and their combined availability for a rain event

bined, 40 GHz, and FSO links are not available, respectively, depending on the 10^{-9} BER criterion.

The availability and specific attenuation of hybrid network for a rain event is shown in Figure 13.22. It can be seen that the availability of 40 GHz link has been reduced as GHz links are more influenced by rain events, and Table 13.2 shows less specific margin for the 40 GHz link. The simulations have been performed using Recommendation ITU-R P.838-1. They have been performed to estimate the availability from the measured data. This time the combined availability remains the same as that of the FSO link. If improvement of availability is required for a rain event, a backup link with lower frequencies should be selected.

It can be seen in Figure 13.23 that despite the 39.49% availability for the FSO link, the combined availability increases to 100% due to 100% availability of the 40 GHz link for the snow event. The simulations have been performed using Oguchi (1983). The simulation with the link budget and propagation models not only allows estimating the availability but also gives insight into its improvement.

13.3.9 Simulation Effects on Analysis

The simulation has provided a great aid to obtaining insight into the real phenomena affecting wireless optical communication. The simulations in Figures 13.1–13.3 help to gain insight into the optical wireless attenuation for different weather conditions of fog, rain, and snow, respectively. These figures show the simulated optical wireless signal behavior at different rates of physical parameters like visibility, rain rate, and snow rate. The specific attenuation has been simulated using the attenuation-predicting models in terms of these parameters. Figure 13.1 also helps to compare the simulated optical wireless signal behavior predicted by different mod-

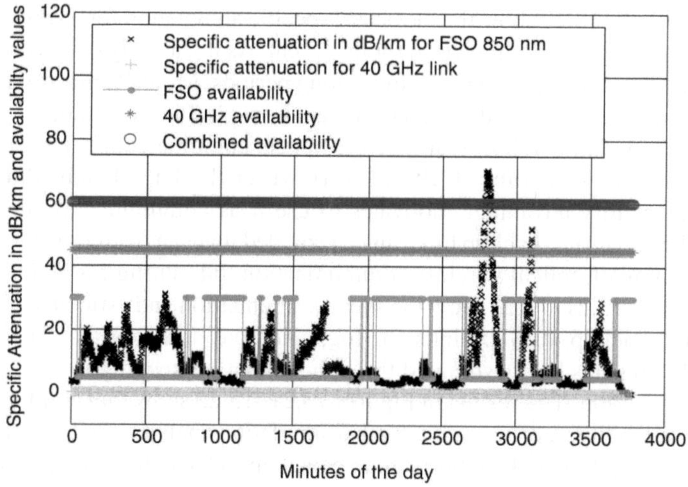

Figure 13.23 Comparison of the specific attenuation and availabilities of FSO and 40 GHz links and their combined availability for a snow event

els. Figure 13.4 shows the measured specific attenuation for wireless optical link wavelengths of 850 nm and 950 nm. However, availability cannot be estimated with such measurements only. Keeping in view the link budget and 10^{-9} BER criterion, the simulation helps to estimate the availability of an optical wireless link as shown in Figure 13.5. The availability has been estimated such that whenever attenuation reaches the level of 3 dB above the receiver sensitivity, which in turn means that BER increases beyond 10^{-9}, we consider that the link is no longer available. In all the availability estimation simulations, these criteria have been considered.

Figures 13.6 and 13.7 show the specific attenuation predicted by different models for fog, and measured specific attenuation. These figures give insight into the accuracy of the specific attenuation prediction model for fog, but these figures do not help in favoring any model over another for the measured data. Figures 13.8–13.12 show the specific attenuation predicted by different models for the recorded fog visibility data. These figures show that despite slight mismatching in measured specific attenuation and model-predicted specific attenuation, the availability estimated through the simulation is same for both cases of measured and model-predicted specific attenuation. But these estimations are for one or two recorded visibility measurements. To estimate the availability for the complete random range of fog visibility, Monte Carlo simulation has been performed and the results are presented in Figure 13.13.

Similarly Figure 13.14 shows the measured specific attenuation and predicted specific attenuation by different models for the recorded rain rate data. Figure 13.15 compares the availability estimates from measured and predicted specific attenuation for rain. To estimate the availability for the complete random range of rain rates, Monte Carlo simulation has been performed and the results are presented in Figure 13.16.

Figures 13.17–13.19 show the measured specific attenuation for a snow event and availability estimated via simulation for this event. As it was only one event and such measurements are not easy to perform for long periods, Monte Carlo simulation has been performed to estimate the availability as shown in Figure 13.20.

All these simulations show that wireless optical communication does not achieve the carrier class availability of 99.999%. However the huge bandwidth potential along with security advantage motivates to use it as communication link. To circumvent the situation a backup link can be provided that can overcome the availability shortcoming of optical wireless communication link during fog, rain, and snow events. Figures 13.21–13.23 show the specific attenuation and estimated availability of FSO and backup 40 GHz links for fog, rain, and snow events. The availability of both links has been estimated through simulation keeping in view the above mentioned criterion. The results in Figures 13.21–13.23 show that combined hybrid network availability improves a lot. Such a simulation analysis can be performed for any other backup link and the best suitable backup link can be selected on the basis of these simulation results.

13.4 Conclusion

Wireless optical communication has the tremendous potential to support the high data rate demands of future communication applications. However, high availability is the basic requirement of any communication link. Due to inherent randomness in underlying attenuation factors, the availability can be estimated through simulation. The measured results show that wireless optical communication link availability is 39.49%, 52.38%, and 24.67% availability for snow, rain, and fog events, respectively. However taking visibility, rain rate, and snow rate as random variables, the availability estimated by Monte Carlo simulation for fog, rain, and snow are 87%, 7.6%, and 0.36%, respectively. The addition of a backup link improves the availability up to 100% for measured results of fog and snow. The simulation with the link budget and propagation models not only allows estimating the availability but also gives insight into its improvement.

References

Acampora A (2002) Last mile by laser. Sci Am, July, vol 287, pp 48–53

Achour M (2002) Simulating free space optical communication, Part I. Rain fall attenuation. Proc SPIE 3635

Akbulut A, Gokhan Ilk H, Arı F (2005) Design, Availability and reliability analysis on an experimental outdoor FSO/RF communication system. In Proceedings of IEEE ICTON, pp 403–406

Al Naboulsi M, Sizun H, de Fornel F (2004) Fog attenuation prediction for optical and infrared waves. Opt Eng 43(2):319–329

Bouchet O, Marquis T, Chabane M, Alnaboulsi M, Sizun H (2005) FSO and quality of service software prediction. Proc SPIE 5892:1–12

Carbonneau TH, Wisley DR (1998) Opportunities and challenges for optical wireless; the competitive advantage of free space telecommunications links in today's crowded market place. In: Proceedings of the SPIE Conference on Optical Wireless Communications, Boston, Massachusetts

Docuet A, de Freitas N, Gordon N (2001) Sequential Monte Carlo methods in practice. Springer, New York

Eldridge RG (1966) Haze and fog aerosol distributions. J Atmos Sci 23:605–613

Kim I, McArthur B, Korevaar E (2001) Comparison of laser beam propagation at 785 and 1550 nm in fog and haze for opt. wireless communications. Proc SPIE 4214:26–37

Kruse PW, McGlachlin LD, McQuista RB (1962) Elements of infrared technology: generation, transmission and detection. Wiley, New York

Mie G (1908) Beiträge zur Optik trüber Medien, speziell kolloidaler Metallösungen, Leipzig. Ann Phys 330:377–445

Nadeem F, Flecker B, Leitgeb E, Khan MS, Awan MS, Javornik T (2008) Comparing the fog effects on hybrid networks using optical wireless and GHz links. CSNDSP July:278–282

Naylor, TJ, Blaintfy JL, Burdick DS, Chu K (1966) Computer simulation techniques. Wiley, New York

Oguchi T (1983) Electromagnetic wave propagation and scattering in rain and other hydrometeors. Proc IEEE 71(9):1029–1078

Recommendation ITU-R P.838-1. Specific attenuation model for rain for use in prediction methods

Recommendation ITU-R P.840-3. Attenuation due to clouds and fog

Reuven Y (1981) Rubinstein simulation and the Monte Carlo method. Wiley, New York

Sheikh Muhammad S, Kohldorfer P, Leitgeb E (2005) Channel modeling for terrestrial free space optical links. In Proceedings of IEEE ICTON

Sheikh Muhammad S, Flecker B, Leitgeb E, Gebhart M (2007) Characterization of fog attenuation in terrestrial free space optical links. J Opt Eng 46(6):066001

Shengming Jiang, Dajiang He, Jianqiang Rao (2001) A prediction-based link availability estimation for mobile ad hoc networks. In Proceedings of INFOCOM, Anchorage, Alaska, vol 3, pp 1745–1752

Shengming Jiang, Dajiang He, Jianqiang Rao (2005) A prediction-based link availability estimation for routing metrics in MANETS. IEEE/ACM Trans Network 3(6):1302–1312

About the Editors

Javier Faulin is an Associate Professor of Operations Research and Statistics at the Public University of Navarre (Pamplona, Spain). He also collaborates as an Assistant Professor at the UNED local center in Pamplona. He holds a Ph.D. in Management Science and Economics from the University of Navarre (Pamplona, Spain), an M.S. in Operations Management, Logistics and Transportation from UNED (Madrid, Spain), and an M.S. in Mathematics from the University of Zaragoza (Zaragoza, Spain). He has extended experience in distance and web-based teaching at the Public University of Navarre, at UNED (Madrid, Spain), at the Open University of Catalonia (Barcelona, Spain), and at the University of Surrey (Guilford, Surrey, UK). His research interests include logistics, vehicle routing problems, and simulation modeling and analysis, especially techniques to improve simulation analysis in practical applications. He has published more than 50 refereed papers in international journals, books, and proceedings about logistics, routing, and simulation. Similarly, he has taught many courses on line about operations research (OR) and decision making, and he has been the academic advisor of more than 20 students finishing their master thesis. Furthermore, he has been the author of more than 100 works in OR conferences. He is an editorial board member of the *International Journal of Applied Management Science* and an INFORMS member. His e-mail address is e-mail: `javier.faulin@unavarra.es`.

Angel A. Juan is an Associate Professor of Simulation and Data Analysis in the Computer Science Department at the Open University of Catalonia (Barcelona, Spain). He also collaborates, as a lecturer of Computer Programming and Applied Statistics, with the Department of Applied Mathematics I at the Technical University of Catalonia (Barcelona, Spain). He holds a Ph.D. in Applied Computational Mathematics (UNED, Spain), an M.S. in Information Technology (Open University of Catalonia), and an M.S. in Applied Mathematics (University of Valencia, Spain). Dr. Juan has extended experience in distance and web-based teaching, and has been academic advisor of more than 10 master theses. His research interests include computer simulation, educational data analysis, and mathematical e-learning. As a researcher, he has published more than 50 papers in international journals, books, and proceedings regarding these fields, being also involved in several international research projects. Currently, he is an editorial board member of the *International*

Journal of Data Analysis Techniques and Strategies, and of the *International Journal of Information Systems & Social Change.* He is also a member of the INFORMS society. His web page is `http://ajuanp.wordpress.com` and his e-mail address is e-mail: `ajuanp@uoc.edu`.

Sebastián Martorell is Full Professor of Nuclear Engineering and Director of the Chemical and Nuclear Department at the Universidad Politécnica de Valencia, Spain. Dr. Martorell received his Ph.D. in Nuclear Engineering from Universidad Politécnica de Valencia in 1991. His research areas are probabilistic safety analysis, risk-informed decision making, and RAMS plus cost modeling and optimization. In the past 17 years that he has been with the University of Valencia, he has served as consultant to governmental agencies, nuclear facilities and private organizations in areas related to risk and safety analysis, especially applications to safety system design and testing and maintenance optimization of nuclear power plants. Dr. Martorell has over 150 papers in journals and proceedings of conferences in various areas of reliability, maintainability, availability, safety, and risk engineering. He is a University Polytechnic of Valencia Scholar-Teacher in the area of probabilistic risk analysis for nuclear and chemical facilities. Dr. Martorell is calendar editor and a member of the Editorial Board of *Reliability Engineering and System Safety International Journal.* He is also an editorial board member of the *European Journal of Industrial Engineering*, the *International Journal of Performability Engineering* and the *Journal of Risk and Reliability*, *Proceedings of Institution of Mechanical Engineers*, Part O. He is Vice-Chairman of European Safety and Reliability Association (ESRA). He has been a member of Technical Committees of the European Safety and Reliability Conferences (ESREL) for more than 10 years and Chairman of ESREL 2008. His e-mail address is e-mail: `smartore@iqn.upv.es`.

José-Emmanuel Ramírez-Márquez is an Assistant Professor of the School of Systems & Enterprises at Stevens Institute of Technology, Hoboken, NJ, USA. A former Fulbright Scholar, he holds degrees from Rutgers University in Industrial Engineering (Ph.D. and M.Sc.) and Statistics (M.Sc.) and from Universidad Nacional Autónoma de México in Actuarial Science. His research efforts are currently focused on the reliability analysis and optimization of complex systems, the development of mathematical models for sensor network operational effectiveness and the development of evolutionary optimization algorithms. In these areas, Dr. Ramírez-Márquez has conducted funded research for both private industry and government. Also, he has published more than 50 refereed manuscripts related to these areas in technical journals, book chapters, conference proceedings, and industry reports. Dr. Ramírez-Márquez has presented his research findings both nationally and internationally in conferences such as INFORMS, IERC, ARSym and ESREL. He is an Associate Editor for the *International Journal of Performability Engineering* and is currently serving a two-year term as President Elect of the Quality Control and Reliability division board of the Institute of Industrial Engineers and is a member of the Technical Committee on System Reliability for the European Safety and Reliability Association. His email address is e-mail: `jmarquez@stevens.edu`.

About the Contributors

Gleb Beliakov received a Ph.D. in Physics and Mathematics in Moscow, Russia, in 1992. He worked as a Lecturer and a Research Fellow at Los Andes University, the Universities of Melbourne and South Australia, and currently at Deakin University in Melbourne. He is currently a Senior Lecturer with the School of Information Technology at Deakin University, and an Associate Head of School. His research interests are in the areas of aggregation operators, multivariate approximation, global optimization, decision support systems, and applications of fuzzy systems in healthcare. He is the author of 90 research papers and a monograph in the mentioned areas, and a number of software packages. He serves as an Associate Editor of *IEEE Transactions on Fuzzy Systems* and *Fuzzy Sets and Systems* journals. He is a Senior Member of IEEE. His e-mail address is e-mail: `gleb@deakin.edu.au`.

Christophe Bérenguer is Professor at the Université de Technologie de Troyes, France (UTT) where he lectures in systems reliability engineering, deterioration and maintenance modeling, system diagnosis, and automatic control. He is head of the industrial engineering program of the UTT and of the Ph.D. program on system optimization and dependability. He is member of the Charles Delaunay Institute (System Modeling and Dependability Laboratory), associated to the CNRS (French National Center for Scientific Research). His research interests include stochastic modeling of system and structure deterioration, performance assessment models of condition-based maintenance policies, reliability models for probabilistic safety assessment and reliability of safety instrumented systems. He is co-chair of the French National Working Group S3 ("Sûreté, Surveillance, Supervision" – System Safety, Monitoring and Supervision) of the national CNRS research network on control and automation. He is also officer (treasurer) of the European Safety and Reliability Association (ESRA) and actively involved in ESRA Technical Committee on Maintenance Modeling and in the European Safety and Reliability Data Association (ESReDA). He is an editorial board member of *Reliability Engineering and System Safety* and of the *Journal of Risk and Reliability*. He is co-author of a several journal papers and conferences communication on maintenance modeling and systems reliability. His e-mail address is e-mail: `christophe.berenguer@utt.fr`.

Héctor Cancela holds a Ph.D. in Computer Science from the University of Rennes 1, INRIA Rennes, France (1996), and a Computer Systems Engineer degree from the Universidad de la República, Uruguay (1990). He is currently Full Professor and Director of the Computer Science Institute at the Engineering School of the Universidad de la República (Uruguay). He is also a Researcher at the National Program for the Development of Basic Sciences (PEDECIBA), Uruguay. His research interests are in operations research techniques, especially in stochastic process models and graph and network models, and in their application jointly with combinatorial optimization metaheuristics to solve different practical problems. He is member of SMAI (Société de Mathématiques Appliquées et Industrielles, France), SIAM (Society for Industrial and Applied Mathematics, USA), AMS (American Mathematical Society, USA), and AUDIIO (Asociación Uruguaya de Informática e Investigación Operativa). He is currently member of IFIP System Modeling and Optimization technical committee (TC7) and President of ALIO, the Latin American Operations Research Association.

Daejun Chang (djchang@kaist.edu) is an Associated Professor in the division of ocean systems engineering, Korea Advanced Institute of Science and Technology (KAIST) since 2009. He leads the Offshore Process Engineering Laboratory (OPEL), whose interest is represented by the acronym PRRESS (Process, Risk, Reliability, Economic evaluation, and System Safety) for ocean and process plants. Since he graduated from KAIST in 1997, Dr. Chang has worked with Hyundai Heavy Industries as a leader of development projects, a researcher for ocean system engineering, and an engineer participating in commercial projects. He was the leader of R&D projects to develop revolutionary systems including ocean liquefied natural gas (LNG) production, offshore LNG regasification, the onboard boil-off gas reliquefaction system, pressure swing adsorption for carbon dioxide and VOC recovery, and multiple effect desalination. Dr. Chang has also participated in development projects with internationally recognized industrial leaders: the compressed natural gas carrier with EnerSea, the methanol plantship with StarChem and Lurgi, and the large-size LNG carriers with QatarGas Consortium. His efforts in ocean system engineering have concentrated on risk-based design: fire and explosion risk analysis, quantitative risk assessment, safety system reliability, production availability, and life-cycle cost analysis.

Kwang Pil Chang is a senior research engineer of Industrial Research Institute at Hyundai Heavy Industries (Ulsan, Korea). He holds an M.S. in Chemical Engineering from the University of Sung Kyun Kwan (Seoul, Korea) and a CRE (Certified Reliability Engineer) issued from the American Society for Quality (Milwaukee, USA). He has extensive experience in optimization of practical offshore production projects and development of new concept processes based on the reliability analysis and risk analysis. He also participated in development of new concept energy carriers: compressed natural gas carrier, Large liquefied natural gas (LNG) carrier, gas hydrate carrier, and LNG-FPSO. His research areas include production availability analysis, safety integrity level assessment, reliability centered maintenance and risk assessment. He has especially focused on application of various analysis techniques to improve reliability or risk based design. He has published several papers in international journals and proceedings relating to reliability and risk assessments. He was a visiting researcher of the Department of Production and Quality Engineering in NTNU (Trondheim, Norway). He is currently an associate member of America Society for Quality and a member of an offshore plant committee managed by a state-run organization of Korea. His e-mail address is e-mail: `envchang@hhi.co.kr`.

Antoine Despujols is Expert Research Engineer at Electricité de France (EDF) Research & Development. He graduated from the French engineering school ESIEE and holds an M.S. in electrical engineering from Sherbrooke University (Canada). He has been working on maintenance management methods, especially on nuclear, fossil-fired, and hydraulic power plants. His research interests include maintenance optimization, physical asset management, indicators, benchmarking, obsolescence management, logistic support, modeling, and simulation of maintenance strategies. He is involved in standards working groups in the International Electrotechnical Commission (IEC/TC56) and European Standardization Committee (CEN/TC319) on maintainability, maintenance terminology, and maintenance indicators. He is member of the board of the European Federation of National Maintenance Societies (EFNMS) and of the French Maintenance Association (AFIM). He is also part-time Assistant Professor at Paris 12 University, involved in a Master degree on Maintenance and Industrial Risk Management. His e-mail address is e-mail: `antoine.despujols@edf.fr`.

Albert Ferrer received a B.S. in mathematics from the University of Barcelona, Spain, in 1978 and a Ph.D. in mathematics from the Technological University of Catalonia (UPC), Barcelona, Spain, in 2003. He worked as Assistant Professor in the Department of Geometry and Topology at the University of Barcelona from 1979 to 1981, and as permanent associate teacher in mathematics of Public High School from 1982 to 1993. Since 1993, he is has been permanent Associate Professor in the Department of Applied Mathematics I of the Technical University of Catalonia (UPC). His research fields are abstract convex analysis, non-linear optimization, global optimization, structural reliability, and fuzzy sets. He has published several papers in international journals, books, and proceedings about optimization, electricity generation, and reliability. He is a member of the Modeling and Numerical Optimization Group at the UPC (GNOM) and of the international Working Group on Generalized Convexity (WGGC). His e-mail address is e-mail: `alberto.ferrer@upc.edu`.

Lalit Goel was born in New Delhi, India, in 1960. He obtained his B.Tech. in electrical engineering from the Regional Engineering College, Warangal, India in 1983, and his M.Sc. and Ph.D. in electrical engineering from the University of Saskatchewan, Canada, in 1988 and 1991 respectively. He joined the School of EEE at the Nanyang Technological University (NTU), Singapore, in 1991 where he is presently a professor of the Division of Power Engineering. He was appointed Dean of Admissions & Financial Aid with effect from July 2008. Dr Goel is a senior member of the IEEE. He received the 1997 & 2002 Teacher of the Year Awards for the School of EEE. Dr Goel served as the Publications Chair of the 1995 IEEE Power PES Energy Management & Power Delivery (EMPD) conference, Organizing Chairman of the 1998 IEEE PES EMPD Conference, Vice-Chairman of the IEEE PES Winter Meeting 2000, Chair of the IEEE PES Powercon2004. He received the IEEE PES Singapore Chapter Outstanding Engineer Award in 2000. He is the Regional Editor for Asia for the *International Journal of Electric Power Systems Research*, and an editorial board member of the *International Journal for Emerging Electric Power Systems*. He is the Chief Editor of the *Institution of Engineers Singapore (IES) Journal C – Power Engineering*. He was the IEEE Singapore Section Chair from 2007 to 2008, and is a R10 PES Chapters Rep since 2005.

Mala Gosakan is a Systems Engineer at Alion Science & Technologies MA&D Operation (Boulder, CO). She holds a Masters in Mechanical Engineering from the State University of New York at Buffalo (Buffalo, NY) and a B.Tech. in Mechanical Engineering from Bapatla Engineering College, Nagarjuna University (Bapatla, India). Her research interests include simulation, human performance modeling and analysis. She has five years of experience working on the Improved Performance Research Integration Tool (IMPRINT). IMPRINT is a stochastic network-modeling tool designed to assess the interaction of soldier and system performance throughout the system lifecycle or for specific missions. Her work involves development, testing, and support of the IMPRINT tool. She has five years of experience of working on the maintenance model within IMPRINT. Her e-mail address is e-mail: `mgosakan@alionscience.com`.

Abhijit Gosavi is an Assistant Professor of Engineering Management and Systems Engineering at Missouri University of Science and Technology in Rolla, Missouri, USA. He holds a Ph.D. in Industrial Engineering from the University of South Florida (Tampa, Florida, USA), an M.Tech. in Mechanical Engineering from the Indian Institute of Technology, Madras (India), and a B.E. in Mechanical Engineering from Jadvapur University (Calcutta, India). His research interests include simulation modeling, reinforcement learning, lean manufacturing, engineering metrology, and supply chain modeling. He has published numerous papers in international journals such as *Automatica, Management Science, INFORMS Journal on Computing, Machine Learning,* and *Systems and Control Letters*. He is the author of a book: *Simulation-based Optimization: Parametric Optimization Techniques and Reinforcement Learning* published by Springer in 2003. His research has been funded by the National Science Foundation (USA), Department of Defense (USA), and the industry. Dr. Gosavi's work in this book was supported partially by the National Science Foundation via grant ECS: 0841055. His e-mail address is e-mail: `gosavia@mst.edu`.

Antoine Grall is Professor at the Université de Technologie de Troyes, France. He is currently the head of the Operations research, Applied Statistics and Numerical Simulation department of the University, and is responsible for the option Operational Safety and Environment in the Industrial Systems academic program. He holds a master of Engineering degree (diplôme d'ingénieur) in computer science, an M.S. in systems control and a Ph.D. in Applied Mathematics from the Compiègne University of Technology (UTC, France). He is giving lectures on applied mathematics, maintenance modeling and systems reliability engineering. As a researcher, he is a member of the System Modeling and Dependability Laboratory of the Charles Delaunay Institute (FRE CNRS 2848). His current research interests are mainly in the field of stochastic modeling for maintenance and reliability, condition-based maintenance policies (performance assessment and optimization, maintenance and on-line monitoring, health monitoring), deterioration of systems and structures, reliability models for probabilistic safety assessment (mainly CCF). He has been author or co-author of more than 90 papers in international refereed journals, books, and conference proceedings. His e-mail address is e-mail: `antoine.grall@utt.fr`.

Joshua Hester is a student of Civil Engineering at the Massachusetts Institute of Technology. At MIT, he has worked with the Buehler Group on developing a mesoscale model of alpha helices using molecular dynamics simulations. He has also worked with the MIT Energy Initiative on implementing an email feedback system to generate environmentally-conscious behavior change on MIT's campus. Most recently, he has collaborated with the IN3 of the Open University of Catalonia in Barcelona, Spain. His e-mail address is e-mail: `jchester@mit.edu`.

Pierre L'Ecuyer is Professor in the Département d'Informatique et de Recherche Opérationelle, at the Université de Montréal, Canada. He holds the Canada Research Chair in Stochastic Simulation and Optimization. He is a member of the CIRRELT and GERAD research centers. His main research interests are random number generation, quasi-Monte Carlo methods, efficiency improvement via variance reduction, sensitivity analysis and optimization of discrete-event stochastic systems, and discrete-event simulation in general. He is currently Associate/Area Editor for *ACM Transactions on Modeling and Computer Simulation, ACM Transactions on Mathematical Software, Statistics and Computing, Management Science, International Transactions in Operational Research, The Open Applied Mathematics Journal,* and *Cryptography and Communications.* He obtained the E. W. R. Steacie fellowship in 1995–97, and a Killam fellowship in 2001–03; he became an INFORMS Fellow in 2006. His recent research articles are available on-line from his web page: `http://www.iro.umontreal.ca/~lecuyer`.

Matias Lee received the Licenciado degree (five-year degree) in computer science from the Facultad de Matemática, Astronomía y Física (FaMAF), Córdoba, Argentina, in 2006. In 2007, he participated in the "INRIA International Internship" program. He was a member of the ARMOR Group, where he worked on Monte

Carlo and quasi-Monte Carlo methods for estimating the reliability of static models. He is currently a Ph.D. student at the FaMAF in Córdoba, Argentina. His Ph.D. thesis is oriented to modeling and analyzing secure reactive systems, where the concept of security is represented by the non-interference property.

Lawrence Leemis is a professor in the Department of Mathematics at The College of William & Mary in Williamsburg, Virginia, USA. He received his B.S. and M.S. in mathematics and his Ph.D. in operations research from Purdue University. He has also taught courses at Purdue University, The University of Oklahoma, and Baylor University. He has served as Associate Editor for the *IEEE Transactions on Reliability*, Book Review Editor for the *Journal of Quality Technology*, and an Associate Editor for *Naval Research Logistics*. He has published three books and many research articles. His research and teaching interests are in reliability, simulation, and computational probability.

Erich Leitgeb was born in 1964 in Fürstenfeld (Styria, Austria) and received his master degree (Dipl.-Ing. in electrical engineering) at the Technical University of Graz in 1994. From 1982 to 1984 he attended a training to an officer for Communications in the Austrian army, (his current military rank is Major). In 1994 he started research work in Optical Communications and RF at the Department of Communications and Wave Propagation (TU Graz). In February 1999 he received his Ph.D. (Dr. at the University of Technology Graz) with honors. He is currently Associate Professor at the University of Technology Graz. Since January 2000 he is has been project leader of international research projects in the field of optical communications and wireless communications (like COST 270, the EU project SatNEx (a NoE), COST 291, and currently COST IC0802 and SatNEx II). He is giving lectures in Optical Communications Engineering, Antennas and Wave Propagation and Microwaves. In 2002 he had a research stay at the department of Telecommunications at Zagreb University, Croatia and in 2008 at the University of Ljubljana, Slovenia. He is a member of IEEE, SPIE, and WCA. Since 2003 he has reviewed for IEEE and SPIE conferences and journals and he acts as a member of technical committees and as chairperson at these conferences. He was guest editor of a special issue (published 2006) in the *Mediterranean Journal of Electronics and Communications* on "Free Space Optics – RF" and also of a special issue (published 2007) in the *European Microwave Association Journal* of on "RFID technology". Since 2007 he prepared the international IEEE conference CSNDSP 08 (July 2008) in Graz as local organizer. In May 2009 he was a guest editor on the Special Issue on Radio Frequency Identification (RFID) of *IEEE Transactions on Microwave Theory and Techniques*. In July 2009 he was a guest editor on the Special Issue on RF-Communications in of the *Mediterranean Journal of Electronics and Communications* (selected papers from the CSNDSP 08).

Adriana Marotta is Assistant Professor at the Computer Science Institute of the University of the Republic of Uruguay since 2003. She received her Ph.D. in Computer Science from the University of the Republic of Uruguay in 2008. She did three internships at the University of Versailles, France, during her Ph.D. studies. Her research interests and activities mainly focus on Data Quality and Data Warehouse Design and Management. She has taught multiple courses in the area of Information Systems, in particular Data Quality and Data Warehousing courses. Adriana has directed two research projects in the topic of Data Quality, supported by CSIC (Comisión Sectorial de Investigación Científica) of the University of the Republic, and has participated in Latin-American projects (Prosul), Ibero-American projects (CYTED), and a Microsoft Research project in the area of bio-informatics.

Adamantios Mettas is the Vice President of ReliaSoft Corporation's product development and theoretical division. He is also a consultant and instructor in the areas of Life Data Analysis, Accelerated Life Testing, Reliability Growth, DOE, Bayesian Statistics and System Reliability and Maintainability and other related subjects. He has been teaching seminars on a variety of Reliability subjects for over 10 years in a variety of industries, including Automotive, Pharmaceutical, Semiconductor, Defense and Aerospace. He fills a critical role in the advancement of ReliaSoft's theoretical research efforts and formulations in all of ReliaSoft's products and has played a key role in the development of ReliaSoft's software including Weibull++, ALTA, RGA and BlockSim. He has published numerous papers on various reliability methods in a variety of international conferences and publications. Mr. Mettas holds an M.S. in Reliability Engineering from the University of Arizona. His e-mail address is `Adamantios.Mettas@ReliaSoft.com`.

Susan Murray is an Associate Professor of Engineering Management and Systems Engineering at Missouri University of Science and Technology (Missouri S&T). She holds a Ph.D. in Industrial Engineering from Texas A&M University, a M.S. in Industrial Engineering from University of Texas at Arlington, and a B.S. in Industrial Engineering also from Texas A&M University. Her research interests include human systems integration, safety engineering, human performance modeling, and engineering education. Dr. Murray has published several papers in international journals and proceedings about human performance modeling, work design, and related areas. She teaches courses on human factors, safety engineering, and engineering management. Prior to joining academia she worked in the aerospace industry, including two years at NASA's Kennedy Space Center. She is a licensed professional engineer in Texas, USA. Her e-mail address is e-mail: `murray@mst.edu`.

Farukh Nadeem obtained his M.Sc. (Electronics) and M.Phil. (Electronics) in 1994 and 1996 from Quaid-e-Azam University Islamabad, Pakistan. His current field of interest is the intelligent switching of Free Space Optical / RF communication links, a field in which he has pursued a Ph.D. since February 2007. He has been the author or coauthor of more than 25 IEEE conference publications. He is actively participating in international projects, such as SatNEx (a network of excellence with work package on "clear sky optics"), ESA project (feasibility assessment of optical technologies & techniques for reliable high capacity feeder links), and COST action

IC0802 (propagation tools and data for integrated telecommunication, navigation and earth observation systems).

Nicola Pedroni is a Ph.D. candidate in Radiation Science and Technology at the Politecnico di Milano (Milano, Italy). He holds a B.S. in Energetic Engineering (2003) and an M.Sc. in Nuclear Engineering (2005), both from the Politecnico di Milano. He graduated with honors, ranking first in his class. His undergraduate thesis applied advanced computational intelligence methods (e.g., multi-objective genetic algorithms and neural networks) to the selection of monitored plant parameters relevant to nuclear power plant fault diagnosis. He has been a research assistant at the Laboratorio di Analisi di Segnale ed Analisi di Rischio (LASAR) of the Nuclear Engineering Department of the Politecnico di Milano (2006). He has also been a visiting student at the Department of Nuclear Science and Engineering of the Massachusetts Institute of Technology (September 2008–May 2009). His current research concerns the study and development of advanced Monte Carlo simulation methods for uncertainty and sensitivity analysis of physical-mathematical models of complex safety-critical engineered systems. He is co-author of about 10 papers on international journals, seven papers on proceedings of international conferences and two chapters in international books.

Verónika Peralta is an Associate Professor of Computer Science at the University of Tours (France). She also collaborates as an assistant professor at the University of the Republic (Uruguay). She holds a Ph.D. in Computer Science from the University of Versailles (France) and the University of the Republic (Uruguay) and an M.S. in Computer Science from University of the Republic (Uruguay). She has extended experience in teaching at the University of the Republic (Uruguay), University of Tours (France), University of Versailles (France), and University of Buenos Aires (Argentina). Her research interests include quality of data, quality of service, query personalization, data warehousing and OLAP, especially in the context of autonomous, heterogeneous, and distributed information systems. She has published several papers in journals and proceedings about information systems and worked in many research projects in collaboration with Uruguayan, Brazilian, and French universities. Similarly, she has taught many courses about data warehousing, data quality, and decision making, and she has been the academic advisor of several students finishing their master thesis. Her e-mail address is e-mail: `veronika.peralta@univ-tours.fr`.

K. Durga Rao works at Paul Scherrer Institut, Switzerland. He graduated in Electrical and Electronics Engineering from the Nagarjuna University, India, and holds an M.Tech. and a Ph.D. in Reliability Engineering from the Indian Institute of Technology Kharagpur and Bombay respectively. He was with Bhabha Atomic Research Center as a scientist during 2002–2008. He has been actively involved in Dynamic PSA, uncertainty analysis, and risk-informed decision making. He has published over 30 papers in journals and conference proceedings. His e-mail address is e-mail: `durga.karanki@psi.ch`.

V.V.S. Sanyasi Rao has worked at Bhabha Atomic Research Centre (Mumbai, India) for the last 35 years. He obtained his Ph.D. in Physics, in the field of Probabilistic Safety Analysis, from Mumbai University, Mumbai, India. He has extensively worked in the area of reliability engineering with emphasis on application to reactor systems, probabilistic safety analysis of Indian nuclear power plants. He has published a number of papers in international journals, and presented papers at various National and International Conferences. His e-mail address is e-mail: vvssrao@barc.gov.in.

Gerardo Rubino is Senior Researcher at INRIA, at the INRIA Rennes–Bretagne Atlantique Center, France. He has also been Full Professor at the Telecom Bretagne engineering school in Rennes, France, in the period 1995–2000. He is the leader of the DIONYSOS team in analysis and design of telecommunication networks (former ARMOR team). He has been Scientific Director at the INRIA Rennes–Bretagne Atlantique Center for four years. His main research areas are in stochastic modeling, and in Quality of Experience analysis. In the former area, he has worked many years in different Operations Research topics (he has been Associate Editor of the *Naval Research Logistics Journal* for ten 10 years) and, in particular, in simulation methods for rare event analysis. He has co-edited a book entitled *Rare Event Simulation Using Monte Carlo Methods* (published by John Wiley & Sons in 2009), and organized several events on rare event simulation. He is currently member of the IFIP WG 7.3 in performance evaluation.

Raul Ruggia is a computer engineer (University of the Republic – Uruguay) and received his Ph.D. in Computer Science from the University of Paris VI (France). He works as Professor at the Computer Science Department of the University of the Republic of Uruguay, where he lectures on information systems, supervises graduate students, and currently directs research projects on data quality management, bio-informatics, and interoperability. Formerly, he worked on design tools and data warehousing areas, participating in Latin-American projects (Prosul), Ibero-American projects (CYTED), and European projects (UE@LIS program). He has also supervised technological projects on environmental and telecommunications domains joint with Uruguayan government agencies.

Carles Serrat is an Associate Professor of Applied Mathematics at the UPC – Catalonia Tech University in Barcelona, Spain. He holds a Ph.D. in Mathematics from the UPC – Catalonia Tech University. His teaching activities include Mathematics, Applied Statistics, Quantitative Analysis Techniques and Longitudinal Data Analysis at undergraduate and postgraduate programs. He also collaborates with the Open University of Catalonia (Barcelona, Spain) as an e-learning consultant. His research areas of interest are related with statistical analyses and methodologies and their applications to different fields, in particular to public health / medicine, food sciences, building construction; survival/reliability analysis, longitudinal data analysis, missing data analysis, and simulation techniques are included in their topics of interest. He has published several papers in international journals, books, and proceedings about survival/reliability analysis and its applications. He is acting as a referee for international journals such as *Statistical Modeling, International Journal of Statistics and Management Systems, Statistics and Operation Research Transactions, Estadística Española,* and *Medicina Clínica.* At this moment, He is currently the Director of the Institute of Statistics and Mathematics Applied to the Building Construction (http://iemae.upc.edu) and Vice-Director of Research, Innovation and Mobility at the School of Building Construction of Barcelona (EPSEB-UPC). His e-mail address is e-mail: `carles.serrat@upc.edu`.

Aijaz Shaikh is a Research Scientist at ReliaSoft Corporation's worldwide headquarters in Tucson, USA. He is closely involved in the development of a majority of ReliaSoft's software applications and has worked on several consulting projects. He is the author of ReliaSoft's *Experiment Design and Analysis Reference* and coauthor of the *System Analysis Reference*. He has also authored several articles on the subjects of design for reliability, life data analysis, accelerated life testing, design of experiments and repairable systems analysis. His research interests include reliability and availability analysis of industrial systems, design of experiments, multibody dynamics, and finite element analysis. He holds an M.S. degree in Mechanical Engineering from the University of Arizona and is an ASQ Certified Reliability Engineer. He is also a member of ASME, SPE, and SRE. His email addresses are e-mail: `aijaz.shaikh@reliasoft.com` and `aij.rel@gmail.com`.

A. Srividya is Professor in Civil Engineering, IIT Bombay. She has published over 130 research papers in journals and conferences and has been on the editorial board and as a guest editor of various international journals. She specializes in the area of TQM and reliability based optimal design for structures. Her e-mail address is e-mail: `asvidya@civil.iitb.ac.in`.

Bruno Tuffin received his Ph.D. in applied mathematics from the University of Rennes 1 (France) in 1997. Since then, he has been with INRIA in Rennes. He spent 8 months as a postdoc at Duke University in 1999. His research interests include developing Monte Carlo and quasi-Monte Carlo simulation techniques for the performance evaluation of telecommunication systems, and developing new Internet-pricing schemes. He is currently Associate Editor for *INFORMS Journal on Computing, ACM Transactions on Modeling and Computer Simulation* and *Mathematical Methods of Operations Research*. He has co-edited a book entitled *Rare Event Simulation Using Monte Carlo Methods* (published by John Wiley & Sons in 2009), and organized several events on rare event simulation. More information can be found on his web page at `http://www.irisa.fr/dionysos/pages_perso/tuffin/Tuffin_en.htm`.

A. K. Verma is Professor in Electrical Engineering, IIT Bombay. He has published around 180 papers in journals and conference proceedings. He is the EIC of OPSEARCH and on the editorial board of various international journals. He has been a guest editor of *IJRQSE, IJPE, CDQM, IJAC,* etc and others, and has supervised 23 Ph.D.s. His area of research is Reliability and Maintainability Engineering. His e-mail address is e-mail: akv@ee.iitb.ac.in.

Peng Wang received his B.Sc. from Xian Jiaotong University, China, in 1978, and his M.Sc. and Ph.D. from the University of Saskatchewan, Canada, in 1995 and 1998 respectively. Currently, he is an associate professor of the School of EEE at Nanyang Technological University, Singapore. His research areas include power system planning and operation, reliability engineering, renewable energy conversion techniques, micro-grid and intelligent power distribution system. He has been involved in many research projects on power system, zero energy plants and buildings, micro grid design, and intelligent power distribution systems.

Valérie Zille is currently an R&D Ph.D. engineer, working in the nuclear industry. She holds a masters of Engineering degree in Industrial Systems at the Université de Technologie de Troyes (UTT, France), and a Ph.D. in Systems Optimisation and Security. Her Ph.D. is entitled "Modelling and Simulation of Complex Maintenance policies for multi-component systems" and she has prepared it within a collaboration between the Charles Delaunay Institute (System Modeling and Dependability Laboratory) of the UTT and the Industrial Risk Management Department of EDF R&D. During her studies, her main research interests were focused on methods and tools for dependability assessments such as Petri Nets, Ant algorithms and Monte Carlo simulation. She has been co-author of a few papers related to her works in international refereed journals (*Reliability Engineering and System Safety, Quality Technology and Quantitative Management*) and conference proceedings and she has made some presentations during international conferences (ESREL, Maintenance Management) and workshops (ESREDA). Her e-mail address is e-mail: zilleval@gmail.com.

Enrico Zio (Ph.D. in Nuclear Engineering, Politecnico di Milano, 1995; Ph.D. in Nuclear Engineering, MIT, 1998) is Director of the Graduate School of the Politecnico di Milano, and full professor of Computational Methods for Safety and Risk Analysis. He has served as Vice-Chairman of the European Safety and Reliability Association, ESRA (2000–2005) and as Editor-in-Chief of the international journal *Risk, Decision and Policy* (2003–2004). He is currently the Chairman of the Italian Chapter of the IEEE Reliability Society (2001–). He is a member of the editorial board of the international scientific journals *Reliability Engineering and System Safety, Journal of Risk and Reliability, Journal of Science and Technology of Nuclear Installations*, plus a number of others in the nuclear energy field. His research topics are: analysis of the reliability, safety and security of complex systems under stationary and dynamic operation, particularly by Monte Carlo simulation methods; development of soft computing techniques (neural networks, fuzzy logic, genetic algorithms) for safety, reliability, and maintenance applications, system monitoring, fault diagnosis and prognosis, and optimal design. He is co-author of three international books and more than 100 papers on international journals, and serves as a referee of more than 10 international journals.

Index

MIX
Papier aus verantwortungsvollen Quellen
Paper from responsible sources
FSC® C105338

If you have any concerns about our products,
you can contact us on
ProductSafety@springernature.com

In case Publisher is established outside the EU,
the EU authorized representative is:
Springer Nature Customer Service Center GmbH
Europaplatz 3, 69115 Heidelberg, Germany

Printed by Libri Plureos GmbH
in Hamburg, Germany